TOWARD SELF AND SANITY:

ON THE GENETIC ORIGINS OF THE HUMAN CHARACTER

ANTHONY M. BENIS, ScD, MD

Research Associate Professor (ret.)
The Mount Sinai School of Medicine
New York, New York

The author was born in Warsaw, immigrating to the United States at an early age. He is a graduate of Princeton University and obtained advanced degrees from MIT. He conducted research at Cambridge University and at the medical center of the University of Nancy, France. Afterwards, he served on the engineering faculty of Columbia University and obtained medical training at the Mount Sinai and Montefiore Medical Centers in New York. He served for many years as director of cardiothoracic intensive care at the Mount Sinai Medical Center. He received career development awards from the National Institutes of Health and from the DGRST of the French government. Author of a number of research papers and review articles in the physical and medical sciences, his present interest is the inheritance of personality traits in European royalty.

Important Note & Disclaimer: The material herein is presented for purposes of information only. The diagnosis and treatment of behavioral disorders should not be attempted without the personal involvement of a licensed health care professional.

TOWARD SELF AND SANITY

ON THE GENETIC ORIGINS
OF
THE HUMAN CHARACTER

ANTHONY M. BENIS, ScD, MD

SECOND EDITION

A.M. BENIS
New York

Originally published in 1985
by

PDI

Psychological Dimensions, Inc.
New York, N.Y.

© 2008 A.M. Benis

SECOND EDITION
1987654321

ISBN 978-0-615-26214-7

LIBRARY OF CONGRESS CATALOGING IN PUBLICATION DATA

Library of Congress Cataloging in Publication Data

Benis, Anthony M., 1939-
    Toward self and sanity. Original edition

    Bibliography: p.
    Includes index.
    1. Personality. 2. Behavior genetics I. Title
[DNLM: 1. Genetics, Behavioral. 2. Personality.
QH 457 B467t]
BF698.B323 1985   155.2'34   84-26313

Copyright reg. 07/06/1982 #TXU 99-426

DEDICATION

To A.G.B.,
wherever you are...

ACKNOWLEDGEMENTS

It is the author's pleasure to acknowledge the assistance of a friend and colleague, Dr. Jacob H. Rand, Director of the Coagulation Laboratory at the Mount Sinai Medical Center, New York, who read the manuscript several times and aided in guiding it to the publisher. The task of producing a typewritten copy was accomplished in good spirits, under less than optimal circumstances, by Ms. Dagmar Martinez. The line drawings were done by Ms. Ann Solomon.

Special mention is due Dr. Robert S. Litwak, Chief of the Division of Cardiothoracic Surgery, Mount Sinai Medical Center, whose dynamism and humanity have made him, to the author, "the most unforgettable character that he has ever met." Innumerable other individuals unwittingly contributed to the manuscript, and the author is grateful to them all.

The author is also indebted to a number of individuals and institutions, cited elsewhere, for their assistance in providing many of the illustrations.

LIST OF PLATES

Following Page 70

Plate

1. A gallery of primates
2. The melancholic orangutan
3. The glare of the baboon
4. Chimpanzee displays coy smile
5. Albino gorilla displays gingival smile
6. Perfectionist chimpanzees "fishing" for termites
7. Chimpanzee smiling in the shower
8. Ring-tailed lemur in narcissistic pose
9. Sea lion in narcissistic pose
10. South American aborigines display adornment and gingival smiles
11. Twelve-year old youth displays gingival smile
12. Queen Elizabeth II and Mrs. Kennedy display smiles of recognition
13. Two-year old child spontaneously assumes narcissistic pose
14. Prince Andrew *à la rose*
15. Somerset Maugham in "narcissistic arms" pose
16. St. Francis of Assisi in narcissistic pose
17. Expressive gestures of a conductor before the orchestra: Leonard Bernstein
18. Self-esteem, adornment and the smile of recognition: Alpine dancers
19. Selket, guardian of Tutankhamun's tomb, in narcissistic pose
20. Alfred Hitchcock with ravens, in narcissistic pose
21. Faces depicting aggressive and narcissistic rages
22. Non-narcissistic grins and laughs

Following page 210

Plate

23. Psychoanalyst Karen Horney: a dutiful NPA– type
24. The four character types of the ancients as depicted by Lavater
25. Perfectionism: Manuscript copy of "Annabel Lee" by Edgar Allan Poe
26. The spirited eyes of the trait of aggression: Archduke Franz Ferdinand
27. The eye contact of dominance and submission: Peter the Great and his son Alexis
28. A perfectionist-aggressive (PA) type haughtily ignores the camera: General Charles de Gaulle
29. A submissive type avoids eye contact with the camera: Somerset Maugham

Plate

30. The countenances of two introverts: Charles I and Cardinal Richelieu
31. Andrew Jackson: an N type
32. Marie-Antoinette: an N type
33. George IV: an N type
34. Catherine of Aragon: an NP type
35. Abraham Lincoln: an NP type
36. Mary Todd Lincoln: an NA type
37. Kaiser Wilhelm II: an N type
38. Friedrich Wilhelm I: an NPA autocrat
39. Henry VIII: a sanguine autocrat of the N type
40. Napoleon I: a pseudoaggressive N type
41. Catherine de Médicis: an N type
42. The Archduchess Sophie of Bavaria: an authoritarian NP type
43. Count Kaunitz of Austria: an eccentric NP type
44. Henry III of Valois: an N type exhibitionist
45. Caligula of Rome: a narcissistic N type who became psychotic
46. Ludwig II of Bavaria: a PA– type who became psychotic
47. Vincent van Gogh: an N–P type who became psychotic
48. Maria Theresa: a dutiful, resolute NP or NPA– type

PREFACE

The reason why the adult no longer wonders is not because he has solved the riddle of life, but because he has grown accustomed to the laws governing his world picture. But the problem of why these particular laws and no others hold, remains for him just as amazing and inexplicable as for the child.

Max Planck (1941)

Man is an abyss of unhappiness and ignorance; he knows neither his body nor his soul, he is aware of being truly Nothing, clothed in a little life; but this knowledge does not make him wise, only unhappy, for philosophy can neither change him nor make him better.

Christina of Sweden (ca. 1670)

PREFACE TO SECOND EDITION

The initial edition of "Toward Self & Sanity" was published in 1985 by Psychological Dimensions, Inc., and an eBook version was first posted in 2004. The present second edition is a printed version of the eBook, with additional updates as of 2008. For the later versions we were tempted to make extensive revisions, having noted deficiencies in the original: wordiness, overblown language, hyperbole and too many literary references. In the end, however, to preserve the spirit in which the book was written (over a period of three months over twenty-five years ago) we have preserved most of the original text, and the index was retained.

Some changes have been made: some of the literary and historical references have been altered or deleted. New material is confined to an ADDENDUM, which follows the text of the original edition. This new section consists of:

- Update as of 2008
- Correspondence with Swiss psychiatrist Manfred Bleuler
- Profiles of the NPA character types
- Comparison of Narcissism and Aggression
- Inadequacy of DSM-IV Criteria for NPD
- The NPA personality test
- Testing of subjects with NPD
- Population genetics: "Out of Africa"
- Article on NPA theory and schizophrenia (1990).

For most readers the interest of this book will lie in practical aspects of the "NPA theory" of character types. These readers can easily skip sections of the book that deal with mathematical or graphical aspects of the theory. We also beg the indulgence of the reader in regard to the literary references: these can also be skipped.

Following Karen Horney, we use mainly masculine pronouns to describe the character types, although the descriptions apply both to males and females. Unfortunately, in the English language there is no good way to avoid this.

There are many roads to self-awareness; but if one ignores the genetic facets of one's personality, then all of these roads may lead one astray. Although the NPA genetic character type is not the only important factor in human behavior, it is, indeed, the only valid starting point. It is in this spirit that, once again, we bid the reader: "Bon voyage!"

AMB
10th November 2008

PREFACE TO ORIGINAL EDITION

Although we consider the ideas set forth here our own, they of course did not germinate in a vacuum. We would like to provide the reader with a few notes so that one may understand the context in which this book came to be written:

I was thirteen years old when I "fell in love" for the first time. She was fourteen and attractive, seductive and brash. I was not only in love, I was madly in love. It did not escape my attention, however, that it was always I who immediately yielded to her ever increasing unreasonable demands, and that when she treated me with scorn, I made no effort to retaliate. And when we went for walks into the woods to pick blueberries, it did not escape my attention that despite her seductiveness and her vivacity, she lacked a quality on which I placed a great deal of importance — tenderness. If this was love, it was a most unpleasant state of affairs. This little liaison soon dissipated, but the memory of it lingered for much longer. I did not know at the time, of course, that what I had experienced as "love" was in actual fact something of a "morbid dependency" between an aggressive dominant type and a submissive type, and I had willingly played the role of the subjugated victim.

I was fifteen years old when Somerset Maugham's novel *Of Human Bondage* fell into my hands. As I read this work during one particular summer, I came to note that the protagonist of the novel, Philip, involved in a morbid dependency of his own, often acted, spoke and *felt* precisely as I would have in the circumstances. Maugham was to become my literary idol, and the merits of any work of fiction that I read thereafter were to be implicitly compared with his. And as time went on, I noted that in other relationships, far from being a subjugated victim I, like Philip, played the role of a dominant subjugator. A most curious state of affairs!

At the age of sixteen I was preparing to enter the university where Albert Einstein was still active as a senior researcher. Knowing that he sometimes mingled with the most junior of undergraduates, I decided that perhaps I should know a bit of his views on life. Thus, I came to buy his little book, *The World As I See It.* I recall reading with some dismay his rather agnostic view of the Creator:

> The fairest thing we can experience is the mysterious. It is the fundamental emotion which stands at the cradle of true art and true science. He who knows it not and can no longer wonder, no longer feel amazement, is as good as dead, a snuffed-out candle. It was the experience of mystery — even if mixed with fear — that engendered religion. A knowledge of the existence of something we

> cannot penetrate, of the manifestations of the profoundest reason and the most radiant beauty, which are only accessible to our reason in their most elementary forms — it is this knowledge and this emotion that constitute the truly religious attitude; in this sense, and in this alone, I am a deeply religious man. I cannot conceive of a God who rewards and punishes his creatures, or has a will of the type of which we are conscious in ourselves. An individual who should survive his physical death is also beyond my comprehension, nor do I wish it otherwise; such notions are for the fears or absurd egoism of feeble souls. Enough for me the mystery of the eternity of life, and the inkling of the marvelous structure of reality, together with the single-hearted endeavour to comprehend a portion, be it never so tiny, of the reason that manifests itself in nature.

These views were in direct conflict with my own thinking of that time, as much of my personal philosophy was vested in the tenets of Roman Catholicism. But this conflict I solved easily, by simply ignoring that it existed.

By the next year I had read Lecomte du Noüy's books, *The Road to Reason* and *Human Destiny,* and I was, like Philip, ready to formulate a "personal philosophy of life" that was to carry me through any difficulty that I might encounter in a potentially hostile world.

The question that I was trying to answer, of course, was the age-old question of "who am I?" This I separated into two parts. The first part of my formulation consisted of an *objective* view of a person as an individual member of the human race, with all the constraints that this implied. This I called "A Look from Without." The second part consisted of a *subjective* view of a person in relation to himself and to other individuals with whom he dealt. This I called "A Look from Within."

The first part of my formulation involved the acceptance of Darwinian evolution by the process of natural selection as fundamental, and this led to my adopting the view of a relative reality that can only be imperfectly perceived by man's limited senses. This, in turn, led me to the conclusion that any attempt by man to understand a concept as complex and nebulous as "the Creator" was fraught with the greatest of uncertainty. Thus, I came to adopt the previously quoted view of Einstein that for a long time had remained shelved to the shelter of the periphery of my thought.

Later I found that my own view of the Creator was somewhat similar to that of Somerset Maugham. In his work *A Writer's Notebook* I found this passage:

> When I consider the vastness of the universe, with its innumerable stars and its spaces measured by thousands upon thousands of light years, I am overwhelmed with awe, but my imagination cannot conceive a creator of it. I am willing enough to accept the existence of the universe as an enigma the wit of man cannot hope to solve. So far as the existence of life is concerned I am not disinclined to credit the notion that there is a psychophysical stuff in which is the germ of life and that the psychic side of this is the source of the complex business of evolution. But what the object of it all is, if any, what the meaning of it all is, if any, is as dark to me as it ever was. All I know is that nothing philosophers, theologians or mystics have said about it persuades me. But if God exists and he concerns himself with the affairs of humanity, then surely he must have sufficient common-sense to take a lenient view, as lenient a view as a reasonable man takes, of the weakness of human beings.

In the second part of my formulation, however, I came to the rather incredible conclusion that the secret to one's life on a personal level was impeccable conduct as an individual living according to the ethics and dogma of my Christian faith. I did not see the obvious contradiction between the two parts of my formulation. Nevertheless, I wrote this up into what was to become a manuscript of some thirty pages, which I pompously entitled *Pensées*. I convinced myself that I wrote this manuscript to gather all my thoughts on paper, but I suspect that the real reason was to impress a certain young lady. I also grandiloquently wrote to my former college advisor, who was also an Episcopalian minister, that I was "working on a new religion."

As the years went by, I retained the first part of my formulation, but the second, my concept of absolute personal perfection, gradually came to crumble like the walls of the proverbial aging fortress. A few years later I came upon a copy of the manuscript, and when I came to read the second section, I instinctively tore out the pages and threw them out.

The second part of my formulation lay idle until I attended medical school and was exposed to the works of Freud, and more important, Karen Horney.

Freud's theory of id, ego and superego was something that neither I nor any of my classmates really understood. Freud was a pioneering giant as far as elucidating the concepts of the unconscious mind, the sexual dynamics of early childhood, the meaning of dreams, and the mechanism of transference (how the patient relates to the analyst) in psychoanalysis. But Freud's tripartite theory of the mind was something that we students thought was simply the object of lip service by psychiatrists, and something that they did not understand any better than the rest of us. I concurred with Horney that Freud's theory was not so much wrong as irrelevant.

In Horney I found someone who was a master at insight by the process of deductive reasoning. She elucidated the concept of "morbid dependency" in her book *Neurosis and Human Growth* and she identified, correctly I believe, three major components of dominant behavior, i.e., *narcissism, perfectionism,* and *aggression.* She was also a careful observer, and I gradually came to learn a rule that is extremely important in the making of observations of human behavior: *Nothing is irrelevant. All behavior, no matter how strange or illogical, is relevant and is begging for an explanation. It is trying to tell us something!*

Next, in the works of the animal behaviorists such as Desmond Morris (*The Naked Ape* and *The Human Zoo*) I found the recurring notion that the understanding of ourselves must come from an appreciation of man's origin in the process of evolution from lower forms, and that this understanding must be thorough and complete to the limits of human intelligence.

Finally, in the poetic songs of Jacques Brel I found the message that life is not just living, but *feeling*... and that to understand one's feelings one must know oneself.

CONTENTS

Acknowledgements   vii

List of Plates   ix

Preface to Second Edition   xii

Preface to Original Edition   xiii

Publisher's Statement   xix

How to Read This Book   xxi

CHAPTER ONE
An Introduction   1

CHAPTER TWO
A Holistic Approach   7

CHAPTER THREE
Alienation from the Human Race   11

CHAPTER FOUR
Alienation from Oneself: A Model of Animalistic Behavior   19

CHAPTER FIVE
A Model of Human Behavior   33

CHAPTER SIX
Character Caricatures   87

CHAPTER SEVEN
Justification of One's Existence   145

CHAPTER EIGHT
Love and Evil: Their Counterparts and Counterfeits   163

CHAPTER NINE
Interactions Between Character Types   183

CHAPTER TEN
Genetics   235

CHAPTER ELEVEN
Disorders of Human Behavior   263

CHAPTER TWELVE
Toward Self and Sanity   317

POSTSCRIPT
Testing the Model   325

APPENDIX A
The Sadist's Choice: A Vulnerable Victim   331

APPENDIX B
Somerset Maugham, William James and Karen Horney   335

APPENDIX C
Is the Trait N Dominant or Recessive?   339

Glossary   341

Chapter References   373

Bibliography   381

Sources of Illustrations   389

Index   391

ADDENDUM   427

Update 1985-2008   429

Correspondence with M. Bleuler   433

Profiles of NPA Character Types   438

Comparison of Narcissism and Aggression   457

Inadequacy of DSM-IV Criteria for NPD   460

The NPA Personality Test   461

Testing of Subjects with NPD   471

Population Genetics: "Out of Africa"   475

Article published in 1990   481

PUBLISHER'S STATEMENT

The publisher is personally proud to have had his firm selected as the publisher of this landmark book by Dr. Benis. Even on a cursory examination, the reader will notice its unusual format, the striking originality of its ideas, and the very broad relevance of these ideas. The questions that Dr. Benis considers are ones that must eventually be faced if we are to progress in our basic understanding of human behavior. Everyone connected with this project believes the author's views merit the most serious consideration by scientists and laypersons alike.

The timeliness of Dr. Benis' work could hardly be more apt. It comes at a time when Darwin's theory of evolution is being questioned by those who espouse "Creationism," and it comes at a time when the mere mention of the term "sociobiology" arouses controversy. It comes at a time when the field of psychiatry is searching for new directions. And it comes at a time when communications between anthropologists, psychologists, psychiatrists and neurologists range from the most dismal to the non-existent, even though they are studying virtually the same behavioral phenomena.

In this book the author develops a theory of human personality that is the most quantitative I have seen to date, and I have been responsible for the publication of many books in personality and related fields. What makes this book especially exciting is that the theory it presents is testable under rigorous laboratory conditions. Certainly the time has come for a new interdisciplinary approach to the investigation of behavioral traits, which makes me enthusiastic about the potential implications this theory has for all of science.

It is important to note that Dr. Benis' book is one that inspires optimism. It cuts across sexual, racial and ethnic boundaries. And more than ever, it reemphasizes that whatever the differences in the fine details of our lives, we are all part of the family of humankind.

If you would like to obtain more information regarding this theory, or if you wish to contribute research to the development of this theory, or if you simply wish to be considered a participant for future research on this theory, you may contact the author.

In conclusion, my staff and I are committed to making further contributions to the dissemination of information in this field. It may be in the form of books, journals, tests, computer information retrieval systems, newsletters, or whatever other format may be best suited for accessing information in behavioral genetics to selected audiences of either pure scientists, applied scientists, medical practitioners, medical researchers, pharmaceutical researchers, industrial managers, scientific administrators, or simply laypersons...

I hope this book generates new insights and new ideas for you, just as the insights and ideas the readers of the manuscript had when they practically "demanded" that this book be published by us.

Robert W. Wesner
President & Publisher

July 1, 1984

HOW TO READ THIS BOOK

This book is intended for all who are interested in the important implications that arise from a theory of basic personality traits that are governed by genes. It is hoped that the format of the presentation is the most direct and intelligible way to introduce this novel view of human behavior.

Although it has been said that "there is nothing new under the sun," we hope that a road to discovery awaits the reader. The discoveries, however, will not come by way of mathematical proofs. It is the reader who will have to open the doors of his or her perception on the basis of personal experience.

We offer the following suggestions:

- For a first reading, approach the opening chapters (Chapters 1–4) with an open mind and even with a bit of indulgence. Save for a second reading your question: "And what if the opposite were true?"

- As you read Chapters 5 and 6, try to categorize individuals with whom you are well acquainted according to the scheme presented. As you develop lists of individuals in the various categories, note carefully the characteristics they have in common and those that are highly variable. Do not attempt, initially, to categorize individuals with whom you are not very well acquainted.

- After perusing Chapters 7 to 9, feel free to interrupt your reading in order to consult some of the references cited, for example the works of Karen Horney or Somerset Maugham.

- If you have no particular interest in quantitative genetics, abnormal psychology or animal behavior, feel free to skim Chapters 10 and 11. Do not get bogged down in these chapters. They should not detract you from a main thrust of the book, namely that there exist discrete human personality types that, with a little practice, you can learn to identify.

- If you have an interest in the area of the personality types of past historical figures, then an exciting adventure awaits you at the biographical section of your local library. Numerous pitfalls await also, but that will make the adventure all the more interesting.

Bon voyage!

CHAPTER 1

AN INTRODUCTION

A wise man once observed that if you study an object of nature intently enough, if you focus upon it long enough all your powers of concentration and attention, there comes a point at which the macrocosm behind the object is suddenly revealed — in somewhat the way in which the vista beyond a keyhole is magnified if one purposefully advances one's eye toward it.

Bernard Rimland (1964)

OUTLINE OF CHAPTER ONE

OBSESSIVE-COMPULSIVE BEHAVIOR

AN INTEGRATED PERSONALITY

A GENETIC MODEL OF BEHAVIOR

Personality types

Sadism

Schizoid & schizophrenic types

"Neurotic" & "normal"

What is this book about?

A.   It is essentially about finding ourselves, to answer the questions, "Who am I?" and "Do I understand what I am doing?"

Q.   I always thought that I was doing quite all right, thank you...

A.   If you do not wish to proceed with coming to terms with yourself, that may be fine with you. But you run the risk of knowing your actions, and the motivations behind your actions, less well than the people with whom you deal. And the rejoinder, "What you don't know won't hurt you," is simply not acceptable. If you do not come to terms with the limitations of your human behavior, with the limitations of your perception of the world in which we live, and with the true motivations for many of your actions, then you are acting irresponsibly — to strangers, to your friends and colleagues, and to yourself.

Q.   What do you mean "true" motivations? I always thought I knew what I was doing.

A.   Unfortunately this may not be true. You may be in a dense fog, and what is worse, you may not even know it. Your motivations are to an enormous extent dictated by instinctual, unconscious forces inherent in your personality structure.

Q.   But don't I know my personality structure? I've spent hours thinking...

A.   One's basic personality structure, according to the model presented here, is determined on the one hand by genetic factors, and on the other hand by environmental influences, primarily those operative during childhood. This structure, once established, becomes relatively fixed and becomes the basis for rigid obsessive compulsive behavior. The conscious mind defends the illogic and irrationality of the individual's behavior so that he appears to have an integrated, cohesive personality. But except for the most superficial acknowledgements that he sometimes behaves irrationally, or for the fleeting moments when he wonders if he is abnormal, or frankly insane, the individual's true personality structure and motivations remain largely unknown to him. As we say, he is in a dense fog.

Q.   How does an individual defend himself to achieve a unified personality structure in the face of this framework of rigid obsessive thought and compulsive activity?

A.   We shall discuss this in some detail. He does so in the presence of an overlay of anxiety, and if the basis of his personality structure is threatened too far, he will undergo personality splits, respond with rages, go into depressions, or finally undergo a psychotic break with reality.

Q.   You said something about a model. What is this model?

A.   The model is an attempt to describe in the quantitative terms of human genetics the occurrence of human character types or phenotypes. Practically all we shall say refers to the model, so we shall only occasionally repeat "according to the model."

Q.   What does the model show?

A.   It predicts the occurrence of discrete character types. These are easily identified in real life and many are identifiable even from photographs. Any individual can, if he wishes, determine his own character type.

Q.   Does the model predict any useful information with regard to the nature or quality of each character structure?

A.   Yes. In particular, it predicts the distribution of *sadistic tendencies* in the human character structure. Some of the personages are practically devoid of sadistic tendencies. Some are at high risk of developing sadistic behavior. In addition, the model predicts degrees of *ambition* and the capacity for *perfectionism* in the various character types.

Q.   How can you possibly do this? Isn't every person a unique individual? Is this some kind of joke?

A.   It is no joke. We are, of course, all unique individuals. However, we are severely limited in our behavior by our intrinsic character types, the bases of which are genetically determined. Just as Gregor Mendel could determine the characteristics of intermediate types of peas and flowers from his knowledge of the basic pure parental characteristics, so can we predict the nature of human character types *once we assume the correct basic variables.*

Q.   What else does the model predict?

A.   It predicts, on both a genetic and environmental basis, the occurrence of *schizoid* or *schizophrenic* types. In addition, the model does not predict any "normal" well-balanced individuals. In fact, all of the character types predicted by the model would be classified according to the principles of "modern psychiatry" as frankly neurotic, psychotic or otherwise mentally deficient.

Q.   Your model predicts no normal individuals?

A.   The model predicts, what many of us have suspected all along, that we are all, at best, "neurotic," and that we have all, including the psychiatrists, been lost in a dense fog of obsessive thought and compulsive behavior. From this moment on, then, we shall no longer use the word "neurotic." According to the model, *the neurotics are we.* It is we, the human race. There isn't anyone else!

Q.   Doesn't this mean that all of modern psychiatry has to be rethought?

A.   We think so. As some psychiatrists have pointed out, "modern psychiatry" has made little advance in the past thirty years. The time has come to reexamine its basic foundations.

Q.   What are the implications of the existence of discrete character types that are readily identifiable by others? Don't these individuals become "transparent" to others?

A.   Yes, to a much greater extent than previously.

Q.   Are there other implications of being able to group individuals according to discrete character types?

A.   The reader will appreciate that if the model is correct, there may be wide and far-reaching implications in the realm of our understanding of human behavior.

Q.   How will this affect my life?

A.   It could affect it dramatically. However, first the model must be verified by many independent observers. As of now, it is just a theory on paper, with observations made by only a single observer. It is possible that it is totally wrong. But if it is essentially correct, then this fact will eventually make itself known.

Q.   Are the results optimistic or pessimistic?

A.   From the point of view that it allows us a "look from within" and that it allows us for the first time to see ourselves as we "really" are, it is optimistic. The results will shock many of us at first, but after the initial shock dissipates, we believe that quiet reason will prevail.

Q.   **Just one more thing.**

A.   Yes?

Q.   **Please try to write in plain language.**

A.   O.K.

CHAPTER 2

A HOLISTIC APPROACH

Peer Gynt. *Just one question: What is it really to "be one's self"?*

Button Moulder. *That's a strange question for a man who just now —*

Peer Gynt. *Tell me what I ask you.*

Button Moulder. *To be one's self is to slay one's self. But as perhaps that explanation is thrown away on you, let's say: to follow out, in everything, what the Master's intention was.*

Peer Gynt. *But suppose a man was never told what the Master's intention was?*

Button Moulder. *Insight should tell him.*

Peer Gynt. *But our insight so often is at fault, and then we're thrown out of our stride completely.*

Button Moulder. *Quite so, Peer Gynt. And lack of insight gives to our friend with the Cloven Hoof his strongest weapon, let me tell you.*

Henrik Ibsen (1867)

OUTLINE OF CHAPTER TWO

LOSS OF FREE WILL

Lack of insight
    Fragmentation
    Empiricism
    Ignoring Darwinian evolution ("Alienation from the Human Race")
    Ignoring one's basic character structure ("Alienation from Oneself")

REGAINING ONE'S FREE WILL

As a human: understanding the limitations of human perception

As an individual: understanding one's basic character structure

Q.   What do you mean by a holistic approach?

A.   By holistic we mean integrated: all the parcels of the important actions and thoughts relevant to one's life must form a synthesis, a cohesive unity.

Q.   A holistic approach to what, exactly?

A.   To free will. There are two things that are most dear to an individual: his body, and his soul or free will. Our body — our health — is very important to us, but otherwise the most important thing that we can lose is our free will.

Q.   How is it that one can lose one's free will?

A.   By lack of insight. And two related processes contribute to this. First, by the *fragmentation* of one's daily life into a series of unrelated acts one can become totally unaware of the illogic of these acts and of one's true motivations behind them. Secondly, by *empiricism,* that is, by the performance of quick, convenient solutions to problems without inquiry into the mechanisms underlying them, one runs the risk of losing oneself in the forest and seeing only unrelated trees.

Q.   How can one regain one's free will?

A.   By casting aside one's prejudices, and looking in the right areas.

Q.   Must I reject my religious beliefs?

A.   That is up to you. After following the reasoning in this book, you may find it expedient to change some of your views. If your religious beliefs have solid foundations and are not based on empty dogma, then you have nothing to fear. But permit us to quote J.A. Thompson:

> To be content with the religious answer — apt to become a soft pillow to the easygoing — is to abandon the scientific problem as insoluble, and there can be no greater impiety than that. It is surrendering our birthright — not for a mess of potage, it is true, but for peace of mind. Therefore, man is true to himself when he presses home the question: How has this marvelous system of Animate Nature come to be as it is?

Q.   Are you, then, irreligious? Atheistic?

A.   Certainly not. True religion means being true to oneself and desiring, in the process, to find truth. If we all do this, then we may all, eventually, arrive at the same destination, albeit by very different paths.

Q.   In which areas should I look in order to gain my free will?

A.   In two areas, which we shall consider in the next two chapters. First, we will look into the implications of man's origins in the process of evolution. Second, we will look into the individual's understanding of his own personality structure, and into the implications of this in his dealings with others and with himself.

Q.   Where will all this lead me?

A.   It may lead you on a journey from which there is no turning back.

CHAPTER 3

ALIENATION FROM THE HUMAN RACE

*The earth has been more or less as we
know it, a fit abode for life, for a period
on the order of two billion years...*

*Evolution is clearly going on around us
today.*

G.G. Simpson (1949)

OUTLINE OF CHAPTER THREE

ALIENATION FROM THE HUMAN RACE: IGNORING DARWINIAN EVOLUTION

ASPECTS OF A SUCCESSFUL MODEL

    Knowledge of morphology

    Understanding of mechanisms

    Tractable mathematical manipulation

FROM DARWINIAN EVOLUTION TO A RELATIVE REALITY

    The limitations of human perception

    Religious models

Q.   What do you mean by "alienation from the human race"?

A.   If one refuses to acknowledge that one is a product of a physical evolution from lower forms, then one is living under a delusion that one is divorced from the implications of being a human being. Thus, one is an individual alienated from the human race.

Q.   But is not the mechanism of evolution still "controversial"?

A.   Not really. There exists an enormous amount of evidence that has been deduced in the fields of anthropology, paleontology, biochemistry and embryology to support the occurrence of the evolution of the human body and of human behavior. Those who disagree often seize on specific areas of uncertainty, to call them "discrepancies" or "contradictions."

Q.   All right. Let's say, for the sake of argument, that I accept the occurrence of evolution as a fact. Where does that bring us?

A.   We are going to use a mathematical model later in this book. Let us see the implications that ensue if we consider man's mind to be a product of evolution.

A model is a description of a combination of elemental parts that may interact. For example, the components of the model in this book are human beings in the present age, thinking, interacting and reproducing according to the laws of genetics that have been elucidated over the past one hundred years.

A model can be a *deterministic* one, namely a model in which the operative mechanisms are specified, in whole or in part. Or it can be a *stochastic* one, namely a "black box" whose behavior can be determined only by its response to perturbations imposed on it. Our model is a deterministic one, and we are definitely interested in seeing if the mechanisms that we assume for the basis of the model are valid ones.

Q.   On what does the success of a model depend?

A.   First, on our knowledge of the physical structure, or "morphology," of the system. Second, on our knowledge of the operative mechanisms. Third, on our ability to manipulate the mathematical description that we formulate.

Q.   How will we know if a model is a successful one?

A.   First, it must describe accurately the known characteristics of the system. Second, it may predict aspects of the system that had remained unknown and would thus be subject to verification. Third, the quantities used in the model must correspond to physically or biologically relevant quantities.

Q.   We are accustomed to acknowledging that "the model is never perfect," but what exactly does this mean? That is, what are the limitations of a model?

A.   Let us take another look at the requirements of a successful model:

- knowledge of morphology

- understanding of mechanisms.

- ingenuity in mathematical manipulation

Considering the last condition first, mathematics is nothing but a logical scheme, a system created by man himself, but sometimes the very success of the model is limited by difficulties in mathematical manipulation. So this is embarrassing, in the sense that man has created a logical scheme, a supposedly consistent scheme, mathematics, and then he has proceeded to show that often he is incompetent to handle it. Happily, this is not true in the case of the model presented in this book.

Furthermore, if the preceding was embarrassing, the worst is yet to come. Taking a look at the first two conditions, namely our understanding of the morphology and mechanisms of the system, here we are often on even shakier ground. In fact, the morphology of biological systems and the mechanisms behind them are *very* imperfectly understood. So man does not even know how his own physiological processes operate, or even how he thinks. He thus encounters the paradox that man does not know how he thinks, yet he must use thought as the process by which he must draw his own conclusions.

With regard to man's understanding of his own morphology, he is put in the embarrassing position of having to dissect himself in order to find out how he is constructed. And in his dissections, he dissects his embryo and finds that it has vestigial gills, showing that his ancestors once lived in the sea. Thus, he comes to the conclusion that he has evolved as a being on this earth through a "slow" (though perhaps "jerky") process of evolution from lower animal forms. So this is the final embarrassment, his realization that his mind is in a very relative state of advancement, hence his conception of the world, dependent on his very limited senses, is a very imperfect one. Yes, the reality that he perceives is a very fuzzy, relative one.

Q.   How do you know that man is not fully mature and thus not fully capable of understanding everything?

A.   First of all, there are some concepts that he just is not capable of totally comprehending, such as the creation and the infinity of space and time. Secondly, his own physical senses are very limited, and his investigations become limited to his ingenuity in constructing machines to investigate natural phenomena. Finally, examine Figure 1 (overleaf). You will note that on the time scale of the age of the earth, which is according to latest estimates about five billion years, man's period of evolution has been short indeed. Man is, in fact, a very, very young baby. We could imagine that one *billion* years from now, the inhabitants of our planet would consider our "sophisticated" models of the present day as the most primitive kinds of fumbling about. So it is really in this sense that we say that our models cannot be perfect.

Q.   You have been using the word "embarrassing." Why do you say "embarrassing"?

A.   Because traditionally man has been pompous with regard to his achievements and to his place in the universe. Consider the scientific researcher who presents the results of his grand model at an international conference. These results are usually presented in the spirit that knowledge is something absolute, hidden away, and is simply there waiting for the researcher to dig up. And when he finds a chunk of this knowledge, he says, "Aha! I've found another one!" and he plunks it down on the table for everyone to see, as if sooner or later all the chunks will be found, and our perception of an absolute reality will be fulfilled.

The realization that our mind is a product of evolution shows that this type of attitude is not warranted. Rather, we should consider that our discoveries, our laws, and even the most complicated theories of Einstein are mere scratches on the surface of an absolute reality that we will never see, but perhaps our progeny could approach billions of years from now.

Q.   Do you mean, then, that models are useless?

A.   No! Quite to the contrary. *Man's ability to recognize his limitations is proof of his intelligence.* It is only if he does not recognize his limitations does he risk the building of models that describe such things as the activities of "gremlins in the forest."

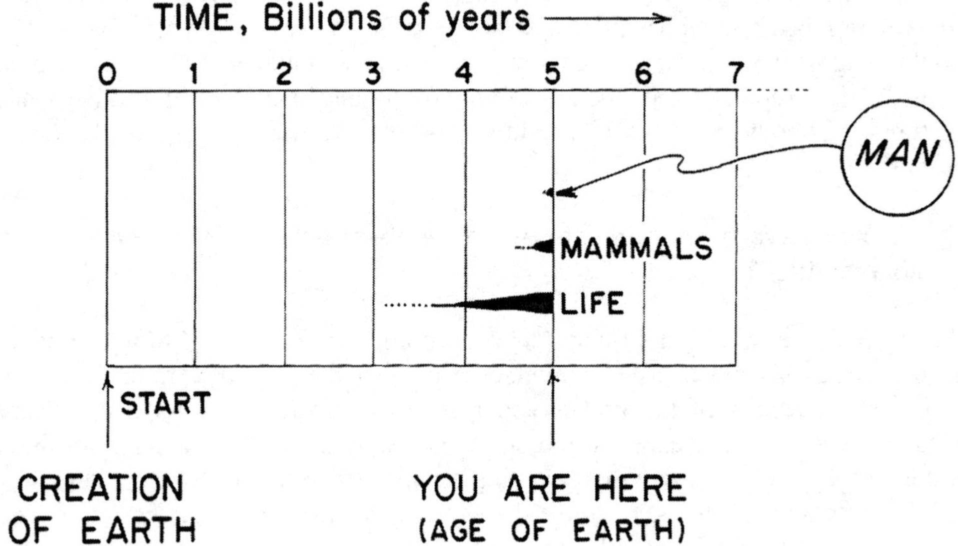

FIGURE 1

The cosmological time scale.

Q.   I see by your petulance that you are referring to religion once more.

A.   As a matter of fact, the role of religion in "man's state of pompousness" is very interesting and paradoxical. It is paradoxical because most religions teach one to be meek and humble, yet the effect of religious dogma is to create the impression that "perfect models" exist. If religion takes on the function of conveniently filling in the gaps in man's understanding of the universe, then to that extent it begins to become counterproductive.

Q.   I see that this chapter is drawing to a close. Where do we go from here?

A.   We will now consider man's role as an individual: his understanding of himself, and his true motivations in dealing with himself and with others. And we shall begin with the premise that man has within him the vestiges of animal behavior, such as one would find in a society of animals living on the subtropical savannah of the Old World.

CHAPTER 4

ALIENATION FROM ONESELF:
A MODEL OF ANIMALISTIC BEHAVIOR

Monkeys reared in total isolation, both from other monkeys and other species including man, find it almost impossible to adapt in later life to any kind of social contact. Placed with sexually active members of their own species, they do not know how to respond. Most of the time they are terrified of making any social contact and sit nervously in a corner.

Desmond Morris (1967)

OUTLINE OF CHAPTER FOUR

ALIENATION FROM ONESELF: IGNORING ONE'S BASIC CHARACTER STRUCTURE

BASIC CHARACTER VECTORS: ANIMAL MODEL

Dominance: Aggression
Submission
Compliance
non-Compliance
Resignation

"PLAYING THE GAME" OF DOMINANCE & SUBMISSION

AGGRESSIVE RAGES

STRESS

SUMMARY OF ANIMAL MODEL

Q.   How are we alienated from ourselves?

A.   We are alienated from ourselves to the extent that we may have only the foggiest idea of the existence of our basic character structures, which are the bases of our unconscious obsessive-compulsive behavior patterns.

Q.   Obsessions and compulsions? Us?

A.   Not only us. Everyone!

Q.   How can this be?

A.   The basic nature of one's obsessive thoughts and compulsive actions is both genetically and environmentally determined. The fact that in any situation involving an interaction with other individuals our actions are dominated, unbeknownst to us, by our basic character structures means that we have, without realizing it, relinquished a good portion of our free will.

Q.   Can we change our character structures?

A.   Since one's character structure has a genetic basis, it has *instinctual origins.* To the degree that one's character structure was mediated by environmental factors after maturity, it may be possible to effect some change, but probably not to any great extent without wreaking havoc to one's general psychic integrity.

Q.   Does this mean that we can effect no changes in our lives?

A.   No. On the contrary. As we learn about the obsessive-compulsive behavior engendered by our own character structures, and about the behavior of other individuals, our relations with others cannot help but change.

Q.   You stated that you were going to begin with an animal model. Why? We are not merely animals.

A.   Let us start with the conceptually simple and then move on to the more complicated. In addition, the animal model will introduce much of the terminology that we will use later in the human model.

Q.   What are the assumptions of the animal model?

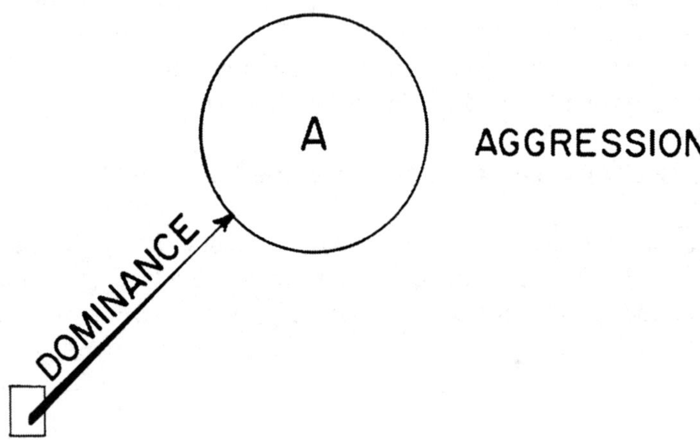

FIGURE 2

Character vector of Dominance based on the genetic trait of aggression.

A.   We assume that our hypothetical animal, imagine for example a monkey-like animal, lives in a competitive society with others of his species. He has the basic quality of attempting to *dominate* his environment by *aggression,* as he must in order to be able to survive in his hostile world. We shall call this behavioral tendency his *basic character vector.* Let us assume that this quality of aggression is determined, or controlled, by the expression of a single gene **a.**

This *state of dominance* of the mature animal may be diagrammed schematically as shown in Figure 2, where the capital letter A denotes the phenotypic behavioral state that is observed. This state, as one can imagine, is typified by the animal's adopting a confident posture, his displaying of teeth, his showing appropriate predatory activities in competition with others of his species and of other species, his being sexually aggressive, and so on.

This mature animal can be reduced to a state of submission in two general sets of circumstances:

- He may be reduced to a *subdued state acutely,* for short periods of time, when he is confronted by circumstances that he cannot dominate, such as a stronger rival or a larger animal of another species. He assumes this subdued state for the sake of his own self-preservation.

- He may be reduced to a *subdued state chronically,* for an indefinite period of time, when he is drawn into a long-term relationship with a group of more powerful individuals, or with a single member of his species whom we will call his companion or mate. This situational state we shall call the *chronic state of subjugation.*

For simplicity we shall denote both the acute and chronic subdued states by the notation A–. Thus, for this mature animal with a *dominant* basic character vector we diagram his states of behavior as shown in Figure 3 (overleaf). Note that we are describing here a basically dominant aggressive animal who adopts situationally determined states of submission according to the environmental circumstances that stress him from time to time, or according to his relations with a subjugating mate or companion. This is indicated schematically by the pathway shown in Figure 3.

Q.   Is the behavior corresponding to the acute subdued state the same as that of the chronic subjugated state?

A.   The animal would be in a submissive state –– there is no question about that. But although the states may resemble each other superficially, the fine structure of the behavioral characteristics would be quite different.

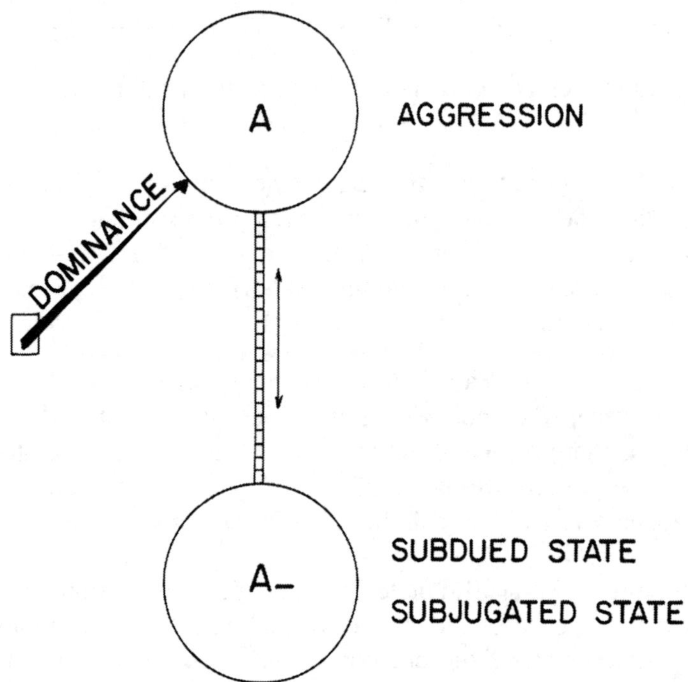

FIGURE 3

Character vector of Dominance showing reversible pathway to a subdued or subjugated state.

Q.   Where do we go from here?

A.   Let us now consider another young animal that is "tamed" during his early developmental period, being reduced to a *state of chronic submission* by what might be fairly rigorous training. This would be a rather traumatic experience for the animal, who as far as his genetic makeup is concerned, is accustomed only to the call of the wild. This state of submission, which we shall for simplicity also denote as A–, is thus the *basic character vector* for this animal.

We also assume that when the animal is returned to his society, he may opportunistically adopt an *energetic state of aggression* A+, which would superficially resemble the state of aggression A of an animal having a basic character vector of dominance.

Therefore, we can diagram the *state of chronic submission,* shown in Figure 4 (overleaf).   Note that the basic character vector here is a phenotypic submissive state A– which is determined both by gene **a** and by the environmental effects of training. Note also the pathway between A– and A+, which allows the animal to assume, opportunistically, the aggressive state A+, for short periods of time according to the situation of the moment.

We now assume that another animal is severely trained, perhaps "flogged" into a nearly completely submissive state A=. He would be reduced to such a state of *compliance* that virtually no amount of enticement would induce him to aggressive behavior. He is basically a helpless animal, always complying with the demands of his environment.

Thus, for simplicity we have defined only two states of submission: *non-compliance* (*n-C*, or A–), as shown in Figure 4, and *compliance* (*C*, or A=). The *n-C* animal, although stunted by his training, is able to act aggressively (but not necessarily "appropriately") for short periods of time. The *C* animal, even under the greatest of stresses that he normally encounters, is unable to do this.

Q.   But should not there exist a continuous distribution, or spectrum, of submissive character states between the *n-C* and *C* states, according to how severe the "training" was?

A.   Yes. But for the purposes of the present discussion, let us divide these individuals into just two groups.

Q.   I suppose that you are now going to introduce another basic character vector.

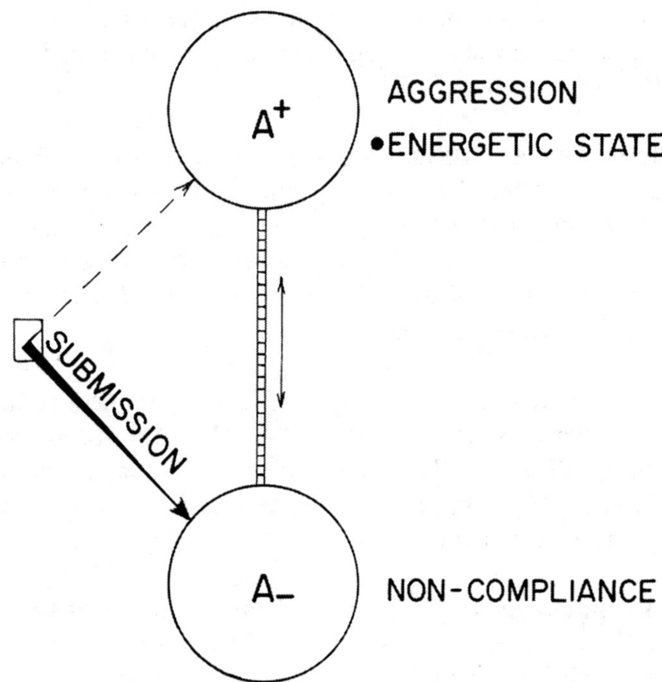

FIGURE 4

Character vector of Submission showing reversible pathway to an energetic state of aggression.

A.   Yes, but first note that the animals whose basic character vectors are dominance and submission are continually interacting with each other so that they must intermittently, and perhaps frequently, change between states A+ and A–. In a single day an animal may have to be in state A+ to his mate but A– to a companion. He may be in state A+ when he goes hunting and then be reduced to A– when he encounters an elephant. And if he is with his mate and encounters a stronger companion, he may be placed in the throes of a severe *intra-psychic conflict,* not knowing whether he should assume state A+ or A–.

Such an animal, without realizing it, is constantly playing a game of dominance and submission. This shall henceforth be referred to as "playing the game." Note that "playing the game" is based on the animalistic quality of aggression.

Q.   What is a result of "playing the game"?

A.   The result can be severe psychic stress imposed on the animal.

Q.   Are there any other general sources of stress?

A.   Yes. Apart from "playing the game," the animal may simply be unable to cope with the constraints or demands of his environment. This may be due either to the fact that his natural abilities are insufficient, or that the demands placed on him by his society (or by an experimenter) are excessive.

Q.   How might such a mature animal respond to stress?

A.   He may go into aggressive rages of attack, which may seem to an outside observer to be illogical, and be diffusely directed. Or they may be directed specifically, and illogically, against his companion or mate.

Q.   How else might this mature animal respond?

A.   The mature animal may respond by eventually "giving up the fight" and resigning himself from most of his active life in the community. He comes to refuse to "play the game" of dominance and submission any more. He becomes detached. We thus denote this state of unaggressiveness or *resignation* by –A, this being our environmentally mediated third *basic character vector.*

Q.   Can the individual be provoked from this state of resignation –A?

A.   Yes. He may be coerced back to his former state A+ by certain stresses, but only temporarily. His state of resignation has become well fixed. This basic character vector of *resignation* may be diagrammed as shown in Figure 5:

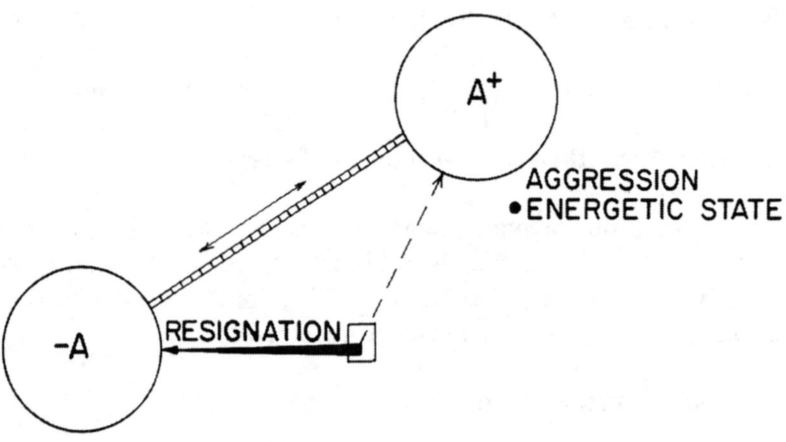

FIGURE 5

Character vector of Resignation showing reversible pathway to an energetic state of aggression.

Q.   So although the detached animal attempts to avoid stress by resignation, he may still be provoked and be subjected to frustration.

A.   Yes. And the detached animal, if sufficiently provoked can, he too, be incited to a rebellious rage of poorly directed aggressive activity. In this model, the individuals of all three character vectors are all subject to stresses as they go about their lives in their competitive society.

Q.   If a submissive individual is further stressed after maturity, can he too adopt the character vector of resignation?

A.   Yes, it is possible. However, in our simplified model we assume that he will more likely be pushed further into submission, becoming more and more compliant.

Q.   Could you kindly summarize all of the above in a single diagram?

A.   The model as presented thus far is illustrated in Figure 6 (overleaf).

Q.   What is the rectangle in the middle of the diagram?

A.   That represents the individual. A solid arrow drawn in the direction of dominance, *or* submission, *or* resignation would indicate that individual's basic character vector.

Q.   How about individuals in the compliant submissive category A=? How do they fare?

A.   In the animal model these placid individuals would not be able to fend for themselves adequately, would barely be able to feed themselves, and their lives would be a passive struggle for survival.

Q.   How about individuals having a severe genetic defect in the expression of gene a?

A.   These individuals would either die *in utero,* or if they survived, they would not be able to fend for themselves and would be put in a "failure to thrive" category.

Q.   Let's say that I observe a group of these animals interacting with each other on the savannah. Would I be able to type each individual according to his basic character vector?

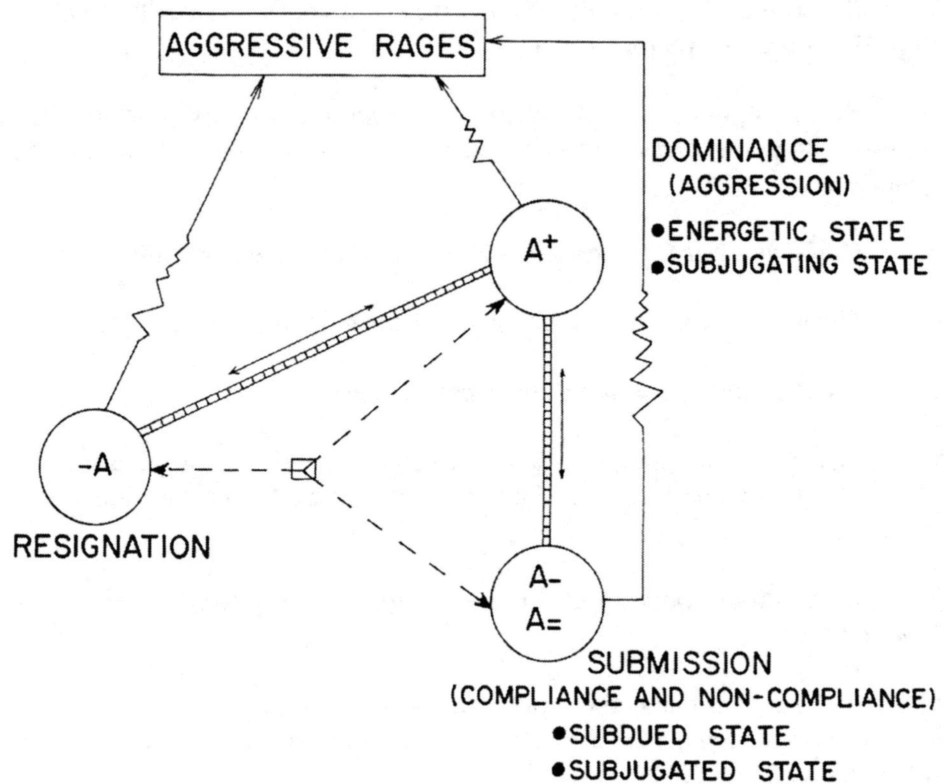

FIGURE 6

Model of animal behavior based on behavioral complex of aggression.

A.   Yes. But only by making careful observations of how each individual reacted in the context of his behavior in the presence of other individuals, his companions and his mate.

Q.   Why might it be difficult to type them?

A.   Animals in the dominant and submissive categories would be "playing the game," and this would have to be carefully observed in order to identify their true vectors. For example, an individual having a submissive basic character vector may be in a dominant subjugating state A+ with respect to an even weaker mate. Similarly, another individual with a dominant basic character vector may be in a subjugated state A– with respect to his mate or an even stronger companion. In addition, resigned individuals –A may resemble individuals who are in a submissive A– state. Finally, all individuals can be incited to aggressive rages, which could be essentially identical in their appearance to the observer and could mask the individuals' basic character vectors.

Q.   What is especially important about individuals who "play the game"?

A.   They may abruptly change from a dominant to submissive state, and vice versa. To an observer, these would appear like dramatic *personality splits.*

Q.   I see that you are beginning to speak in human terms. No doubt you are going to use the animal model as the basis for the human model. Could you summarize the assumptions of the animal model?

A.   The hypothetical animal model has the following characteristics:

- The animals have an innate tendency to attempt to dominate their environment by the mechanism of aggression. This tendency to the state of *dominance* A may be considered to be rigid and compulsive, or instinctual. It comprises the individual's basic character vector.

- The behavioral complex of aggressive behavior is assumed to be dependent on the expression of a single genetic allele **a.**

- Individuals of a second basic character vector (*submission*) are animals who are genetically identical to dominant animals, but have been rendered submissive by unfavorable conditions existing during their phase of nurturing and growth.

- Submissive animals may have been rendered compliant (C) or non-compliant (n-C) to their environment. The n-C animals may be elevated to an energetic dominant state of aggression A+. True C animals are compliant to almost all other individuals and are essentially helpless in their society.

- Dominant A and non-compliant submissive A– animals "play the game" of dominance and submission, and can alternate between the character states A+ and A– according to situational requirements.

- Individuals of a third basic character vector (*resignation*) are detached animals –A who had a dominant character vector A at maturity but then renounced their aggressive behavior. They do not normally "play the game," but they may occasionally be provoked back to the energetic state A+ for short periods of time.

- A dominant animal may be in a dependent chronic state of subjugation with respect to a mate, companion, or group.

- Animals may experience frustration that can stem from two sources. First, the natural abilities of an animal may be insufficient for him to cope with the stresses of the environment. Second, the stress of "playing the game," leading to intermittent and conflicting changes in the character state, may lead to frustration.

- Animals of all basic character vectors can be incited to aggressive rages by frustration, as well as by acute situational conflicts.

- Animals lacking the expression of gene **a** fail to thrive.

Q.   Does the above-described animal behavior actually exist in real life?

A.   Some of the behavioral characteristics are certainly accurate for some non-human primates. However, we have introduced the model of this hypothetical animal mainly for didactic reasons, in order to sharpen a bit our faculties for the observation of human behavior.

Q.   Where do we proceed from here?

A.   We now proceed to . . .

CHAPTER 5

A MODEL OF HUMAN BEHAVIOR

Thus, from the war of nature, from famine and death, the most exalted object which we are capable of conceiving, namely, the production of the higher animals, directly follows. There is grandeur in this view of life... and that, whilst this planet has gone cycling on according to the fixed law of gravity, from so simple a beginning endless forms most beautiful and most wonderful have been, and are being evolved.

Charles Darwin (1858)

OUTLINE OF CHAPTER FIVE

HUMAN QUALITIES
    Intellectualization
    Ambition: narcissism & aggression
    Perfectionism

BEHAVIORAL COMPLEXES
    Narcissism
    Perfectionism
    Aggression

BASIC CHARACTER VECTORS: HUMAN MODEL
    Dominance
    Submission
        Compliance
        non-Compliance
    Resignation

CHARACTER TYPES
    Dominant
        N type
        A type
        NA type
        P type
        NP type
        NPA type
        PA type
    Submissive
        Compliant: NA= & NPA= types
        non-Compliant: NA– & NPA– types
    Resigned: NP–A & N–A types
    Non-aggressive withdrawn
        Juvenile onset: N–, N=, N–P & N=P types
        Maturity onset: –N & –NP types

PERSONALITY SPLITS

ABJECT STATES

EXHILARATED STATES

SUMMARY: DIAGRAMS OF HUMAN MODEL

Q.   How do we now proceed with the human model?

A.   We recognize that man is a being of an order higher than an animal living in the wild. We ask ourselves, "How is he different?" This leads us to the consideration of four qualities:

1. He has a higher order of intelligence and is capable of intellectualization and planning.

2. He has the self-esteem to dominate his environment non-aggressively, with an expectation of a reward from his fellow humans.

3. He has the drive for perfection, by persistence and repetitive action.

4. He retains, however, his animal-like instinctual drive of aggression.

These qualities may be summarized by the terms:

- intellectualization

- self-glorification

- perfectionism

- aggression.

Let us first consider his *capacity to intellectualize.* We recognize that man has intelligence far superior to the lower animals. But we recognize also that intelligence, for example as measured by the intelligence quotient (I.Q.), seems to be a continuously variable function rather than a discretely occurring one. Furthermore, if we look at the intelligence levels of some of the higher mammals — apes and porpoises for example — we are drawn to the conclusion that man's intelligence is different from these animals, but it is by no means a discrete attribute which humans have and animals lack. We prefer to look at man's capacity for intellectualization, problem solving and planning as the background, or foundation, on which other character qualities are overlaid.

We consider next man's tendency to *self-glorification,* or to non-aggressive achievement, where one is drawn to the image of an individual achieving glorious deeds, not only for himself but also for the benefit of others in his community. He stands before them, upright and proud, basking in the limelight of public attention and... *smiling.* In fact, we do indeed have the vision of the Greek god Narcissus sitting on the bank of a pool, smiling at his reflection in self-admiration until finally being transformed into a flower. Thus, we conclude that the human smile, far from being a frivolous gesture, is rooted in man's instinctual desire for recognition before his fellow men and serves to acknowledge their praises and good feeling toward him for a job splendidly well done. Thus, *narcissism,* far from being a dirty word, is the essence of man's non-aggressive striving for achievement and personal recognition in his complex world.

Moving on to man's quest for *perfection,* one notes the necessity for a concerted effort in man's drive for accomplishment, through persistence and repetitive action. Man's tendency to perfection would be an empty drive, however, if he did not have the *ambition* to strive for accomplishment. Thus, persistence and repetitive action would be meaningless without ambition.

Q.   What is ambition?

A.   Ambition is the striving for glory and power, rooted respectively in narcissism and aggression. The striving for perfection mediates the drives for glory and power, which by themselves would be coarsely formulated and diffusely directed. One could say that the striving for perfection gives the strivings for glory and power their redeeming social value.

Q.   What role does aggression play in the human model?

A.   As you may have guessed, it is essentially the same role that we presented in the animal model. However, in the animal model aggressive states were regarded as "natural," and necessary for an animal to survive in his society. In the human model, the instinctual tendency to aggression raises its ugly head, not only in overtly aggressive acts which are easy for all to see and to attempt to deal with, but in pervasive, insidious, far-reaching, and unfortunately very common sadistic behavior.

Q.   Sadistic behavior? Who? I thought sadism referred to the bizarre sexual practices of a few very rare individuals.

A.   Wrong. Sadistic behavior in the broad sense refers to behavior based on aggression in which an individual having achieved dominance over another individual abuses his privilege. At best he neglects his partner, and at worst he inflicts physical or psychic pain on him. Sometimes sadistic behavior can be diffusely directed at many individuals in the outside world. Very often, however, it is done in the context of "playing the game," and the dominated individuals become either hapless or willing victims.

Q.   So humans also "play the game"?

A.   Only those individuals who have a measure of the trait of aggression. The model will show that not all individuals have aggressive tendencies, hence these do not "play the game." Consequently, these non-aggressive individuals are virtually devoid of sadistic behaviorisms.

Q.   How is it that some individuals will behave in a sadistic manner in their relations with others?

A.   It appears that sadistic behavior in man is a consequence of his animal past. It remains in our model a genetically determined vestige in some of our present character structures. The individual, when given the chance to assume a position of dominance or power will use it, as he will simply follow the instinctual obsessive-compulsive undercurrents of his character structure. Since the aggressive behavior underlying his sadistic acts is written in his genetic code, he will deep down feel that it is a natural part of him. He will, in human terms, *derive satisfaction from his acts,* even though on an intellectual plane he may realize that his behavior may be in conflict with the rules of the society in which he lives.

A good portion of our discussion will be devoted to the questions of the real nature of the sadistic behavior of man: how it is camouflaged and rationalized so that it may go unnoticed; who is at risk to be sadistic or to be the victim of sadistic behavior; and how sadistic behavior might be controlled in man's society once its true nature is exposed.

Q.   I think that I am ready for the human model, but I do not understand how you are going to incorporate the complex, rather vague concepts of narcissism, perfectionism, and aggression into a genetic model.

A.   The concepts are complex, true enough, but they are not all that vague.

Let us consider first the one that we understand the best, *aggression.* Aggressive behavior is based on compulsive, fairly stereotyped acts that ensue when individuals encounter each other. The aggressive individual may be thought to carry the unconscious motto: "Every day and in every way, I am always the most powerful."

Without going into any great detail with regard to the posturing, gestures, self-grooming and other mannerisms that occur, we will just mention that aggressive or submissive individuals who "play the game" have a certain level of activity, nervousness, edginess or lability, as they are constantly testing each other for relative strength.

An important part of "playing the game" is the function of eye contact. Very briefly, when two competitive individuals encounter each other, the individual establishing dominance fixes his eyes on the other, who is to become subdued. The latter, after a period of tentativeness, looks away, masking his averted eyes by a subtle pattern of agitated but undirected activity. Once dominance is achieved, the dominant individual feels free to avert his eyes from the deferential submissive individual, while the latter may then look pleadingly to him for recognition. That is, the slave is not entitled to any personal attention from the master.

In encounters having a sexual connotation, during the courting stage, mutual eye contact is the rule. Once dominance is achieved, if the dominant individual no longer is interested in continuing the relationship, then he will avert his eyes from his partner, and the latter will once again seek to reestablish eye contact with him to gain his attention. This is inevitably the state of affairs in a "morbid dependency," which will be discussed in a later chapter. In other situations, eye contact by aggressive types can also be used as a means of intimidating other individuals, in the form of an icy stare or glare.

When encounters between individuals become tense, then the "flight or fight" behavioral sequence may be invoked. This is a complex physiologic response mediated by the sympathetic nervous system and by catecholamine hormones (e.g., adrenaline) released into the circulating blood from the adrenal glands. A hallmark of this response is the *aggressive-vindictive rage,* which takes the form of a directed attack on another individual and carries the motto: "I will get you in retribution for what you did to me!"

Given that the above modes of aggressive behavior are so complex, one wonders whether they could be dependent on just a single gene. This is not obvious, but we cannot ignore the fact that many human individuals simply do not have these qualities of aggressive behavior. They do not "play the game," they do not become involved in the "morbid dependency" of a subjugated state, and they cannot be incited into an aggressive-vindictive rage. Thus, it is reasonable to assume that, in the human, aggressive behavior is a *discrete character trait,* and may in fact be controlled by a single gene or group of genes assorting in tandem. Hence, in the human model, we assume that a single gene **a** controls the expression of the behavioral complex of aggression.

Q.   How about the behavioral complex associated with narcissism?

A.   Once more the clue presented to us is the existence of unaggressive individuals who are exhibitionistic and seemingly ambitious. They have a glorified opinion of themselves with regard to their abilities, flaunt their physical attractiveness and irresistible sexual wiles, and speak incessantly in the first person of their future glory and fame. There is often an instinctual need for self-adornment and for the use of expansive arm gestures.

The motto of narcissism seems to be: "I am the grandest and most beautiful, and I shall be the most successful," and "I deserve honor and recognition for my wonderful qualities."

Whereas the aim of the behavioral complex of aggression is that of domination by power over all rivals, by brute force if necessary, that of narcissism lies in the realm of direction, motivation, creativity, preparing oneself for the work ahead or applying one's intellectual faculties to the planning of future achievements. Consequently, a key difference between the goals of aggression and narcissism becomes evident:

- An *aggressive* individual furnishes to himself his own reward, whether it be power over others, sexual conquest, or victory in overt combat. He may enjoy somewhat the recognition that he receives for his exploits, and he may acknowledge the recognition with a grin or with a laugh. On the other hand, he may scorn any recognition that he receives.

- The *narcissistic* individual, however, seeks future achievement by planning, which requires intellectual capacities rather than brute force. His reward comes not from himself, but from his expectation of recognition and honor when, having accomplished his splendid piece of work, he steps into the limelight before his compatriots, accepts their glorification of him and responds to the recognition with a *radiant smile*. Thus, it may be said that whereas the behavioral complex of aggression consists essentially of "moving against people," that of narcissism consists of "moving toward people."

We have advanced the hypothesis that the warm, radiant human smile is rooted in the narcissistic behavioral complex, representing the acknowledgement of the individual's recognition by others. The narcissistic individual has, in fact, so much invested in the anticipation of recognition in the limelight, that the converse, embarrassment in the limelight, due to any deficiency whether it be large or small, is of particular sensitivity to him. Since *blushing* before others is a physical sign of embarrassment, we must not ignore its possible significance in individuals showing evidence of narcissistic behavior.

Finally, we note the response of the narcissistic individual when he is frustrated from attaining his achievement and the attendant limelight. The response is the *narcissistic rage*, which is an explosive but unaggressive rage in defense of his position and subsequent withdrawal. It carries the motto: "I'm not going to have anything to do with him any more!" or "All of you ungrateful people simply do not recognize my fine qualities, so I am going elsewhere where others will appreciate me!" This is more often than not delivered out of earshot of the offending individuals, hence differs markedly from the directed attack of the aggressive-vindictive rage.

The narcissistic rage sometimes resembles a childish tantrum. In fact, we propose that a classic tantrum of childhood, characterized by hyperactivity, diffusely directed accusations of others, and the random destruction of objects within reach, is rooted in the behavioral complex of narcissism.

To continue the model, then, we assume that the narcissistic behavioral complex is, like that of aggression, controlled by a single gene **n**.

Q.   How about the behavioral complex of *perfectionism?*

A.   The motto of the behavioral complex of perfectionism seems to be: "Do it slowly, keep repeating it again and again until it is just right." However, in the absence of ambition — in the absence of instinctual narcissistic and aggressive drives — the drive of a "pure perfectionist" would have an aimless undirected quality to it.

A clue to the nature of the behavioral complex of perfectionism lies in the repetitive mannerisms of autistic and schizophrenic children. These individuals have a severe disability with regard to establishing social relationships with others. Their behavior pattern is often ritualistic and stereotypic, with an emphasis on repetition. Their speech may be an echolalic, parrot-like repetition of words. Autistic children may endlessly line up toys, or make patterns of objects they have collected. In addition, they may become obsessive in achieving a "perfect" arrangement of objects, for example furniture, and may become extremely distressed if the arrangement is altered.

The model of human behavior presented here does, in fact, predict the occurrence of such autistic-like and schizoid individuals. It is proposed, then, that the behavioral complex of perfectionism is controlled by a single gene **P**. Evidence will be presented later that many autistic, schizoid and schizophrenic individuals are those in whom the components of ambition, i.e., narcissism and aggression, have been suppressed by genetic or environmental factors in childhood or after maturity, thus revealing in the individual a primitive state of aimless perfectionism.

Q.   Why are you using small letters for the genes of narcissism (n) and aggression (a), but a capital letter for the gene of perfectionism (P)?

A.   There is evidence, to be presented later, that the gene **P** is expressed as a Mendelian dominant, while **n** and **a** are recessive.

Q.   Are the behavioral complexes of narcissism and perfectionism really distinctly human qualities?

A.   We do not believe so. Anyone who has seen a chimpanzee smile, or who has observed baboons grooming each other by careful repetitive actions, will allow for the possibility that these traits may have their real origins in the lower animal forms.

Q.   Can we go on now to the human model? Feel free not to have to repeat the qualities of the animal model that are common to the human model.

A.   Let us start with the basic character vector of *dominance*. We assume that there may be concurrent expression of the genes for the behavioral complexes of narcissism N, perfectionism P and aggression A. As in the animal model, dominant individuals having a component of aggression may adopt situationally determined acute or chronic states of submission according to environmental circumstances.

Our model may be schematized with the aid of the Venn diagram shown overleaf in Figure 7. Note that the model predicts the possibility of seven discretely different dominant character phenotypes, namely:

- narcissistic, N
- perfectionist, P
- aggressive, A
- narcissistic-perfectionist, NP
- perfectionistic-aggressive, PA
- narcissistic-aggressive, NA
- narcissistic-perfectionistic-aggressive, NPA.

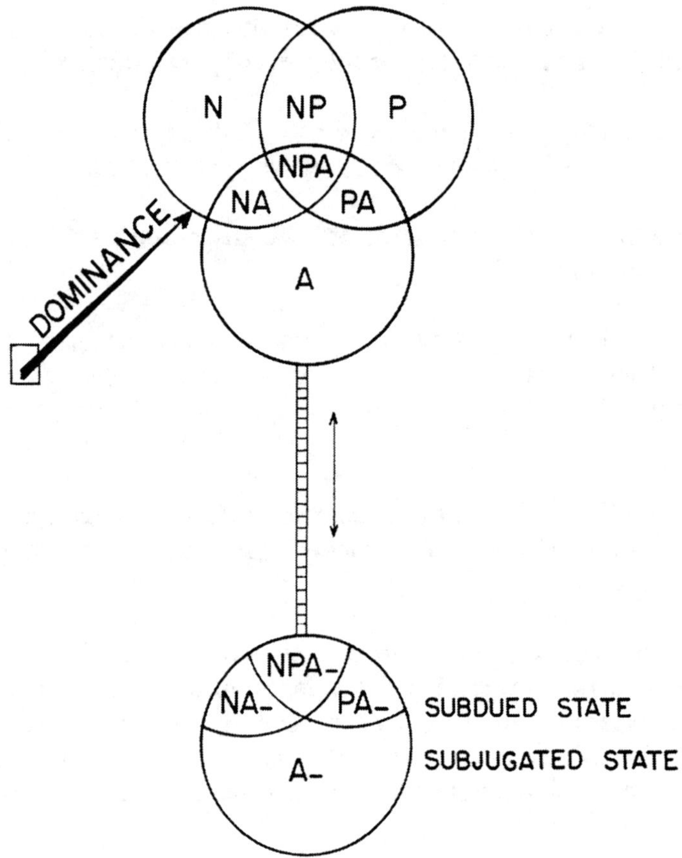

FIGURE 7

Character vector of Dominance based on the genetic traits of narcissism, perfectionism and aggression.

We shall discuss later the behavioral characteristics to be expected for these dominant character types. Note that the types having an aggressive component are susceptible to *personality splits* to an A– subdued or subjugated state.

Here we make a fundamental assumption (to be verified by observation) that a dominant individual's behavior while in state XA– resembles that of an X personage in his natural state. (Here, the symbol X may represent either N or P, or both, or neither.) For example:

- An individual in the NA– state would resemble an N (narcissistic) personage. Another individual in the NPA– state would somewhat resemble an NP (narcissistic-perfectionist) personage.

- An individual in the PA– state would resemble a P (perfectionist) personage. This latter "pure perfectionist state" could appear rather dramatically as an acute personality split to a schizoid state of profound incapacitation and perhaps immobilization. If a PA individual is drawn into a chronic PA– state, where he is subjugated by a stronger partner, he would have a schizoid personality for much of the time and might be severely limited in social relations.

- Finally, an aggressive individual in a prolonged A– state would also tend to be relatively limited in interpersonal relations, perhaps with an acute onset. It is difficult to imagine an aggressive A individual in a chronic A– state of subjugation, since this individual has all of his ambition vested in his aggressive drive, and loss of this drive would threaten his very psychic integrity.

In general, dominant XA types find their personality splits to subdued or subjugated states to be very unpleasant and stressful occurrences. Such stressful states cannot remain stable for any great length of time, and the individual will react in rages, depressions, or conflicts with his mate or companion. This will be considered in more detail in Chapter 7.

Q.   You have considered acute personality splits in aggressive individuals XA who enter a subdued state XA–. How about individuals with a narcissistic component? Do they split to an N– state?

A.   Not in the above sense. Individuals of character types N or NP do not "play the game" and are not usually susceptible to acute suppression of

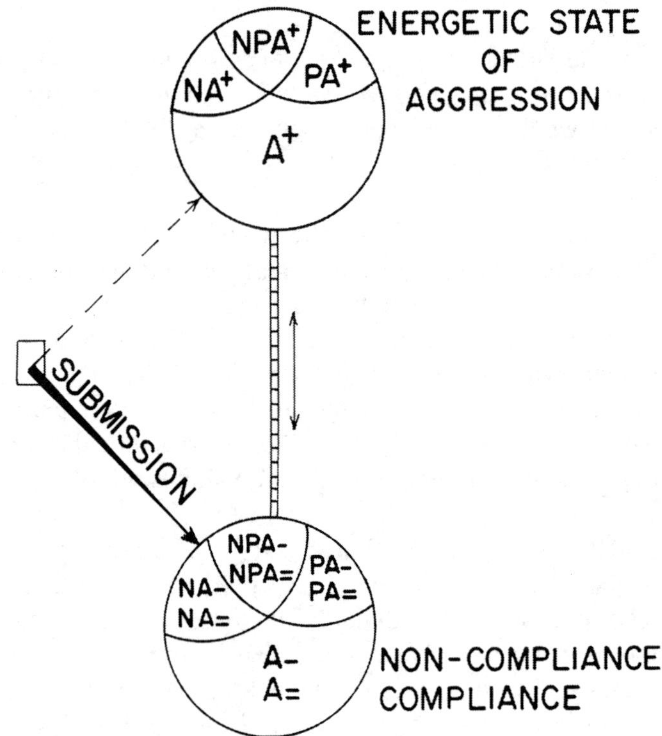

FIGURE 8

Character vector of Submission showing reversible pathway to energetic states of aggression.

narcissistic ambition.   However, in such an individual a state of chronic suppression of narcissistic ambition, lasting days to years, can occur either before or after maturity. We shall refer later to such states, in particular with reference to autistic (schizoid) and schizophrenic states in N and NP individuals.

Q.   What is next?

A.   Let us go on now to the basic character vector of *submission*, which is diagrammed in Figure 8. As in the animal model, we assume, for the time being, that individuals having a basic character vector of submission are those who have been exposed to unfavorable conditions during their developmental period, especially in early childhood. They are submissive in the sense that their aggressive component of ambition has been stunted. We again divide these individuals into two main groups:

- *non-compliance, n-C* , and

- *compliance, C.*

Of these individuals, the most motivated would be the non-compliant NPA– and NA– ones. They would superficially resemble personages of the NP and N character types, respectively. The PA– type would be a perfectionistic schizoid individual, while the A– type would be a non-perfectionistic schizoid individual.

All of the non-compliant individuals are capable of assuming the energetic state of aggression (i.e., XA– $\rightarrow$ XA+). Of special interest here will be the schizoid individual PA– who periodically assumes the energetic state of aggression PA+ to perpetrate sadistic acts. We believe that we have identified here one type of psychopathic thrill killer, the quiet loner who always seems to surprise the neighbors when he is arrested as a psychopathic murderer.

Of the compliant types, the NA= and NPA= types would be severely withdrawn individuals of the "shrinking violet" type. We suggest that the A= and PA= types would usually be classified as mentally ill (see Chapter 11).

In addition, we propose that under unusual conditions of nurturing, there could be produced individuals whose narcissistic behavioral complex was not permitted to flourish, either with or without suppression of the aggressive behavioral complex. This would produce categories of withdrawn individuals (N–, N–A–, N–A=, etc.), borderline autistic individuals (N–P), schizophrenic individuals (e.g., N=P), and even animal-like individuals (e.g., N=A, N=PA).

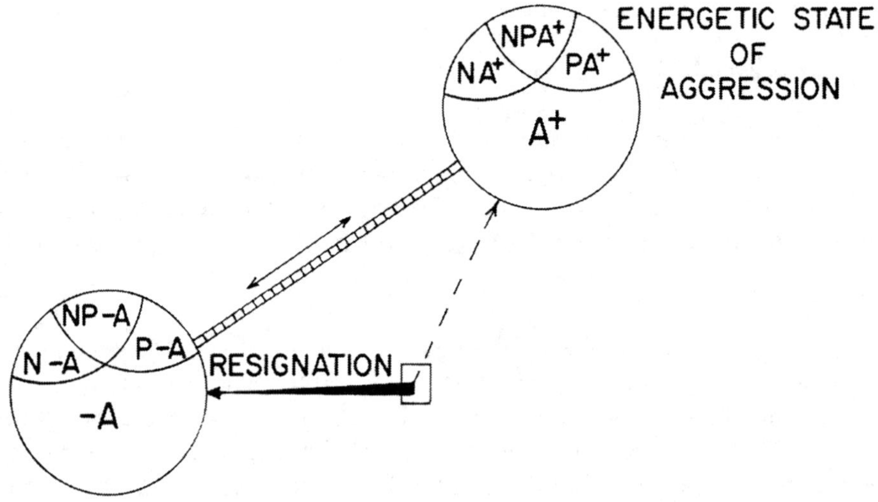

FIGURE 9

Character vector of Resignation showing pathway to energetic states.

Finally, we note that the submissive types are capable of the aggressive-vindictive rage. This may come as a shock to these individuals' friends and acquaintances, since the baseline behavioral state of these individuals is a quiet submissive one.

Q.   I suppose that you are going to introduce the third basic character vector for the human model.

A.   Yes. Again we assume that somewhere along the late developmental period, or at maturity, the stressed dominant individual refuses to "play the game" and gives up the struggle with the demands being made on him. He gives up. He resigns. He does not care very much what happens, or at least he will not fight for his goals. He desires independence. He develops a philosophy of splendid detachment.

As expected, the basic character vector for *resignation* is diagrammed as shown in Figure 9. The healthiest individuals are the NP-A and N-A ones, some of whom were formerly aggressive individuals in childhood or adolescence. They superficially come to resemble the NP and N character types respectively. The -A and P-A individuals are former A and PA types, respectively, who suffer *schizophrenic breaks* in late childhood, adolescence or at maturity (see Chapter 11).

Q.   You mentioned that all of these individuals are under stress and can enter into depressions. Where do depressions fit in?

A.   This will also be discussed in Chapter 11. Reactive depressions, or *situational abject states* as we will call them, may occur in any of the personages. They are the result of hopelessness engendered in the individual, as a result of a failure in his ability to fulfill the special needs of his particular basic character structure. Like rages, the individual may not be aware of the real reasons for his abject state, or why it may seem totally out of proportion to the failure incurred.

And of course, the reasons for his almost continual anxiety, as he goes about in his competitive society, are also usually a total mystery to him.

Q.   Individuals of the submissive character types *C* and *n-C*, and of the resigned types, seem to be quiet individuals who would tend to resemble each other. Is this not true?

A.   Yes, but only very superficially. Individuals of the submissive character types are those who had their aggressive component stunted in early childhood. They, however, *do* "play the game," although usually playing the part of the dependent individual.   Hence, they have a quality of apprehension, mild agitation or overt nervousness when stronger individuals are in the vicinity. The resigned types, having renounced "the game," are usually more placid in such circumstances.

The basic difference between the two character vectors may be summarized by the statement that, in general, a submissive type seeks to place himself "beneath people," while a resigned type seeks to "turn away from people." The undercurrents of intra-psychic activity are very different for the two vectors. The submissive type has a general feeling of inferiority and sometimes deep feelings of shame or guilt. The resigned type has a self-confident feeling of splendid detachment and independence.

Q.   I gather that we are going to look for these discrete characters in real life. Will the descriptions of the character types hold equally for males and females?

A.   Surprisingly enough they hold equally for males and females, as well as for the different races, nationalities, ethnic groups, social classes, and classes of sexual preference.

Q.   On the basis of the model, what can we reasonably expect find in the character structure of each one of the types?

A.   Let us consider them one by one, as outlined in the beginning of this chapter:

DOMINANT TYPES

The Narcissistic Type (N)

He would convey the impression that "I will be the greatest, the most glorious and the most beautiful, and in fact I think that I already am!" He is self-admiring. He has great ambitions with regard to future accomplishments, but does not recognize his limitations. His eyes are set more toward himself and toward the limelight of recognition to be attained in the future than toward the actual tasks with which he must deal. His voice is soothing and clear, and may be directed toward the horizon where all can hear it. His pride is also invested in the beauty of his physical body when in the presence of others. With his not so subtle body postures, mannerisms, and self-congratulatory laughter, he flaunts himself.

Lacking perfectionist qualities, he may have difficulties in organizing himself. He lacks persistence or staying power. Perfectionism by careful repetitive action is alien to him. He has no aggressive-vindictive qualities, does not "play the game" and does not become involved in the dependency of subjugation. He does not split his personality to a subdued state. He cannot be incited into an aggressive-vindictive rage.

He cannot tolerate any serious criticism of his qualities, or interference in his ambition to attain the limelight. When frustrated he may, on the one hand, be incited into the narcissistic rage of defense and withdrawal, or on the other hand, he may undergo a depression to an abject state of hopelessness.

Aristotle in his *Ethics* gives us this character sketch of the narcissistic personality:

> Conceited people, on the other hand, are fools ignorant of themselves, who make themselves conspicuous by being so; they try for positions of honor under an impression of their own abilities, and then, if they get them, prove failures. They rig themselves up in fine clothes and pose for cffcct, and so on; thcy wish what good fortunc thcy have to be known to the world, and talk about themselves, as if that were the road to honor…

> …So boastful people, if their object is reputation, pretend to the qualities that win praise or congratulation, and, if their object is gain, pretend to qualities useful to their neighbors, their own lack of which cannot easily be proved, as, for example, a skill in prophesying or in healing…

Note that we have discovered the personality type of the *proselytizing evangelist* or the *self-anointed prophet*. The important qualities of such a personality are as follows. First, he does not recognize his limitations, and his glorious goals become frank delusions of grandeur. Second, lacking perfectionist qualities, he does not occupy himself with the fine details of the means by which his goals are to be attained. For this he requires disciples and other followers to rally around him for the Cause. Third, he is self-anointed; thus in glorifying his Cause, he is in inverted terms glorifying himself. Finally, lacking aggressive qualities, he is not only sexually unaggressive but also essentially defenseless against attack. If brought before a tribunal he can only repeat in direct or inverted terms that he is the anointed one. And if maltreated and physically abused, he can but turn the other cheek. If led to the gibbet or burned at stake, the anointed one will die the death of a passive martyr, all the while hearing inner voices reassuring him that in the glory of his death lies the glory of his purpose in life.

Aggressive type (A)

He conveys the impression that "I am the strongest. I come first. Period!" He cannot help but be an extrovert and be overtly arrogant. To him arrogance is power, since in arrogating to himself qualities of special importance, nay of omnipotence, he can self-righteously use the brute force of a steamroller to attain his goals. His hallmark is that of seeking the vindictive triumph in overt intimidation, or indeed in any manner that is available to him. He should be able to do anything to anyone, but no one — but no one — can have any claim on him. He is loud. He cannot be missed. He is first.

He "plays the game" but compulsively must dominate all others. He will avoid at all costs his own subjugation in a personal relationship or in a "morbid dependency"; rather he will be the subjugator. He may split his personality for short periods of time to the subdued state, but will be most unhappy there and will emerge fighting. Having so much invested in aggressive dominance, he will scorn anyone and anything reminiscent of weakness or of submissive tenderness. When opposed, he is the master of vindictive retaliation, and will thirst for the thrill of the vindictive triumph, with strict accountability until retribution is obtained. When attaining subjugation over others he may become sadistic, but his sadism is in the realm of brute force: it is out in the open for all to see. When cornered, with his back to the wall, he will fight and may be at his best.

He has little in the way of narcissistic qualities, hence he has less invested in the anticipation of future accomplishment than in the maintenance of a position of power, where no one can have any claim on him. He has little in the way of perfectionist qualities; therefore, his actions are coarse, with a premium on speedy, goal-oriented, self-obtained satisfaction. He is thus openly hedonistic. Perfectionism by quiet, careful repetitive action is alien to him.

When frustrated, he has in his armamentarium the aggressive-vindictive rage, which may be activated with a hair trigger. With repeated defeats he may undergo a depression to an abject state of hopelessness.

Narcissistic-Aggressive type (NA)

He would synergistically combine the mottos, "I am the most glorious" and "I am the most powerful." Thus, lacking the behavioral complex of perfectionism, his life is ruled by the unbridled ambition of the behavioral complexes of narcissism and aggression. We surmise that these two traits, as expressed separately in the N and A personages, would be found together in this individual.

The synergistic drives for both power and glory must produce an extroverted, hyperactive individual. Given the pride invested in "I am beautiful" and "I am powerful," one would expect this individual to be domineering, sometimes with an almost megalomanic drive for ambition, in power, glory and sexual domination. This individual lacks concerted perfectionist qualities, hence his search for power and glory is likely to be superficial, spasmodic and lacking in direction.

When reduced to the subdued state NA− this individual strongly resembles the self-flaunting unaggressive narcissistic personage N. Of course, he "plays the game," and with his hyperactivity and tendency toward "hypersexuality" he would involve himself in many compulsive dependencies, usually as the subjugator but sometimes as the subjugated individual. As is often the case in the dependency of subjugation, he may become overtly sadistic, especially in frustrating and in playing on the emotions of his subjugated companions, of which there may be several at one time. And he too, if opposed, seeks retribution in the self-justified vindictive triumph.

This individual when frustrated can be incited to a narcissistic rage, an aggressive-vindictive rage, or a combined narcissistic-aggressive rage (NA rage). In the latter, the aggressive-vindictive component usually appears first, followed by the narcissistic rage of withdrawal as the individual leaves, slamming the door behind him.

Finally, as in all character types, he is liable, when frustrated to a point of relative hopelessness, to enter an abject state of depression. In such a hyperactive individual, with ambition unchecked by the lack of perfectionism, we are drawn to propose that he is in some cases the *hysterical, cyclothymic,* or *hypomanic-depressive personality* of the psychiatric literature.

Perfectionist type (P)

Here the individual would have the inner drive, "Do it carefully, perfectly, over and over no matter how long it takes," and "Arrange it neatly; arrange it into a nice pattern and keep it like that." However, lacking either narcissistic or aggressive ambition, the individual cannot do much else. As mentioned earlier, we propose that this behavioral pattern characterizes the *autistic child,* who we believe is an N=P type, and the *childhood schizophrenic,* who we believe is usually a PA= type. In addition, the perfectionist phenotype would characterize some mature individuals undergoing schizophrenic breaks with reality (e.g., −NP and P−A types).

Narcissistic-Perfectionist type (NP)

This individual has narcissistic qualities that are mediated by the behavioral complex of perfectionism. Thus, he has narcissistic ambition ("I will be glorious") but also perfectionistic qualities ("Do it well..."). This individual would be well motivated, but a perfectionist plodder. He would be slow. He would do it over and over again. He would chip away at something ever so slowly, so he will not go too fast, or too far all at once. He must be something of a loner. He must be somewhat subdued. He fills up his time with activities, doing, redoing, starting, not quite finishing, polishing, giving him a "workaholic" quality. His time may be filled with ambitious goals, but slow repetitive actions, so that he is always late for appointments. We propose that he may be one of the so-called *obsessive-compulsive* individuals of the psychiatric literature.

He has his eye on the limelight, to be sure, with a vision toward the future. But the self-flaunting aspect of the previously described narcissistic (N) personage is mediated by this individual's necessity to be perfect in all ways. Thus, the conceit of the narcissistic individual is moderated into persistent achievement in quiet modesty. In fact, this individual feels that he must do whatever his friends, family and society demand of him as a perfect person. He simply must try to do to the best of his ability what he feels he *should* do. And he feels that he should do everything. And do it perfectly. Thus, to his friends, to his family and to himself, he is a prisoner of his own sense of duty. He is a quiet achiever but at the same time a prisoner of perfection.

Although the behavioral complexes of narcissism and perfectionism appear to be acting in concert, it is apparent that the demands of modern society can lead to great conflicts in this character type. Narcissism gives him the vision of a glorious future and the possibility of the limelight, while perfectionism is the root of his excruciating, painstakingly slow progress. In addition, no human being can possibly accomplish in the time of twenty-four hours all that he expects of himself, not to mention all the demands that others place on him.

Lacking aggressive qualities, he is defenseless against aggressive types. He does not "play the game," does not split his personality to a subdued state, and cannot be induced into an aggressive-vindictive rage. He is usually not sexually aggressive in a predatory manner. Since his love relations cannot be based on a dominant-submissive relationship, he loves on the basis of his narcissistic and perfectionist qualities. In particular, he loves because he *should* love, and he is tender because he *should* be tender. In fact, he will do anything that a devoted mate should do, simply because his inner nature tells him, "I deserve it to myself and to everyone that I should do everything perfectly."

In essence, his attitude in its compulsive rigidity becomes equivalent to a "deal with life." If he pays attention to all the details of life, if everything is neatly in its place, if he has thought of everything in advance, then nothing should go wrong. And when things do go wrong, whether they are by any stretch of the imagination his fault or not, he blames himself for not having foreseen the difficulty. This may send him to the doldrums of a melancholic abject state for days or months, especially if the cause of the failure actually was some deficiency on his part.

He may be incited to the *perfectionistic-narcissistic rage* when frustrated by others. This begins as a few moments of inwardly directed seething, and then may break out into the narcissistic rage of defense and withdrawal.

Finally, we note that in the NP character structure we expect to find a quiet, unaggressive individual who should be content with perfecting the various aspects of his life with a minimum of conflict with others. However, in a poorly adjusted individual his inner voices may demand of him an incessant repetition of bizarre acts, so that he may attempt to satisfy some poorly-understood need for order or completeness in his life. If these acts come into conflict with the norms of his society, then he may become a compulsive gambler or bank robber, a kleptomanic collector, or even a demented ritualistic murderer. Hence, such an individual, although lacking the aggressive behavioral trait, could certainly be branded as "aggressive" by his society.

The risk of the NP type of succumbing to a psychosis of schizophrenia will be discussed in Chapter 11.

Narcissistic-Perfectionistic-Aggressive type (NPA)

This individual obviously has a concurrent combination of all the behavioral complexes: glory, perfection and power. We would expect the following qualities: narcissistic ambition, perfectionistic attention to detail and sense of duty, and the aggressive need for triumph through power. It is not difficult to surmise that this person would be outgoing, active, and sexually vigorous.

This individual, having aggressive qualities, "plays the game" and splits his personality to a subdued state when dominated. His love is based on his narcissistic qualities, on the dynamics of the dependency of subjugation, as well as on his sense of perfectionist duty.

This character would be the most susceptible to be incited to rages, since both his vanity based on narcissism and his pride based on omnipotence are subject to being wounded. This is the narcissistic-perfectionistic-aggressive super-rage, or NPA rage (see Plate 21). This often begins with some violation, reasonable or not, of the individual's sense of perfection, order or justice. That is, all rules should be followed exactly; everything should be exactly in its place. After a few milliseconds of seething, an explosive rage bursts forth, first having an aggressive-vindictive quality to it and being personally directed. Gradually it becomes directed to the horizon, and may finally end as a narcissistic rage of defense and withdrawal. After the rage subsides, this individual's sense of duty may require him to apologize. We believe, in fact, that this individual is the source of the *explosive personality disorder* of the psychiatric literature, and is described in the writings of Aristotle:

> Quick-tempered people get angry quickly and with the wrong person, or for the wrong reasons, or more than is right. But they soon get over being angry; indeed this is the best point about them. It is because they do not control their anger; they are so quick-tempered that they retort bluntly and then have done.

Finally, we note that here we have found the personality type of the *extroverted autocratic tyrant* of the historical literature. This imperious personality is known on the one hand for his sense of duty to his people, and on the other hand for his towering rages. Indeed, there is no other character type that comes even close to resembling him. Unfortunately, as we shall see, despite his apparent sense of duty, the NPA autocratic tyrant often finds himself with blood on his hands.

Perfectionist-Aggressive type (PA)

Here we have an individual who attempts to streamline both his aggressive and perfectionistic qualities into a cohesive unity. How can he possibly do this? His aggressive behavioral complex tells him to forge ahead, to gain power over all others, to achieve vindictive triumph after triumph, to intimidate, and to be first. However, from his behavioral complex of perfectionism comes the subliminal voice, "Do it carefully, slowly, perfectly." Thus, perfectionism modulates aggression, and here we realize that we have discovered the *passive-aggressive personality* of the psychiatric literature. However, this character is not so much passive-aggressive as perfectionist-aggressive.

In a life situation where he has attained a relative equilibrium, or has achieved success in circumstances where he is not threatened, this type may appear as a quiet or fairly outgoing, well mannered, somewhat stern individual. He may, like all other types, go through life uneventfully, although his relationships with others may be rather tenuous and distant. He may appear as a relatively content, careful, reliable worker who is persistent and pays attention to detail.

This individual, despite his basic passive-aggressive tendencies, may feel a strong need to interact with people, according to the dictates of the gregariousness that modern society demands of him. He may find himself, in fact, in a life situation where he is constantly interacting with others in more or less stressful circumstances. He may then appear in one of two different forms. In one form, he is an individual who sees himself as a perfectionist worker, but whose aspirations are constantly being thwarted by the imperfections or malevolence of others. He, thus, becomes the "chronic complainer" or "chronic criticizer." In a second form, he appears as a fairly gregarious individual with a cutting, sarcastic sense of humor — the "sardonic wit."

If this individual aspires to high ambition, his lack of the narcissistic trait means that his ambition must be vested entirely in his aggressive drive. Since his perfectionist trait does not permit the overt use of force, he must use it quietly, unobtrusively, insidiously, obstructively. To a casual observer it may not be noticeably visible. He becomes a manipulator, quietly accumulating relevant information, pondering over it, collating it, putting it into place, using it for insidious, obstructive attacks on others and using it to batten his defensive perimeter. As he becomes manipulative, he becomes more and more mistrustful of others, including his superiors, his colleagues and his subordinates. And in the general mistrust of others he becomes suspicious, cynical and paranoid. Hence, we believe that any study of the *paranoid* personality type might begin with the study of this character structure.

Continuing with a poorly adjusted individual who has developed a sense of motivation, what emerges is a moody, brooding character who isolates himself within his self-created magic circle of manipulation, suspicion, cynicism and paranoia. And whatever power he attains, and whatever recognition he actually receives from others, do not mask the fact that he is essentially a loner. If circumstances of life allow him to become a political or military strategist, a concentration camp director, a prison guard, a lieutenant in charge of a group of civilians, or a world leader with a nuclear trigger within the reach of his finger, then the road that he paves may lead directly to hell. To the extent that such an individual considers himself above others, he becomes their master and they become his slaves without rights. And since we gradually come to the realization that aggression is the root of sadistic acts in man, in a master-slave situation, we gradually realize that the perfectionist-aggressive individual in a position of power has all the potential for becoming a perfectionist-sadist, whether his acts be directed at a single person, ten million people, or an entire ethnic group.

Here we discover an important facet of sadism in man. Diffusely directed aggressive activities whose aim is, to subjugate people into submission and to cause them physical or psychic pain is rooted not only in the open aggression of an A, NA, or NPA type, but also in the perfectionist aggression of a PA type. Thus, we might ask ourselves if the danger to the world lies not so much in an arrogant steamroller as in a cynical, brooding loner. It should not escape us that here we have identified a personality type of another *tyrannical despot* who has time and time again throughout history caused the deaths of countless innocent people. And we must examine whether he is inherently a self-destructive type.

No less harrowing, on a smaller scale, is a PA individual who has been subjected to unfortunate circumstances during his years of nurturing and growth, and finds himself a suspicious loner at maturity. One of his inner voices is telling him, "Be strong — Be powerful — Achieve triumph over others." Another voice is telling him, "Direct your efforts — Choose carefully — Do it slowly, perfectly, over and over again until it is just right." It does not take much imagination to surmise that here we have identified one type of a *ritualistic criminal,* the quiet individual who chooses his victims carefully and is driven to perpetrate a series of stereotyped criminal acts. This is an individual who cannot, of course, show remorse. He is, in fact, following the inner dictates of his very soul.

The reader will appreciate that the implications of sadistic trends in the human character are of utmost importance. Chapter 8 considers this topic in detail.

The PA character structure, we propose, may also be the source of other "character disorders" that have been described by psychiatrists. First, it is probable that the condition known as *folie à deux* is based on the symbiotic involvement of two individuals having the PA character structure. This condition is a psychotic disorder in which two schizoid persons, usually members of the same family, mutually share similar paranoid delusions. Second, it is probable that many litigious individuals described under the category of *compensation neuroses* are of the PA character type.

On a personal level, the PA type, of course, "plays the game," becomes involved in subjugation dependencies and is subject to incitement from a quiet state to the *perfectionist-aggressive rage.* This begins as a period of outwardly directed seething, and then with the slightest provocation bursts into a directed aggressive-vindictive rage (see Plate 21).

When dominated, this individual enters a PA– state of perfectionistic schizoid behavior, since he lacks any narcissistic component of ambition. He must thus attempt, almost at all costs, to maintain a position of dominance. Finally, if he is dislodged from his position of dominance, and hopelessness sets in, then he is subject to a deep abject state of schizoid depression.

The PA type can be a member of an interesting symbiotic relationship, called "the power behind the throne," and this will be discussed later in Chapter 9.

SUBMISSIVE TYPES

Compliant types (*C*)

This introverted individual carries the mottos, "I am the most unselfish, the most sympathetic and the most loving," or "I will do anything, but anything for you so long as you protect me for the rest of my life."

He has, without realizing it, renounced his aggressive tendencies in early childhood, hence has been reduced to a shadow of his human genetic potential. He may be a self-conscious, painfully shy "shrinking violet" and be easy prey to any aggressive type. He may have a strong feeling, or a vague uncomfortable suspicion, that all was not well during his very early childhood. Somehow, he feels and acts as if, in the deep recesses of his mind, he were ashamed of something that he is, guilty of something that he did wrong or something he should have done right, or was somehow, somewhere deeply humiliated before others.

Being defenseless, and easily frightened, he fears any demands to be made upon him, particularly those forcing him to any position of responsibility. In fact, the word "responsibility," "ambition," "career," or "success" is taboo to him, and may send shivers down his spine. If he finds himself in a hierarchal structure, he wants to stay right where he is. He certainly does not want to move up to a position of greater responsibility, and he will invoke the "Peter Principle" in his defense. That is, he will say that he does not want to exceed the limits of his capabilities, which despite intensive rationalization, must have become painfully obvious to him.

Although his aggressive qualities have been suppressed into profound submission, they are nevertheless latently present. He does "play the game" but instinctively feels himself helpless at the bottom of the "pecking order." In his helplessness, his only salvation in life is to offer to all comers helpfulness, love, sympathy, compassion and self-sacrifice.

In love relationships the "shrinking violet" flourishes into a "clinging vine," and he "falls in love," in the form of the morbid dependency, with almost any strong individual. And if he obtains a commitment for protection, it must be total. He must have everything done for him, while in return he offers little else than the promise of his total abandon to "true love."

In his insatiable desire to achieve safety in the promises of protection and love, he becomes vulnerable to abuse by others, and in fact, does come to feel that he is abused. He, thus, finds himself in the position of offering himself to be abused in order to fulfill the needs of his character structure and to find satisfaction in his life. Hence, it is in this character structure that we find the roots of *masochism,* that is, the finding of satisfaction in life through being abused by others. He may, in fact, be overtly sexually masochistic.

As with all of the character types, frustration of the most serious kind is engendered when the premises of the basic character vector are threatened. For this individual, the worst threat is that of the loss of his protective master, boss or subjugator. This will be defended vigorously in the form of a claim of fidelity from the master. Since he, the slave, has been so faithful and loving, the master must respond in turn. The dynamics of such a "morbid dependency" will be discussed in more detail in Chapter 9.

If frustration mounts to a breaking point, the individual may, finally, incite himself to an *aggressive-vindictive rage.* This will surely surprise his onlookers, who are used to seeing a very quiet, shy individual. It will also surprise and frighten the individual himself, who may not have known that a spirit of aggression lurked deep in the catacombs of his character.

If the hopelessness of the situation comes to the fore, then a deep abject state of depression will ensue. In this individual, the abject state is characterized by *suffering,* and the suffering provides a source of "positive feedback" to the unconscious motivations behind his basic character structure. That is, the suffering becomes a further reinforcing alibi for the individual's not mobilizing himself, and for his continuing demand that he be rescued by the master — or by anyone else — without any positive effort at all on his part.

The above descriptions would apply to individuals of the NPA= and NA= types, who would superficially somewhat resemble the NP and N types, respectively. That is, the NPA= type would tend to be a quiet, meticulous perfectionist worker, while the NA= type would tend to be more labile, more aware of his physical and sexual attributes, and less a perfectionist worker than a task-oriented "doer." The A= and PA= types, as discussed previously, would be a *chronic schizophrenics* (see Chapter 11).

Non-Compliant types (*n-C*)

The character structure of the non-compliant type is similar to that of the compliant type. That is, this individual is typically not a forceful person and may even be a shy, self-conscious introvert. We noted that the *compliant type* views his life from the depths of the "pecking order" and is quite satisfied to stay there, somewhat contentedly awaiting retirement and death, provided that he is well taken care of and protected. However, whereas the *non-compliant type* also views life from the bottom of the "pecking order," he has one eye open toward the top and aspires to achieve power, glory and domination, even if ever so briefly.

He does, in fact, retain his narcissistic behavioral complex, even though his aggressive capacities have been stunted (NPA– and NA– types). He is, therefore, in the constant throes of an intra-psychic conflict. He desires narcissistic glory, but he lacks the aggressive power to support his ambitious drive. In competitive society he may be so unsure of himself, and fraught with stage fright, that at the same time that he craves the limelight, he also fears it. In overtly stressful circumstances, he begins to glance about in search of protection or escape, like the classic coward who dies a thousand deaths.

He can ascend to an energetic XA+ state for only a brief period, and he is quite apprehensive and vulnerable there if he is in competitive surroundings. If he encounters an aggressive type in competition with him while in the energetic state, he becomes agitated, uncomfortable, easily intimidated, and is usually easily defeated. He may find himself behaving like a frightened rabbit. For one so low to have aspired so high, the agony of defeat is nevertheless severely stressful. Not only does he come crashing down to his original XA– state of submission in an abrupt personality split, but he may also enter an abject state of depression because of his failure. With repeated failures, he may become more and more wary of any aggressive ventures, or even of assertive behavior toward others of lower status. He may develop phobias or "panic attacks" with respect to situations in which he has previously felt himself to be humiliated.

Finally, he comes to temper somewhat his desires for glory. He will convince himself that somehow, in this life anyway, he was not meant to accomplish those glorious deeds of his visions. He becomes an opportunist and a dreamer. If opportunity knocks and presents itself to him, with little danger of his being overcome or humiliated in the presence of aggressive types, then he will abruptly split his personality to the energetic state and seize the opportunity by the horns. If he is lucky, he will descend to his baseline submissive state of his own accord, with the victory intact. To this individual "nature abhors a vacuum," and if a vacuum presents itself, he will step into the situation. If thrown into a lifeboat with stronger types, he will be quite content to stay submissive. But if the others are weaker, then he will take command, and command he will. In the land of the blind, the one-eyed man is king!

The love relationships of this character type are, as one may have guessed, often based on the "morbid dependency," and he may adopt the role of subjugator or subjugated, or even both simultaneously in a love triangle. Using the most blunt of terminology, he may assume the role of either a hardened sadistic master, or of a suffering masochistic lover desirous of nothing in life except to lose himself completely in a warm, tender, sentimental love of subjugation to another individual.

As with any individual possessing the trait of aggression, if the individual adopts the dominant role in a relationship of subjugation, or to the extent that he comes to consider himself above any person or group of people, then aggressive-sadistic trends may come to the fore. These may show themselves in several ways:

- First, the aggressive sadistic behavior may be overt, especially if the individual is incited to one of his rare aggressive-vindictive rages. If goaded into a barroom fight, this type can become a wild panther, and he may not stop until no one is left standing and until every bottle in the barroom is broken.

- Second, if the circumstances present themselves, he may become paranoid and quietly manipulative, hence similar in behavior to a PA type.

- Third, he may become a "situational sadist" if he is tied to a relationship with a companion or mate from which, because of life circumstances, he cannot extricate himself.

- Finally, his ventures to the energetic state of aggression, if frequent and short-lived, may become transient states of exhilaration, or thrills. If these thrills are channeled into aggressive behavior, then our model gives us a quiet, introverted individual who becomes a psychopathic thrill killer. If the individual happens to be a PA– or A– type, then we have a psychopathic criminal who has a deeply schizoid baseline personality.

As may be predicted, NA– and NPA– individuals would have behavioral characteristics somewhat similar to N and NP dominant individuals. In addition to being incitable to aggressive-vindictive rages, the NA– and NPA– types would also be capable of narcissistic or combined NA rages.

What finally emerges in the non-compliant submissive character type is a basically introverted, active, even hyperactive, individual who may be capable of assuming practically any psychic state of any of the other character types. He may be meekly submissive or even masochistic. He can be sadistic. He can go into the wildest of rages and be depressed into the deepest of abject states. He can split his personality to aggressive behavior, and back to the submissive state in a flash. He can see the "morbid dependency" from both ends, even simultaneously. He may go into flurries of introspective narcissistic-perfectionist work activity of the NP type, or of hypersexual activity of the NA+ type. He may feel like an NPA+ king, or he may be drawn by circumstances to the psychopathic aggression of a PA+ individual. He seems to have the widest range of psychic states and the widest range possible of emotions. He has the ability to empathize with practically any of the other character types but not necessarily to sympathize with them. For example, he may be able to *feel* what his subjugated partner is feeling, but not necessarily be able to offer his love and sympathy to him.

But life goes on, and little by little, the non-compliant type loses his energy and vitality and becomes compliant or resigned. The sweet verve of youth dissipates itself, little by little, and he comes to await the sweet kiss of death.

RESIGNED TYPES

These types are mature individuals having an aggressive component in their character structure, who have given up the struggle of "playing the game" and have entered a state of detachment. We denote the state of suppressed aggression after maturity by −A. We identify two groups of resigned types: 1) former dominant types having the trait of aggression, and 2) former non-compliant submissive types. Such individuals carry the mottos: "I am self-sufficient; I am independent of everyone and everything," and "I don't need anyone else, thus no one can hurt me."

Whatever the cause for the detachment, whether it was rooted in inheritance, in a single traumatic event such as a disfiguring accident, or in a spectrum of stressful circumstances, the individual has, in essence, taken to the hills. He has felt his psychic equilibrium to be in jeopardy, and in defense he has rationalized a philosophy of resourcefulness and splendid inner independence. He becomes an onlooker of life, or to the extent that he considers himself to be superior to others, he may adopt the attitude of a detached overseer.

NP−A type. Considering first the NP−A type, he develops a personal philosophy of non-involvement with all things both great and small, perhaps a philosophy of equilibrium or communication with nature. His philosophy may be based on achieving peace through religion or through non-involvement with the environment. As he would consider his involvement with the environment a desecration of the natural order, so does he resent any intrusion of the environment into his life. He will resist acceptance of any philosophy or any way of thinking that may lead to irresolvable problems or to unforeseen conflicts. He may deny the evidence of Darwinian evolution, thereby assuming ultimate independence from the world around him in effectively denying that he is a member of the human race (Chapter 3).

He develops his own magic circle of detachment and bitterly resents any unwanted intrusion into it. There is an undercurrent of anxiety with regard to being intruded upon or of being drawn into circumstances that would impose themselves on him. He is constantly scanning the horizon for the approach of events casting their shadows before them, and he is always prepared for escape if the coercion becomes too great. If an escape route is not available, then his anxiety level will rise. He may be literally claustrophobic in constrained situations.

His whole life becomes geared to the maintenance of his detachment. He will seek work in a non-hierarchal structure where the fewest demands are made on him. He will, at all costs, avoid making demands on others. He may become a physician, a taxi driver, a free-lance writer, a nun, a lighthouse keeper or a vagabond. If coerced, he will leave his job abruptly, and the search for another suitable one may take an interminably long time. In sports he is an avid spectator, or if he is active he will row a single scull or be a cross-country jogger.

His personal life is often a mystery to his companions and co-workers. Often no one really knows where he lives or what he does. He may keep an unlisted telephone number and no identifying sign on his door. In a hotel, a "do not disturb" sign appears immediately on his door. He lives and travels alone. And to all observers he appears to live alone and like it.

He is, of course, unaggressive. He is friendly, cordial and good humored. He is reliable, helpful and has a real sense of integrity. He is a person who is well liked by others and is considered to be dependable.

In order to maintain his detachment he must continually be on guard so that friendships do not become overly constraining. Relationships with a sexual connotation become interludes with the understanding that real involvement is not around the corner. The prospect of marriage is frightening unless his prospective mate is able to show the promise of supporting his detached status. His aversion to close friendships and to the expression of giving oneself to another person cannot help but lead to an emotional numbness. He becomes bland and phlegmatic. As he goes through life and is exposed to more and more, he responds by taking less and less. His life has become peaceful, placid with not a conflict in sight, but it has become shallow.

When threatened or goaded, he can be activated to an aggressive state, but like the non-compliant submissive type, he is not comfortable there. He will be deeply disappointed with himself that others were able to penetrate his aura of placidity and goad him into "playing the game" once more.

When coerced to the breaking point, he may, once in a lifetime, erupt in a vociferous rage of rebellion, which carries the motto, "To hell with you all! I am not going to do all these things for you any more! I'm getting out of here!" — which will be seen to be a narcissistic rage in disguise.

When not coerced, this NP–A type is a quiet, resourceful worker who has a strong resemblance to the NP dominant type, "the quiet achiever." However, the resigned type's residual aggressive component is subtly evident by his occasional aggressive language and gestures. Like all types expressing the aggressive gene, he is prone to the undercurrents of sadistic behavior. In the detached individual this may occur in the form of passive obstructionist behavior reminiscent of the PA type, in the form of non-involvement in situations where disaster is imminent, in the morbid interest in natural disasters such as the following of the progress of a hurricane, or finally in simply standing back in detached amusement and watching the faults and foibles of others as they fritter and flounder about, fumble and fail in their frenetic ventures of futile human folly.

N–A type. Turning to the N–A resigned type, according to the model he is lacking in perfectionist qualities. He is less a quiet worker than a narcissistic individual who is well capable of task-oriented accomplishments, but less of directed efforts requiring the planning and execution of many interrelated details. In addition, when activated to the NA+ state, he may resemble the hyperactive, hypersexual "bird of prey" type. This, we propose, would be the *hypersexual resigned type* of the psychiatric literature. In such an individual, his only bridge to involvement with others is on a sexual plane, and of course the affairs can lead to no stable relationship.

Finally, as is true for all character types, the resigned type is vulnerable. If his magic circle comes to be repeatedly penetrated, he may feel the walls of life closing in on him, and he may descend into the profound depression of an abject state.

NON-AGGRESSIVE WITHDRAWN TYPES

Q.   You mentioned previously that the narcissistic component of ambition can also be suppressed in an individual. How does this enter the model?

A.   We noted that of the two components of ambition, aggression is the more labile one. That is, individuals having the trait A easily undergo personality splits as they "play the game" of dominance and submission. The second component, narcissism, is not so labile, and individuals do not ordinarily alter their narcissistic drive to undergo personality splits over short periods of time. In defining the character vectors of submission and resignation, we simplified matters somewhat by considering alterations in the drive of aggression only.

Considering the chronic suppression of narcissistic ambition, again as the result of environmental influences, we focus on the N and NP types. These types do not have the trait of aggression, hence the character vectors of submission and resignation are not relevant to them. However, in analogy to the states of submission and resignation, we consider two categories of individuals in whom the N trait has been suppressed, namely juvenile onset and maturity onset. In both categories these individuals have a moderately to severely stunted component of ambition, hence we shall refer to these individuals as *non-aggressive withdrawn types*.

- In the *juvenile onset category* we obtain types N–, N=, N–P and N=P. The N–P type is a not uncommon meek, nervous, melancholic individual. We believe that he corresponds, in some cases, to the *asthenic personality disorder* of the psychiatric literature. The N=P type is best known as the *autistic child.*

- In the *maturity onset category* we obtain the types –N and –NP. These are dominant individuals of the N and NP types, respectively, who suffer "schizophrenic breaks" at maturity, and we shall discuss them further in Chapter 11.

Q.   Well, that was quite a cast of characters! They all seem to be nutty as fruit cakes! Are you sure that you are not exaggerating?

A.   Not really. But we are talking about individuals in a competitive environment and under stress. If the individual is in a life situation where there is little stress imposed on him, then many of the qualities that we have described would simply not be seen.

Q.   And you actually found all of these characters in real life?

A.   Yes. And more detailed caricatures of the individuals that we found will be presented in the next chapter.

Q.   And you found no "normals"?

A.   No.

Q.   Not a one?

A.   Not a single one…

Q.   Hmmmm… Well, that makes me feel better.

A.   Me too!

Q.   But aren't some of the characters "better" than others?

A.   No, that is a value judgment that we would never pronounce. There are saints and sinners among all of the different character types. The model is simply telling us in more direct terms what we already knew: that all of us, because of our basic genetic structures and because of our own particular childhood experiences during the attainment of adulthood, are capable of many different useful and pleasurable activities during our lives. However, we are all very vulnerable to the foibles inherent in our character structures. It is evident that there are some activities in life for which we are all, according to the limitations of our particular character structures, definitely not suited.

Q.   Could you summarize this chapter with a few concise diagrams of the human behavioral model?

A.   They are shown overleaf in Figures 10 and 11.

Q.   Where are the *non-aggressive withdrawn types* in Figure 10?

A.   Since N and NP types do not normally undergo acute personality splits, the withdrawn types (N–, N=, –N, N–P, N=P and –NP) are not on this diagram.

For a society of such *non-aggressive types* the comparable diagram is given in Figure 11. Note that the pathway indicates that at maturity a dominant individual can undergo a "schizophrenic break" to a withdrawn state, and that this process may be a reversible one.

Q.   You stated earlier that a non-compliant submissive individual can become compliant. Where is the pathway showing this in Figure 10?

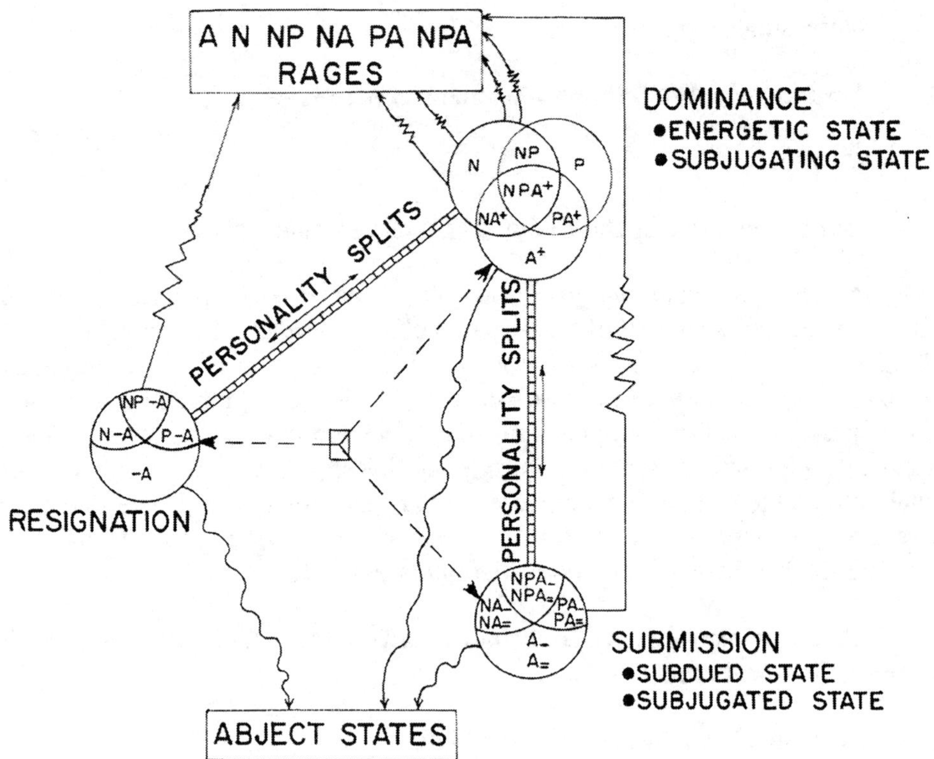

FIGURE 10

Human model based on behavioral complexes of narcissism, perfectionism and aggression. Character types having a measure of trait A can undergo *personality splits*.

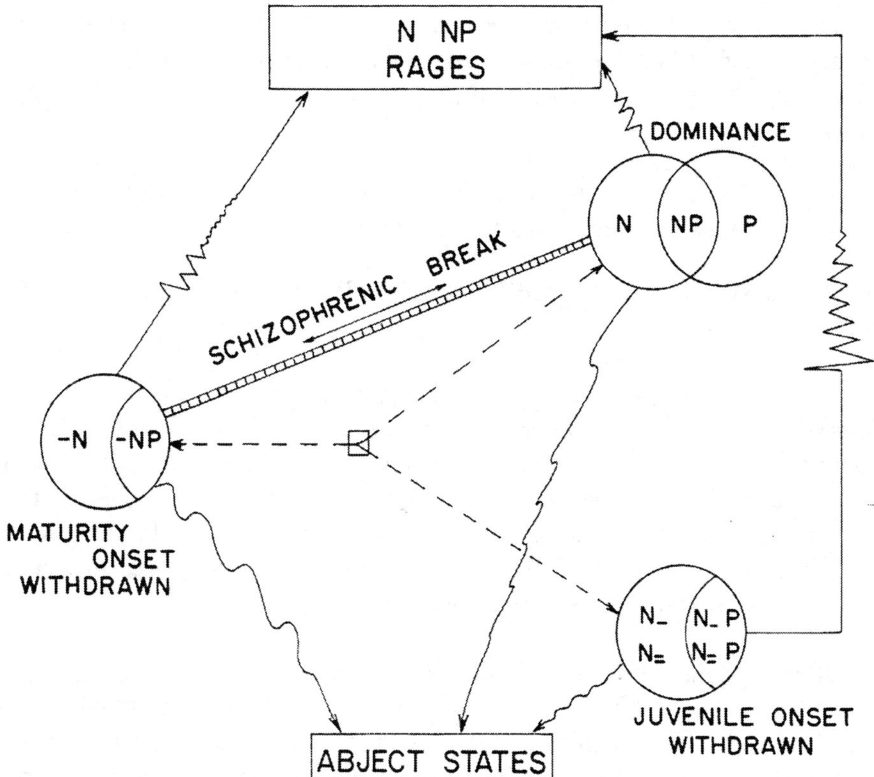

FIGURE 11

Human model based on behavioral complexes of narcissism and perfectionism. Loss of expression of trait N after maturity is shown by the *schizophrenic break* to a withdrawn state,

A.   For simplicity the submissive types, *n-C* and *C*, have been lumped together in the figure, so no pathway is shown. The character vector of submission is determined before maturity, but we should keep in mind that a non-compliant type can nevertheless become compliant after maturity. Such a personality change could, in fact, be a dramatic one. This will be discussed later in the context of one class of schizophrenia having an onset after maturity.

In like manner, referring to Figure 11, we point out that the character vector associated with the N– or N= state is also determined before maturity. Nevertheless, an N– or N–P individual can decompensate into a more deeply withdrawn N= or N=P state after maturity, again in the context of schizophrenia having an onset after maturity.

More detailed diagrams of the various classes of schizophrenic withdrawal will be presented in Chapter 11, on pp. 276-277.

Q.   The concept of "personality splits" is interesting. Could you give an example?

A.   Consider a businessman employed in a hierarchal structure. He is an aggressive individual, rules over his subordinates with a vigorous hand, and normally has poise and confidence. If, however, he receives a surprise visit from the president of the firm, he becomes submissive, and in his subdued state will appear nervous and ill at ease. To an onlooker, he will appear like a completely different individual. And he will be in a profound state of stress, especially if some of his subordinates are also present.

Another example is the *n-C* submissive individual who is vacationing with a group of tourists. He finds that he is by far the dominant individual. He will abruptly split his personality to a narcissistic-aggressive energetic state and become the tour guide. He, too, will be in some state of stress, but the subjective feeling will be one of well being, rather than dejection. Again, to an outsider, he will appear like a very different individual.

Q.   Is there anything else that you want to add to the model?

A.   Yes. To complete our model, all individuals can assume a non-aggressive *situational exhilarated state,* when the individual's life situation is beautifully matching the requirements of his basic character vector. These exhilarated states are seen to be the converse, or mirror image, of the abject states, which occur when the life situation seems hopeless in relation to the requirements of the basic character vector.

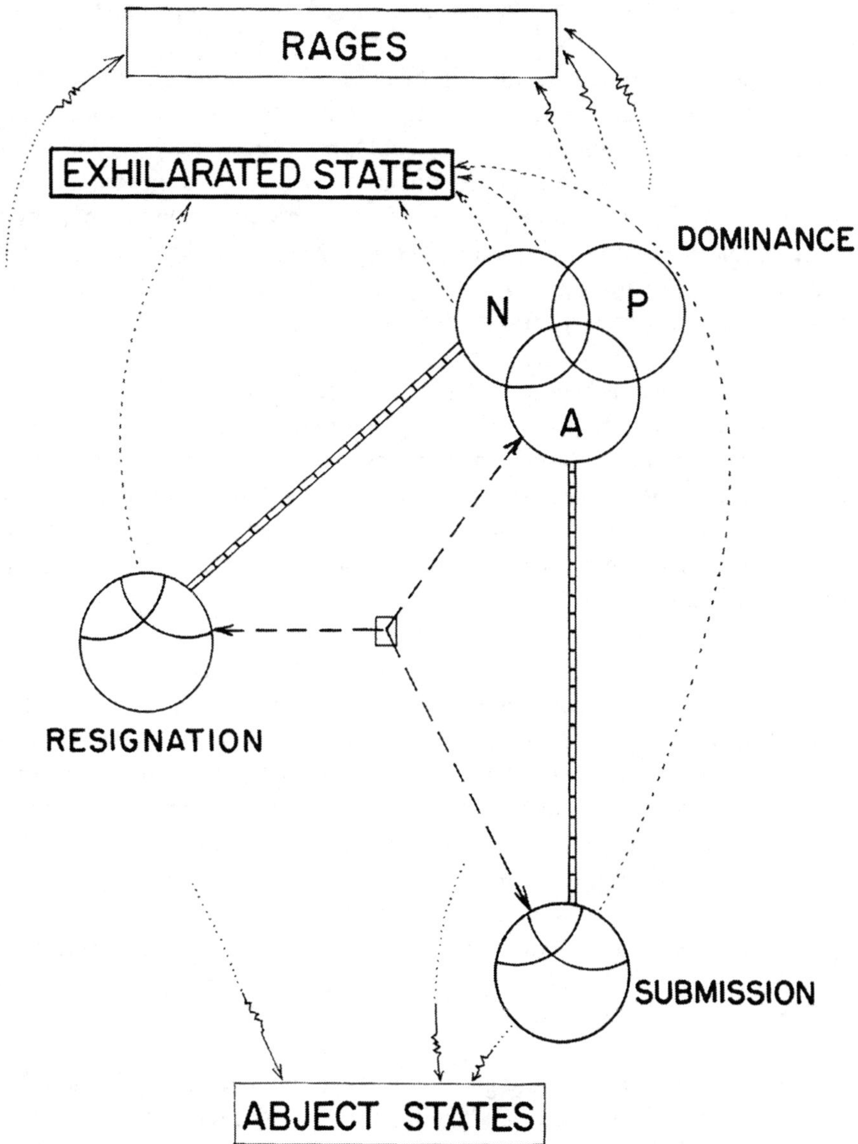

FIGURE 12

Model of human behavior including exhilarated states.

The *exhilarated states* were not indicated in Figures 10 and 11. If we include the exhilarated states, then the schematic diagram of our model of human behavioral states evolves into Figure 12, shown on the previous page.

As shown in Figure 12, any individual may assume an exhilarated state, just as any individual can be depressed into an abject state. An *exhilarated state* is characterized by euphoria and laughter, which may seem curiously inappropriate. For example, in dominant individuals an exhilarated state is characterized by an accentuated drive of ambition and relative hyperactivity. Conversely, an *abject state* is characterized by dejection and weeping, which may also be apparently inappropriate. For dominant individuals the abject state is characterized by a depressed drive of ambition and relative lethargy.

Since the baseline state of the narcissistic-aggressive NA personage is a hyperactive one, and sometimes a frankly hypomanic one, we might consider the full-blown *manic state,* as described in the psychiatric literature, to be simply this individual's exhilarated state. However, there is evidence that this is not so. In fact, the manic-depressive disorders are, with little doubt, distinct "affect disorders" and have their own distinct mechanisms of inheritance, as will be discussed in Chapter 11.

Finally, on a point of terminology, we have reserved the term *personality splits* in reference to dominant and submissive individuals having an aggressive component XA, who "play the game" between XA+ and XA– states, and in reference to resigned individuals who are occasionally provoked from the X–A state to the XA+ state. The rages, exhilarated states, abject states, and depressions of narcissistic ambition are not "personality splits" but are rather considered to be separate classes of behavioral states.

Q.   Why is it that these personages have not been discovered until now?

A.   There are several reasons:

First, we have all been taught that human character type, like intelligence, follows a continuous distribution, something akin to the infinitely variable colors of an iridescent rainbow. Hence, little effort has been directed toward deducing the "primary colors" of this rainbow, even though they have, literally, been staring us in the face since time immemorial. Second, like the animals in the wild, human individuals may camouflage their basic character vectors, especially the *n-C* submissive type and XA aggressive types who "play the game" and split their personalities. Third, the NP, PA and resigned types may confuse the picture, as these often-quiet individuals may resemble submissive types. Finally, there may have been religious objections to the concept that our basic character structures, unknown to us, have been stamped from only a few basic molds.

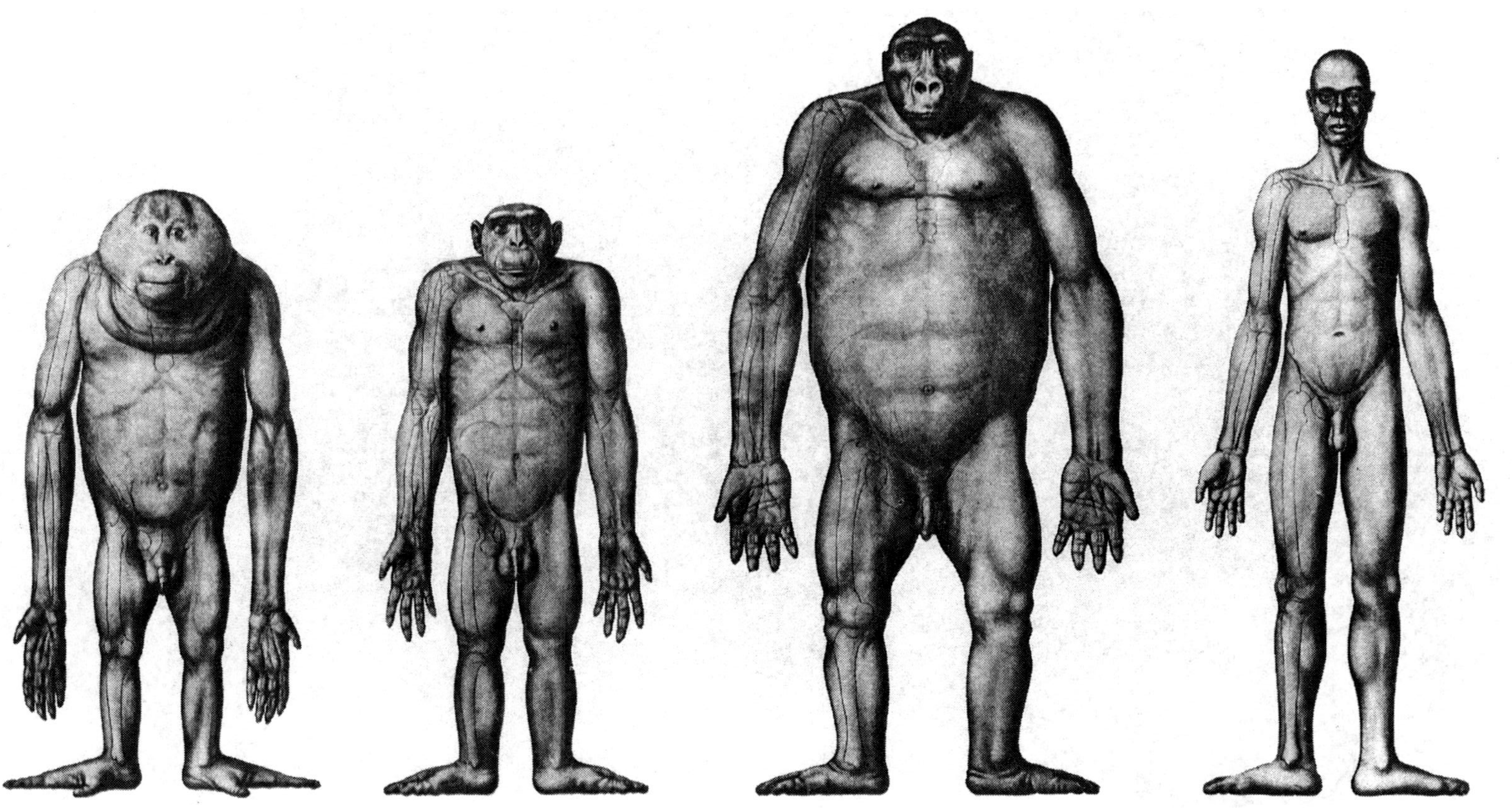

PLATE 1. A gallery of primates, shown to scale in an upright position with the hair omitted. From left to right: orangutan, chimpanzee, gorilla and man. [*A.H. Schultz*]

PLATE 2. The aloof, melancholic orangutan, whose name means "person of the jungle" in the Malay language. The author hypothesizes that orangutans are primarily NP types. [*American Museum of Natural History*]

PLATE 3. The piercing glare of the baboon, who plays an active "game" of dominance and submission. It is proposed that baboons are primarily PA types. [*T.W. Ransom & the National Geographic Magazine*]

PLATE 4. A chimpanzee displays a coy narcissistic smile of recognition.
[*N.Y. Zoological Society*]

PLATE 5. The gingival narcissistic smile in an albino gorilla. It is proposed that gorillas are primarily NP types. [*D. Hosking*]

PLATE 6. Chimpanzees "fishing" for termites, exhibiting attributes of perfectionism. It is proposed that chimpanzees are primarily NPA types.
[*M. LaFarge & the Princeton University Press*]

PLATE 7. In the limelight: a chimpanzee exhibits a gingival narcissistic smile. [*G. Lill & Hallmark Cards*]

PLATE 8. A lemur exhibits an arm gesture of recognition. Lemur-like mammals were among the earliest primates. [*The Zoological Society of San Diego*]

PLATE 9. A sea lion strikes a narcissistic pose of recognition. [*R. Harcztark*]

PLATE 10. Adornment and gingival narcissistic smiles in South American aborigines. [*Jesco, Brasilia & Macmillan Co.*]

PLATE 11. The gingival narcissistic smile in a twelve-year old youth. [*Collection of author*]

PLATE 12. Queen Elizabeth and Mrs. Kennedy appear at Buckingham Palace. At the highest levels of society the narcissistic smile provides a bond between leaders and the populace. [*UPI*]

PLATE 13. A two-year old child spontaneously assumes a narcissistic pose. [*Collection of author*]

PLATE 14. Prince Andrew *à la rose*. A narcissistic gesture of adornment. [*AP*]

PLATE 15. Somerset Maugham in a "narcissistic arms" pose.
[*The Rank Organisation*]

PLATE 16. From Bellini's *St. Francis in Ecstasy,* showing "narcissistic arms" pose. [*Frick Collection*]

PLATE 17. Leonard Bernstein on the podium. According to the model musical expression finds its basis in the traits A or N, or both. [*E. Hausner, N.Y. Times*]

PLATE 18. The behavioral complex of narcissism at work: dance, adornment, self-esteem and a smile of recognition. [*Culver Pictures*]

PLATE 19. Selket, guardian of shrine within Tutankhamun's tomb, showing "narcissistic arms" pose. [*F.J. Maroon*]

PLATE 20. Alfred Hitchcock in "narcissistic arms" pose. A sanguine N type, he combined qualities of arrogance and diffidence. [*P. Halsman, LIFE*]

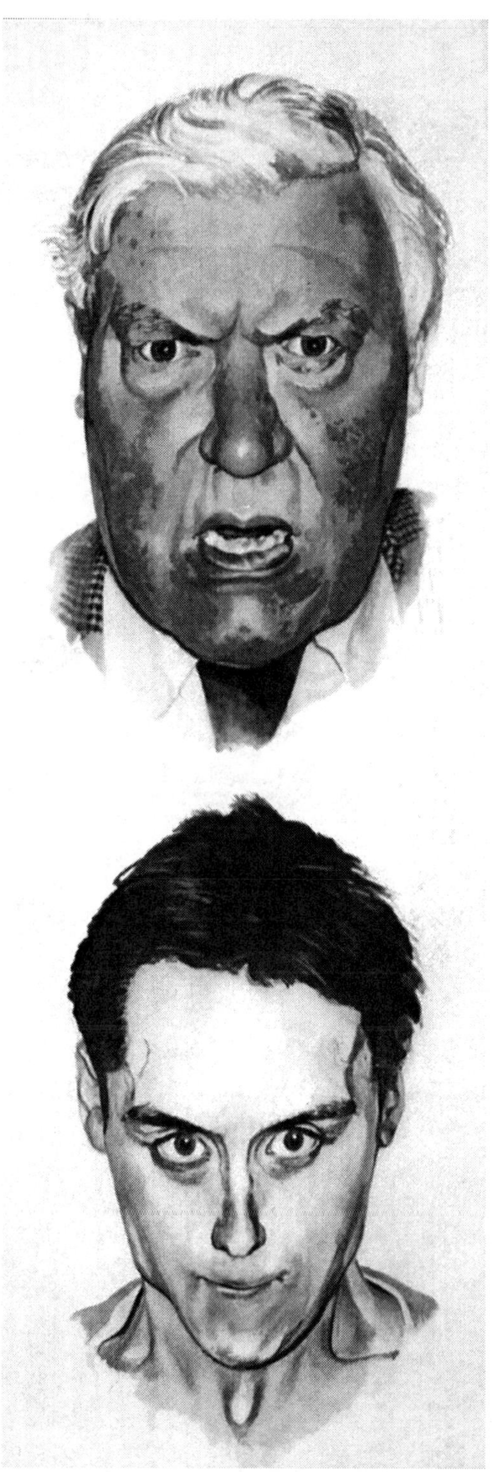

PLATE 21. Faces in Rage. *Bottom:* Ashen-faced incipient aggressive rage in the physiognomy of a non-sanguine PA type. *Top:* Red-faced explosive narcissistic rage in the physiognomy of a sanguine N (or NPA) type.
[*Anthony Moore & Elsevier Ltd.*]

PLATE 22
Non-narcissistic grins and laughs.

CHAPTER 6

CHARACTER CARICATURES

One reads that no one exactly resembles anyone else, and that every man is unique, and in a way this is true, but it is a truth easy to exaggerate: in practice men are very much alike. They are divided into comparatively few types.

Somerset Maugham (1938)

We must consider at least three subdivisions of the "expansive type": the narcissistic, the perfectionistic and the arrogant-vindictive type.

Karen Horney (1950)

OUTLINE OF CHAPTER SIX

INVERSIONS OF CHARACTER TRAITS

CHARACTER CARICATURES

Dominant
    N type
    P type
    A type
    NP type
    PA type
    NA type
    NPA type
Submissive
Compliant: NA= & NPA= types
    non-Compliant: NA– & NPA– types
Resigned
Non-aggressive withdrawn
    N– type
    N–P type

PSEUDOAGGRESSION & PSEUDONARCISSISM

LACK OF AMBITION

IMPRINTING

SUMMARY: LIST OF CHARACTER TYPES

ATTRIBUTES OF CHARACTER TYPES
    Complexions
    Smiles
    Appearance in photographs
    Gestures
        Aggressive
        Narcissistic: "narcissistic arms" pose
    Handwriting

Q. I am interested to hear what you found. Is there anything that I should know before you describe the characters in greater detail, one by one?

A.   We will present what we believe to be the essential aspects of the character structures as we have observed them. However, they will emerge as *caricatures* stripped of their real life situations; the latter would, of course, add many positive aspects to their apparently grotesque lives. Some of the descriptions, in their adjectives, adverbs and metaphors, will seem to be of a pejorative nature, and seem to be casting value judgments, but this is not true. Each individual's value to himself and to society is determined not only by his character structure but by his real life situation. How he adapts may depend on sheer luck — there is no question about that — but also on his free will to the extent that he has gained it, and on his use of the more subtle qualities of the human spirit that are completely beyond the scope of our model of human behavior.

Q. Anything else?

A.   Yes. The psychodynamics underlying each separate character structure are complicated. We will note that each personage, as he goes about his complex life, is at best only very vaguely aware of the instinctual forces underlying his character structure. Often he is totally unaware of them. He manages to maintain an integrated, cohesive sense of unity, but just barely, in the presence of a pervasive overlay of continual anxiety. And in the defense of his psychic unity he is profoundly selfish.

Q. All of these characters are selfish?

A.   Yes. They are selfish and arrogant. And by arrogant we mean in the literal sense of arrogating to themselves special qualities that are required to maintain a psychic unity in the face of their instinctual drives, which themselves may sometimes be conflicting. Each individual maintains within himself a complex spectrum of rationalizations and alibis, so that his conscious psyche emerges like a rose, apparently flawless and intact.

Q. Anything else?

A.   Even in his highest anxiety or lowest abject state the individual cannot, or will not, recognize the true reason for his frustration. The reason, of course, is that he is, by his life situation, attempting to violate the tenets of his basic character structure.

He may, however, from time to time, seize on one fact or another in his life that seems to be illogical and compulsive. Or, perhaps, others may point this out to him (It is always easier to see the illogic in the lives of others!). He may, then, consciously or unconsciously bend over backwards to try to change a few of the most superficial manifestations of his rigid character structure. These behavioral adaptations we call *inversions* and we must be prepared to note their presence in various real life characters, as "the exception that proves the rule."

For example, a male N type, wary of the social implications of adorning himself too flamboyantly, may in fact be rather modest in his dress. An NP character, reacting to the jokes of his colleagues or to his own self-criticism regarding his neatness, may deliberately keep a disheveled desk. An NPA type, wary of his explosive rages, may try to keep a tight lid on his emotions, and the rages may in fact only rarely emerge. Another NPA type will exhibit an inverted modesty, praising others in long public speeches when deep in his own mind he is praising himself. Or he may accost a colleague saying, "Let me show you this work that I'm doing. I don't think that it merits the Nobel Prize, but..." A perfectionist-aggressive PA type may realize his general brooding nature and his poor relations with people, and will choose one or two family members or friends for lavish attention, or he may choose to devote himself to a cause, and this fact will be flaunted for all to see. And so on.

In addition, the individual is subject to overt pressures of his peers and of his society to conform to accepted standards. Thus, he may be subject to act in a manner that may seem "out of character." For example, an NP type may tease others, use vulgar language or speak in a loud voice if his real life situation demands this of him.

Q.   Are you not, then, saying that inversions allow an individual to assume any modality of behavior whatever? Do not inversions allow the individual to become the well-rounded personality of his dreams, the *homo universalis* of yore?

A.   No. Quite the opposite. It is true that the individual would like to believe that he is capable of any mode of behavior that his heart desires. But this is, in fact, nothing more than a fantastic dream.

First of all, the individual's behavior is limited by the *one immutable principle* to which our model directs our attention: individuals lacking the A trait cannot exhibit the aggressive-vindictive rage, and individuals lacking the N trait cannot exhibit the narcissistic rage. This one fact imposes enormous constraints on the behavioral modalities of any individual. And as we come to realize that all behavioral pathways eventually lead to the rages, we come to avow that each individual treads on a limited number of narrow paths in his life.

Secondly, if we examine the various inversions of basic character traits that individuals utilize, we can easily uncover their superficial nature. It is, in fact, their superficiality that draws us to say that a particular inversion appears to be "out of character" for the individual, or is the "exception that proves the rule."

Finally, if an individual in a last ditch effort to preserve his psychic integrity undergoes a profound inversion of his basic character structure, then he may be trading his birthright for a mess of potage. If he adopts the modalities of *pseudoaggression* or *pseudonarcissism* (discussed later), he may be trading one behavioral modality for another one that is just as uncompromising and coercing. He may find himself constantly bucking the norms of his society, and he may find that his relations with others range from the tenuous to the nonexistent.

To summarize, the individual's use of inversions is simply an example of a more general principle:

> *The individual will seize upon any life situation that he can possibly use* as *a mold for the requirements of his own particular character structure.*

And by conforming to certain aspects of the pressures exerted by his society, and by using selected inversions of his basic character structure he attempts to convince himself, and others, that he is an integrated human being.

Q.   And is he not?

A.   No. Western society, which man has constructed by himself and for himself, has far outstripped his ability to deal with himself, as an individual interacting with others of his species and with his environment. Furthermore, he does not have the slightest idea of what is going on when he enters into relationships involving love and sex, and the divorce rates in Western countries are not so moot evidence of this fact.

Q.   Why do you keep referring to "Western" society?

A.   We must examine later the question as to whether the distribution of character types is different in various cultures. And from this information will come ideas and ideals with regard to man's future on this planet.

Q.   May we have the character sketches now?

A.   Here they are, as outlined in the beginning of this chapter:

DOMINANT TYPES

The Narcissistic Type (N)

He is the epitome of unbridled narcissism. Hence, his complexion is often sanguine and his smile sublime. If he is at all handsome or beautiful, or even if he is not, he will adorn himself in finery. The female, especially, will not be able to resist wearing colorful clothes or painting her face in a manner that may startle onlookers. If the opportunity arises to flaunt his body in the nude, or semi-nude, he will not be able to resist it. Thus, he displays immense vanity with regard to personal appearance, for example in matters of cleanliness and hair style, but there his punctiliousness stops. In other matters, we shall see, he is flamboyant and extravagant, but as he rides his tiger through life he never seems to achieve any true sense of order or stability.

In placid circumstances this individual is a friendly, unaggressive, perennially optimistic, often charming individual who becomes radiant like a sunflower when flattered. Rather than be embarrassed by adulation, he will revel in it. Whatever his station in life, he will not be able to resist the temptation to mount the podium if the opportunity presents itself. If he has even mediocre talent, he will be a compulsive jokester, a mimic, an amateur singer, or will play some kind of musical instrument. If he has real talent or physical beauty, he will be inexorably drawn to a career in the field of fashion modeling, in the performing arts or in show business. Alternatively he may be drawn to any one of the professions where public oratory is possible. He may thus be a politician, a social activist or an evangelist. As an engineer, a businessman or a physician his genuinely friendly manner may mystify those with whom he comes in contact. They may think his extroverted affectations to be somewhat strange, and may consider him to be either incredibly naive, incredibly conceited or simply groutheaded.

If he inverts his mannerisms of self-importance, he shows no trace of conceit. He presents himself as a dedicated individual with a soothing, self-assured manner. The male may have a starry-eyed appearance and may be accused of being elfish, effeminate or "flaky." The female may be described as "angelic" or "dreamy." If his voice is loud, which it often is in the male, it is a voice of hollowness rather than forcefulness, as if it were a clarion call being delivered into the void around him

Although he is not predatory in a sexual manner, he is extremely vulnerable to flattery and may seek adulation. Hence, he is prone to attracting sycophantic parasites and may even develop intimate physical relations with adulators of the same sex. He may present himself as a sexually promiscuous person but in a naive and ingenuous manner. More than sexual pleasure, he needs the constant reassurance of the opposite sex that he is indeed the wonderful person of his dreams. To others, therefore, he appears to have the amorality of innocence, rather than of vice.

As an unaggressive individual, the N type may, at first blush, be confused with the NP type. However, the NP type tends to be poker-faced or melancholic, and tends to work by himself in a perfectionistic manner. The N individual, on the contrary, tends to be buoyant, self-assured, jocular, expansive or even charismatic, and tends to seek people out in order to enlist their aid in his projects. Thus, the NP individual will work quietly to solve a particular problem, while the N individual will organize a conference so that he may lead a discussion on how his ideas may be implemented.

In competitive society, it may be difficult to avoid him, for although he is not aggressive he is obtrusive. He may precede his visit with a letter, definitely written in the first person, describing his extraordinary talents, the fine things that he has done in the past, despite all odds against him, and the grand things that he hopes to accomplish, nay *will* accomplish. He stands proud, tends to flaunt his body, and projects his resonant voice not only to his partner in conversation, but also to anyone within earshot, or even beyond. His objectives may be grandiose and far beyond the limits of reasonableness in relation to his present status in life.

Any new activity or new person arriving on the scene arouses his great interest, and he will not be able to resist imposing himself on others, ostensibly to see if he can help. And he wishes to help because he believes that he has the qualities and talents to help. On close examination, though, it is he who desires assistance, and his favors done to others reflect this ulterior motive, especially with regard to long-range assistance in his career. And in the background, deep in his unconscious mind, there lives a pervasive suspicion that perhaps all those glorious deeds will never come to fruition. Hence, he requires constant reassurance from others of his worth to them.

What emerges in this character type is an individual who has everything invested in the attainment of the fruits of his ambition, namely in the attainment of the limelight. However, lacking perfectionist qualities, he lacks the ability to apply himself to the finer details of the tasks to be done. This may not be obvious to others at first because he does indeed have boundless energy, and he may seem to have endless "persistence" when it comes to adorning himself or his abode to "perfection," often to the point of garishness. And, it is true, he may be an enthusiastic author when it comes to writing his autobiography or other epic, soaring literary efforts that glorify himself in direct or in barely camouflaged inverted terms.   Nevertheless, despite such occasional activities of self-glorification, where he appears to be motivated, dynamic and goal oriented, on closer examination he lacks the persistence to create a meaningful synthesis of the mass of interrelated details that surround him. He is vague. He may be grossly, illogically imprudent. He has infinite pride in being recognized for his nebulous "creativity." But he is like a Don Quixote who leaps onto his horse and tries to gallop off in four directions at once.

For the narcissistic type the ends justify the means, and in fact, he would rather not have to bother with the means at all! We see, then, that he judges people not even so much for their potential for future accomplishments as for their already proven abilities to deliver the goods to him. And, of course, his estimation of others varies in direct proportion to the reassurance, deference and outright flattery that they give him.

Having so much invested in attainment of the limelight, once he arrives there it may be difficult to ease him out. His showmanship may take precedence over virtually all other aspects of his character. He may, then, neglect the accepted modalities of human decorum, as he talks endlessly in the first person, and in so doing he inevitably begins to neglect the truth.*

Finally, he is vulnerable in the limelight and may be self destructive there. If he is a sports figure, he may risk his life in attempting a dazzling but dangerous play. And it is he, of course, who will perform the most courageous of courageous acts if they are done in the presence of others, but will be strangely immobile if his noble deed would remain unknown to them.

At work he flaunts himself with his unctuous manner, his ostentatious handwriting and his mellifluous voice. In meetings he arouses resentment in others when he speaks of himself in overt or inverted terms. If others protest, he will not know what they are talking about, for he *is* the anointed one. And if he is seriously criticized or otherwise frustrated, he will explode into the red-faced narcissistic rage of defense: "I, in fact, am the only one who does anything worthwhile around here!" and of withdrawal: "If you people don't appreciate me, I'm leaving!"

He does not "play the game" and is not overtly vindictive. If he does raise his voice in frustration to reprimand someone, his bark is but a hollow bellow, and everyone knows it. He is usually a person who would not hurt a fly. His love

---

* "Friends, Romans, countrymen, lend me your ears! He comes not to bury Caesar. Nor does he come to praise him. He comes to praise himself!"

relations are not based on subjugation but on his conviction of his own beauty, his sexual wiles, and his irresistibility. And in his self-glorification he shows his Achilles heel, for he may become less able to interact meaningfully with those around him, and more and more a prisoner of his own inflated image of himself.

Nevertheless, an individual of the N type may, despite the constraints of his character structure, achieve great success in life. Although lacking the behavioral complex of perfectionism, his narcissistic drive for achievement may allow him to arrive at real accomplishments, especially if his perennial optimism can attract others to help him bring to fruition those dazzling visions of future triumphs.

Those of his acquaintances who do not know him well will be among those who like him the least. They will titter behind his back for his unctuous manner, his conceit, his extravagance and his frivolousness. They may ridicule him for his affected charisma, his perpetual smile, his weakness for empty-headed adulators of either sex, or for his outlandish cosmetics and dress. But those who come to know him more than casually may come to be enchanted by his kindness, his accessibility, his ingenuousness and his impeccable manners.

Individuals of the N type are fairly common in Western society. Illustrious examples may be found in personages such as Lorenzo the Magnificent of the Medici, Marie-Antoinette, Napoleon III, Catherine the Great, Theodore Roosevelt and Marc Chagall.

In the works of Somerset Maugham we find examples of females of the N type in the characters of Louise in *The Narrow Corner,* and of Maugham's beloved Rosie in his later novel, *Cakes and Ale.*

As an example of an illustrious N type, the reader is referred to Castelot's perceptive biography of the Empress Josephine:

> It is true — and it had often embarrassed him — that Josephine, with a Creole's naive and unthinking immorality, often talked shamelessly of her former lovers. Without doubt, she was fickle, light, flirtatious, "even somewhat amorous," as he was to say on St. Helena, and where love was concerned "made a few zigzags." Again, her education often left something to be desired. She seldom, if ever, read anything and holding a pen tired her out. Only the pleasure of adorning herself, ordering a dress, matching a ribbon for her hair could arouse her from indolence.

She was indeed ignorant, but she had nevertheless managed to acquire and retain a certain amount of knowledge which she knew how to use. She was said to have no brains. But she appeared to have enough, or at least to be very clever at making the most of what she had. Her rival, Mme. de Vaudey, said she had "only a quarter of an hour's wit a day"...

To be convinced that he was right to make his wife an empress, Bonaparte only had to recall with what grace, charm and distinction she received people. One might perhaps reproach her with being almost too welcoming, too easy of approach. But those were the defects of her virtues. And he was pleased, moreover, to find her "usefully captivating."

It was enough also to watch her walk with a "suppleness in her movements, a lightness that gave something aerial to her step without excluding the majesty of a queen."

...Then there was the point of her beauty. Was Josephine beautiful? She was more than pretty. Her countenance "was affected by all the impressions of her mind without ever losing that charming gentleness which was its main character." When his wife's "bright and gentle" eyes were turned on him, Bonaparte was often touched as in the old days. He loved her "long, silky" chestnut hair, which she dressed so prettily in the morning with a red madras kerchief "which gave her the most attractive Creole air." He loved her skin, whose "transparent satin" amazed him. He loved that body which had lost none of its suppleness, those arms and bosom like those of a girl. He loved the perpetual mobility of her features, which constantly assumed new expressions. Above all, he loved that gentle, silvery, caressing voice, whose tones were so enchanting that "one stood still simply for the pleasure of hearing it," that voice which caused him to say to Bourrienne, after Marengo, when the populace were shouting their enthusiasm:

"Do you hear the noise of that continuing acclaim? It is as sweet to me as the voice of Josephine."

Perfectionist type (P)

Individuals of this type were not found as functioning members of society. As mentioned earlier, we propose that this type is most closely approximated by the *autistic child* (N–P) and other *schizophrenic individuals*.

Aggressive type (A)

His complexion tends to be sallow rather than sanguine. Lacking the narcissistic trait, he tends not to adorn himself. But he cannot be missed. He has a loud voice and a brusque demeanor. He is rough around the edges. He is arrogant, aggressive and often cannot help being callous. He is a steamroller who cannot be stopped, or if he is halted, it is only for a moment. He must be first, and no one can have any true claim on him. If he does submit to others, it is he who is magnanimously doing them a favor.

He cuts corners in almost every aspect of his life, in his relations with others, in time, in space, in his poor handwriting and in the realm of the truth. Perfectionist traits of careful, directed activity and of a sense of duty to others are alien to him. He will be happy to say that doing something over and over again to make it better and better will only make it worse and worse. To him it is obvious: perfectionism is the enemy of progress.

He is not by nature a contemplative person; hence, he has only the slightest understanding of the forces that propel him. And it is only for the most fleeting moments of introspection that he wonders why he must so often act like the callous boor that he so often is.

He may not realize it, but his *modus operandi* is a claim of omnipotence. If in his present situation he is not the lord and master, then he tells himself that he soon will be. He "plays the game," and to the extent that he considers himself the master of all, his eye contact with others is often poor. Why should he, the all-powerful master, waste his time with eye contact on mere weaklings?

If he is seriously criticized, he will bristle and shoot from the hip. His argumentative reply will not finish until he has achieved a vindictive triumph. He must have the last word.

Just as he does not know how to give thanks or give compliments, he does not know how to apologize. Whatever the issue at hand, truth becomes secondary to an instinctive urge telling him that power must prevail. In fact, in a competitive society he is a dynamo whose meaning, and satisfaction, in life is the vindictive triumph over all individuals, whom in the final analysis he considers to be weaker than he.

But satisfaction in life through the vindictive triumph over weaker individuals is, the reader will recall, our definition of sadism, so in the wrong time and the wrong place he is capable not only of the most callous disregard of the rights of others, but also the worst of cruel, sadistic acts (Chapter 8).

In better circumstances, his aggressive tendencies are out in the open for all to see. His friends, family and acquaintances will come to recognize and predict his vindictiveness and his accountability for retribution like an elephant that never forgets. They will come to be accustomed to his vindictive rages, much like modern city dwellers become accustomed to the sonic boom. And despite his self-devotion, his hedonism, his arrogance, his brashness and his brusqueness, they may like him. This may be because they have sensed that in admiring him, and in letting him know that they admire him, they give him his only link to tenderness in interpersonal relations — personal recognition for his accomplishments. And they learn that this is the only way to gain his fleeting "smile."

Finally, he may be admired by both the strong and the weak for his qualities of getting things done, and getting them done fast. "Damn the torpedoes! Full steam ahead, come hell or high water!"

As a "player of the game" he must be the dominant figure, and so it must be in love relations, which are, of course, usually based on the subjugation of other weaker aggressive types (XA types) or submissive types. And if his daily life is one modest vindictive triumph after another, then his true vindictive rages may be few and far between. He may then be less of an Attila the Hun or a Genghis Khan than a brusque, abrasive dynamo who, despite his being a prisoner of the primeval forces that drive him, charges through life as a useful functioning member of society.

For contemporary examples of the A type we need only to look to some of our aggressive competitors of the various fields of athletics and sports.

Narcissistic-Perfectionist type (NP)

We will describe the NP type with the understanding that the baseline temperament of these individuals may vary widely, from the "phlegmatic" to the "nervous."

If one were required to describe this individual as an introvert or extrovert, one would tend to say that he is a quiet, friendly, sincere, somewhat sheepish introvert. He is polite, dependable, calm, careful, conscientious, considerate and cooperative. He is modest, pleasant and cheerfully reserved. He may have a pensive, bashful quality. He is unselfish, tolerant and sympathetic. He is neat and scrupulously attentive to detail. His desk is usually neat, his affairs are in order, and his dress, although not necessarily fashionable, is impeccable. He does not have an unbridled need to adorn himself. He may have a fetish for order, symmetry, neatness, with a place for everything and everything in its place. His handwriting is usually highly legible, with every letter clearly visible. It may have a calligraphic quality. A piece of work imperfectly done or lying unfinished causes him no end of smoldering grief. He may be a compulsive housecleaner.

He has a strong sense of duty: dutifulness to his family, to his friends, to his country, perhaps to his god, and to himself. In his duty to himself he must accomplish what he must, and this may sometimes be one detailed task after another. Thus, he may develop a "workaholic" quality to his life, always seemingly busy and always late. He may be a contemplative procrastinator. He is, therefore, punctilious but not necessarily punctual.

He is almost universally liked, or regarded to be a quiet person who is benignly tolerated. In American parlance, he is often a "nice guy" or a "fine gal." He may be somewhat prudish or even pious. He seems to be rather aloof, ignoring others in a benign way as he quietly goes about his business of pursuits for his own self-satisfaction. However, he is easily approachable and almost always tries to be helpful. He may present himself as a straightforward, uncomplicated person, even a "Simple Simon," but this is misleading and it is only on closer examination that we see the stringent demands that he places on himself.

His natural facial expression is a poker face or deadpan look of perfectionist restraint. As he plays his game of cards, he holds them close to his chest, and he does not particularly wish to discuss how his game is proceeding. In fact, he secretly has everything invested in the finality of success and in the recognition of his success by others. Thus, if recognition does come in any shape or form, whether it is a colleague to bring him good cheer, or actual recognition for something well done in an actual limelight, then he will smile. And his gingival smile, whenever it breaks out through his poker-faced visage, is a sight to behold! It is a warm, radiant, captivating smile, a kind of sheepish smile, appearing suddenly like the sun breaking through the clouds, or of a Cheshire cat suddenly appearing in the mist. It is often so breathtakingly, sincerely radiant that one is led to believe that this smile by itself is enough to give the human race its redeeming social value. And it does not take much insight to realize that this smile is the smile of narcissism bursting forth through the clouds of perfectionism.

As the quiet, somewhat bashful person that he sometimes is, he may surprise others when he mounts the podium as a public speaker. Somewhat stilted at the outset, he slowly gathers momentum and may gradually become radiantly charismatic, especially if he is attempting to persuade his audience.

This individual may have so much pride invested in the recognition of his work as a "conscientious achiever," that embarrassment in the limelight is of special sensitivity to him. In fact, of all the character types, he is the one to blush the most readily. Indeed, in the Caucasian he sometimes has a pinkish or ruddy complexion. If he is male and of the appropriate body habitus, he may have a cherubic, "blushing boy" quality.

He may be quietly ambitious, but lacking qualities of the behavioral complex of aggression, this individual is unaggressive and unassertive. He may, in fact, become annoyed when others continually point this out to him. If he is in a position where he should be aggressive, for example as a baseball manager or football coach, he may sometimes adopt a loud voice, but on close inspection he is simply an unaggressive person with a loud voice. Perhaps the most aggressive act of which he is capable is to interrupt someone while he is speaking.

Being unassertive, he finds it difficult to give clear-cut orders. Rather he expects that others will know what to do by his excellent example. He goes forth, and he expects others to follow. But, of course, as they often do not, he may have serious problems maintaining discipline. He cannot berate a subordinate, and he is the master of the short, two-sentence reprimand delivered in an almost apologetic manner.

He does not "play the game" of dominance and submission. His eye contact with others is almost universally good, but does not have the intensity of the NPA type. His gestures are reserved, as is his language. Ostentatious behavior is alien to him. His laugh, even if loud, is reserved. His humor has a pleasant impish or elfish quality to it, and he is not the type to play practical jokes of the kind that might cause someone physical discomfort or harm. He enjoys social situations, but is somewhat stilted in demeanor in them.

This usually quiet, unaggressive individual easily bends to the will of others, right? Wrong! He is persistent. He is obstinate. He is recalcitrant. He is downright stubborn! He has a will all of his own, and if he wants to do something in his own orderly way, then wild horses will not be able to budge him from his position.

When criticized he will immediately, quietly, and logically defend his position. If abused, he may scowl, but he may often turn the other cheek, simply not being able to defend himself against aggressive behavior. His only defense may be recalcitrance in the face of the demands of others. He may withdraw himself from the offending individual and ignore him, or pretend not to understand his wishes. He may use various modes of passive resistance, or make lame excuses for his recalcitrance. If pressed further he may quarrel, even vigorously, but he will not fight unless he is cornered. If he does, at long last, punch someone in the nose, it will be in the context of escape or of a narcissistic rage of vanity rather than an aggressive rage of vindictiveness.

He may appear as a somewhat negativistic "passive resister" or "chronic criticizer." If his sense of order is intruded upon repeatedly, he may become more and more negativistic and more and more recalcitrant to the point of resemblance to a PA type. Under greater stress, his negativistic recalcitrance can degenerate either into a catatonic state of near immobility or into a harried nervous state of agitation. In such a latter state he may closely resemble the n-C submissive type, or even a hypomanic NA type.

He tends to suffer in silence and to be self-berating. He tends not to be vindictive, neither in a vindictive rage nor in calculated vindictiveness, nor even in frustrating others or begrudging them what is their due. He holds no grudges and sometimes appears to have infinite patience and understanding. He may tease others good-naturedly, even persistently, but this is probably a Western cultural habit, and he is essentially devoid of sadistic trends. If his vanity or his pride in perfection is wounded, then he may respond in the perfectionist-narcissistic rage, as has been described earlier.

He is often drawn to activities in which painstaking, repetitive action is required. He may be an artist, a musician, a poet, a craftsman, a collector, or a do-it-yourself tinkerer. He is often the dutiful writer of long, careful letters to his family and he may keep a diary.

As an example of an NP type, we present the following account of Dieter's father, the parent of an autistic child, taken from Bosch's monograph:

> Dieter's father had worked his way up via elementary school and a trade apprenticeship in evening classes to the position of civil engineering technician and got his engineer's diploma at the late age of 46. He was a do-it-yourself addict and was engrossed in technology. Outside his job he was reportedly somewhat out of touch with the

rest of the world. He had a habit of composing long, typewritten reports about his son with dates carefully set out in the margins and many underlinings of points that he thought important. Despite the fact that he could have telephoned us, he preferred to communicate by means of painstakingly prepared letters. There is no doubt that he went out of his way to cater for his boy's special technical interests, which were further encouraged and influenced by his own do-it-yourself activities. While Dieter was with us at the hospital and he was away on holiday, he wrote his son a letter which, apart from the first sentence dealing with the nice place the parents were staying at, was exclusively given over to a description of all the locomotives complete with type number and colour that the couple had seen during their journey to the resort.

In the context of mating, the NP type is not aggressive and usually not predatory, although he will certainly acknowledge that he is a sexual being. If he is promiscuous, the overtones of aggression or exploitation are absent. The subjugated love of the aggressive type is unknown to him, as is sadomasochistic sexuality. His love relations are based on his narcissistic-perfectionist qualities: his love is tender rather than unabashedly passionate. He tends not to "fall in love" easily because the decision to devote himself to a mate must be a perfect decision, and such a perfect decision, like all of his decisions, is not easily made on the spur of the moment. If he enters into a long-term relation with an aggressive-vindictive personage, for example with an NA type, he may suffer in silence like the "brave wife" or the "henpecked husband," all the while berating himself for not being perfect enough to make the relationship an ideal one.

Finally, in his deal with life: "I will be perfect, so life will be perfect to me" he is especially vulnerable to any failure intruding into his existence, whether it be a flat tire on his automobile, a natural disaster or the loss of a loved one. Such failures may register not only as deep disappointments, but as hopeless reversals in what he perceives should be a natural order in life. The loss of a loved one, in particular of a family member, is an incomprehensible, unbearable crushing blow to him. If he does not throw himself on the pyre, then he will enter a melancholic abject state of deep mourning. He will wear black, literally or figuratively, for months or years and often will take his grief to the grave.

His requirements for narcissistic glory, but especially for perfectionist order, may lead him to abject depression, which may surprise even those who thought that they knew him well. The news of his fate will be met by the stunned disbelief of his neighbors, friends and other family members. They will say, shaking their heads, "We never had the slightest inkling that something was not right... Everything seemed to be in such perfect order..."

In the final analysis, his lack of free will has a poignant quality. He may at heart be so innately sympathetic, so intrinsically devoid of evil... so "good," that it is sometimes pathetic to see him go through life, whatever his real accomplishments, a prisoner of his own demands on himself.

The NP individual has certainly made his mark in history. Both male and female monarchs of the NP type have often emerged as just, strong-willed rulers. However, they sometimes appear as well-meaning "lambs among the wolves" (e.g., Catherine of Aragon or Tsar Nicholas II).

To conclude our discussion, we refer the reader to Herndon's *Life of Lincoln:*

> His penmanship, after some practice, became so regular in form that it excited the admiration of other and younger boys. One of the latter, Joseph C. Richardson, said that "Abe Lincoln was the best penman in the neighborhood"...
>
> He was a very sensitive man — modest to the point of diffidence — and often hid himself in the masses to prevent the discovery of his identity. He was not indifferent, however, to approbation and public opinion. He had no disgusting egotism and no pompous pride, no aristocracy, no haughtiness, no vanity. Merging together the qualities of his nature he was a meek, quiet, unobtrusive gentleman...
>
> He was unusually considerate of the feelings of other men, regardless of their rank, condition, or station. At first sight he struck one with his plainness, simplicity of manner, sincerity, candor, and truthfulness. He had no double interests and no overwhelming dignity with which to chill the air around his visitor. He was always easy to approach and thoroughly democratic. He seemed to throw a charm around every man who ever met him. To be in his presence was a pleasure, and no man ever left his company with injured feelings unless most richly deserved...
>
> Lincoln's melancholy never failed to impress any man who ever saw or knew him. The perpetual look of sadness was his most prominent feature...

There is, in short, little doubt that "Honest Abe" Lincoln was an individual of the NP type. With regard to his melancholic nature, some say that this was rooted in the death of his beloved Ann Rutledge (probably another NP type). For years after her death, Lincoln was known to have revered the entire Rutledge family. The tendency of the NP type to assume a melancholic nature will be revisited in Chapter 11.

Perfectionist-Aggressive type (PA)

As with all the character types, the PA type includes a wide spectrum of individuals who may be enormously successful, and greatly admired, in their societies.

With regard to the physiognomy of the PA individual, his complexion tends to be dull rather than sanguine. It is often swarthy, pale, pallid, dusky or sallow. In a fair individual it may be milky white. His countenance is usually one of a deadpan poker face. In his choice of dress and cosmetics he tends not to adorn himself in an outlandish manner. He may present himself as a "strong silent type," exhibiting a courteous reserved charm and a well-mannered gallantry. Thus, in unstressed circumstances he may have the quiet personal magnetism of the NP type. He may be greatly admired for his proud bearing, his undemonstrative low-key manner, and his dry humor, often providing others a welcome relief from the madding hysteria of modern society.

The individual of this character type may present himself as something of a high-strung extrovert. More often than not, he tends to keep to himself and may lead a fairly quiet life as a somewhat wary, withdrawn but coolly efficient perfectionist achiever. But with his aggressive drive frustrated to a muted level by his perfectionist behavioral complex, what may emerge is a laconic individual who displays an imperfectly concealed deep dissatisfaction with life. This will not go unnoticed, and his acquaintances will sometimes accuse him, behind his back, of having an air of haughty superiority.

If this individual is constantly interacting with others in a fairly competitive or stressful setting, he may become the classic passive-aggressive personality, who will not accede to the desires of others except in the most grudging manner. If he is a bit more of an extrovert, then he too may appear as a "chronic complainer" or "chronic criticizer."

Alternatively, this moody personage may find that an effective link with humanity can be made only through the medium of dry humor. In this case, he appears in public as something of a gregarious individual who relies, compulsively and almost exclusively, on his sardonic wit to gain the favor of others. Such an individual may display a real talent for creative, albeit sarcastic, humor, and he may be immensely popular with his colleagues for this quality.

In the description that follows, we will present a caricature of one particular PA subtype when he is thrown into the throes of a stressful competitive society, into a position of leadership, or when he is involved in a love relationship.

When this individual is encountered in a stressful setting, one finds oneself picturing him alternately as either an introvert or extrovert. Basically, he appears as a moody extrovert in whom one senses an undercurrent of deep hostility toward the outside world.

He tends to isolate himself, to be introspective and to be basically unfriendly. He is usually coldly calm, but when he is involved in a stressful ambitious venture, a directed task, or a situation of conflict with people, he may become mildly or severely agitated. His pallid face will blanch even more. His normally even voice will rise and have a cutting quality to it. It may barely hide a snarl. His eyes will flash and his brow will furrow. And in his voice one senses not only anger, but also the contempt and disdain of others, and a thinly veiled threat of vindictiveness.

His face in calmer circumstances usually shows the deadpan look of an austere poker-faced perfectionist. Although he is usually at ease before a group of people, to the extent that he believes that he dominates over them, he does not smile easily. What usually emerges is an odd grimace or a grin. He is usually aware of the fact that he cannot at will break out into a natural smile, and in compensation he has developed a short, repetitive laugh accompanied by a closed or half-open mouthed grin, which for him takes the place of the smile that his modern society demands of him. If he tries to smile for a photograph, the result is usually a broad "false smile," a non-symmetric frozen grin, or a sardonic smirk, and often he will not even try.

His handwriting may reflect either his perfectionist or aggressive tendencies. The handwriting may be neat and well formed, especially in the female. In contrast, it may be a slurred illegible scrawl, signifying at the same time his instincts of repressed physical aggression and his disdain of others.

In relations with people he pictures himself as a superior person and would not see himself as isolated. However, his view of life may be seen by his usual physical posture, which is sometimes slightly stooped, with the head dropped slightly down and his eyes directed to the ground before him. He may have, in fact, an amazing capacity to ignore people around him in familiar surroundings, almost to the point of obliviousness. When encountering a subordinate, or in a social setting, his normally stooped posture may change to one of erect, pompous rigidity. His basic unfriendliness may be reflected in obligatory but only cursory greetings to acquaintances or colleagues.

In a stressful social situation where strangers are present he gives the impression of being decidedly uncomfortable and will adopt either a posture of defensive rigidity or will be somewhat agitated. When meeting people in public, he has the roving eyes of suspiciousness, as if scanning the horizon for the possible approach of enemy aircraft. He instinctively evaluates a new arrival for his strengths and weaknesses, as his eyes move in frequent saccades of critical examination. What he is looking for, of course, is the new arrival's particular weakness, his soft underbelly, for in his isolation he senses himself vulnerable, and the only way to conquer his vulnerability is to dominate all others and to be stronger than them.

To the extent that he feels above his fellow humans, he does not make good eye contact with them. If challenged, however, he is the master of staring down his opponent to submission with a steely-eyed intimidating glare or glower. It is a glower that says, "You had better do what I, the perfectionist, expect, or else my vindictive rage may be activated with a hair trigger." His general demeanor, then, may be austere and dour, and may have a sinister aspect to it. His lips may take the form of a fixed smirk. He is the past master of the short derisive laugh.

Is he sarcastic and pessimistic? Yes, he certainly is. But he may be more than that. He may be deeply cynical, and one senses that behind the cynicism of the disasters to come lies the wish for a self-fulfilling prophecy.   And in his cynicism and his brooding nature, which exudes the gloom of the basic hopelessness of the human condition, we find the roots of his defensive paranoia and his offensive sadistic potential.

In a position of power, being aloof and suspicious, he is almost unapproachable to his subordinates or colleagues. His door is usually closed, and telephone calls are made in an atmosphere of secretiveness and suspicion. To the extent that he is secretive and mistrustful of others, so are others mistrustful of him. And if they do not trust him, they are intimidated by him, and they are ill at ease with him.

He is an individual who is above all self-righteous, and absolutely convinced that his position of dominance over others is not only right, but inevitable. He may develop a sense of invulnerability. When the going gets tough, and the weaklings scatter in all directions, he will stand firm, tall and proud, with his back to the wall if necessary, and he will survive over, nay conquer, all of his adversaries.

Is he a perfectionist? Yes, he certainly is. He has a sense of duty, but he lacks a broader sense of devotion to others that is often present in the NP and NPA types. His perfectionism lies in the area of pacifying his aggressive tendencies, so that they operate smoothly with maximum efficiency, which means that they are insidious and almost unseen. His *modus operandi* may be summarized by the expression "manipulation with care."

As he discovers, consciously or unconsciously, that his relationships with people are distant, he may attempt to invert his aloofness from time to time. As a perfectionist, an inner voice tells him that he should be warm, loving, sympathetic and gregarious. He will, then, from time to time descend to the level of the lower minions to mingle with his subordinates and colleagues, banter with them, "smile" with them, and attempt to prove to them, and to himself, that at the core he is a down-to-earth humane person. And he must, almost invariably, have at least one "exception that proves the rule" in a cause, or in a person to whom he shows, openly, care and devotion.

He, of course, "plays the game," and here we will only briefly mention the implications of this tendency rooted in his aggressive drives. If he is subjugated in marriage or in a love relationship, then he may find himself in a helpless schizoid state and may be reduced to achieving dominance only in occasional vindictive rages. In the dominant role he is, of course, much more successful, provided that his partner is not completely helpless or too tolerant of abuse. Finally, his love relationship may assume a stable symbiotic form in which he adopts the role of what we call the "power behind the throne." As the "power," he protects his mate, who is his sole source of security in life, with all the perfectionist-aggressive talent that his personality type can muster. This type of relationship, the dynamics of which are not consciously perceived by either individual, will be discussed in greater detail in Chapter 9.

The possibility of harrowing sadistic behavior by a PA type has been mentioned previously and will be considered further in Chapter 8. It goes without saying that a motivated PA type is no less committed to success in life through the vindictive triumph over others than is the aggressive (A) type. The difference lies in the muted way that he goes about it: by manipulation, by conniving, by holding grudges, by begrudging others amenities that they desire or recognition that they deserve, by withholding information from them, and in short, by an always ongoing process of perfecting a multitude of interrelated offensive and defensive actions, so that the final result is a fine tapestry of interwoven vindictive triumphs.

As is the case with all of the personages, this type is one who arouses pathos, for he too is a prisoner of his character structure. He is a prisoner of his aggressive tendencies, for if they were to be suppressed he would be left with only an aimless perfectionistic drive, with no ambition to perfect, and he would thus become schizoid or schizophrenic.

In Maugham's novel *Christmas Holiday* we find a masterful characterization of a non-sanguine PA type: Simon is a young reporter who dreams of revolution and of becoming the head of a Gestapo-like police force. Charley is his former school companion:

> Charley had a sudden fear that his visit to Paris was going to be a failure and he awaited Simon's arrival with a nervousness that annoyed him. But when at last he walked into the room there was in his appearance at least little alteration. He was now twenty-three and he was still the lanky fellow, though only of average height, that he had always been.

> He was shabbily dressed in a brown jacket and gray flannel trousers and wore neither hat nor great coat. His long face was thinner and paler than ever and his black eyes seemed larger. They were never still. Hard, shining, inquisitive, suspicious, they seemed to indicate the quality of the brain behind. His mouth was large and ironical, and he had small irregular teeth that somewhat reminded you of one of the smaller beasts of prey. With his pointed chin and prominent cheek-bones he was not good-looking, but his expression was so high-strung, there was in it so strange a disquiet, that you could hardly have passed him in the street without taking notice of him. At fleeting moments his face had a sort of tortured beauty, not a beauty of feature but the beauty of a restless, striving spirit.

> A disturbing thing about him was that there was no gaiety in his smile, it was a sardonic grimace, and when he laughed his face was contorted as though he were suffering from an agony of pain. His voice was high-pitched; it did not seem to be quite under his control, and when he grew excited often rose to shrillness...

Narcissistic-Aggressive type (NA)

This individual is driven by the instincts of narcissism and aggression, without mediation of these traits by that of perfectionism. Thus, the behavioral qualities found in the N and A personages may also appear in essentially unaltered form in the NA individual. Those qualities will not be repeated in detail here.

This individual is an extrovert, there is no question about that. He usually has a moderately loud voice, but more than loud he is talkative, or outright loquacious. His loquacity is characterized by bursts of rapid speech and has a gossipy quality to it, often with the use of slang expressions, sexual innuendo, and many references to current fads. He is vivacious, flamboyant, flashy, theatrical and somewhat agitated. He may have a hyperactive, labile, mercurial quality and may employ openly seductive body language, accompanied by agitated gestures. He may be in a state of perpetual motion, literally being unable to sit still. Others may comment that "he runs around like a chicken with his head cut off." We might say that he has flounce and he has bounce.

He may be described as charming, but his charm seems to lack any real depth, as if it were a highly polished artificial veneer. He seems in posture and in manner to be self-assured, but his self-assurance, like his other qualities, seems to be superficial, as if it could be popped like a balloon. In the same way, he may be described as attractive, but more in the literal sense of being capable of attracting the opposite sex, rather than having the attributes of profound beauty. He does not hesitate to talk about any subject, in public or in private, with little regard as to whether it might cause embarrassment to his companion in conversation. In fact, he is practically unembarrassable. He may be "nosey." To use American parlance, "he has a lot of verve and a lot of nerve."

With his agitation and vibrancy, his eye contact with others is often not good, and this is, once more, true to the extent that he feels himself to be above others. His laugh may be loud, penetrating and in the female shrill, but may seem to be forced. He smiles easily. The female can, with especial ease, flash a charming smile at will and hold it indefinitely, as if for a photograph.

As was true for the A type, he is basically aggressive and must dominate over all others, in little matters and in big, in every way and throughout every day. If challenged, or even if not challenged, he may be arrogant, brash and belligerent. If a stressed relationship is to be terminated, whether with a friend, a colleague, an acquaintance or a mate, it must be terminated in the context of a vindictive triumph. And the snarling, vicious, directed vindictive rage is certainly in his repertoire but may rarely be seen.

In his perpetual motion he fills up his time with trips, visits, classes and activities but is not really a "workaholic" in the sense of devoting himself to any real work. He is a dilettante. He is at his best in task-oriented activities. He tends to lack any deep sense of duty. Lacking in perfectionist qualities, he has little staying power. His most reliable attributes are his impulsiveness and his inconstancy.

He is, if you listen to him, *intensely ambitious* and he may say so in those words, but there is a great discrepancy between his ambitions and his real accomplishments. Sometimes he seems propelled by the jet stream of his own self-generated hot air. He is extravagant. There is a definite proclivity to travel, preferably by airplane, and it seems as if someone else is always paying for his trip. He is an intensely social person, a partygoer, and indeed the classic "life of the party." And as he becomes older, lacking the staying power for satisfying his deep needs for accomplishment and affection, he senses that something is wrong, that perhaps he is not "normal." But more often than not he will acknowledge that of course he is not normal — he is exceptional!

His relations with the opposite sex may be characterized as troublesome, turbulent and tormenting. Although the range of variability in human behavior in the realm of sexuality is enormously wide, there is no question that many of these individuals place a premium on sexuality in their lives. Some of them could well be placed into the categories of satyriasis and nymphomania. He is a coquette, a tease, a bird of prey, a vampire. There may be a continual, compulsive search for sexual partners, ostensibly to satisfy a conscious desire for a stable relationship. Instead, there occurs only one turbulent, tormenting relationship after another, and he views the past in perplexed amazement as he recounts how many partners fell in love with him, how many wanted to marry him, and how many he crushed by leaving them abruptly. Why, oh why, he wonders, cannot he *really* fall in love? Why, oh why, does he end by treating his partners so scornfully, so . . . sadistically?

The answers to these questions are not difficult to find, and lie, of course, in the narcissistic-aggressive character structure itself. His narcissistic self tells him that he is a grand person, thereby laying a framework of conceit. His aggressive, hedonistic instinct tells him that he must dominate his partner and get his pleasure while he can. And in dominating his partner, he cannot give of himself to him. He, thus, cannot provide his partner with the one thing that his partner craves for the most — his tenderness. And as his partner clings to him more and more, all the more there is established a master-slave relationship. The master becomes alienated. His eyes become averted from the slave. And he begins to treat the slave with all of the tormenting, sadistic means that are available to him in the repertoire of muted, and not so muted, aggressive human behavior.

The slave for his part is usually a weaker dominant type having an aggressive component, or a submissive type who becomes literally enslaved in a chronic state of subjugation. Here, it is truly fitting to call the relationship a "morbid dependency," the slave usually being totally incapable of extricating himself, however illogical the situation. The affair ends when the NA master moves on, sometimes sadistically playing off a former partner against his new interest, setting the stage for a new turbulent relationship. And this may be repeated again, again and again.

If he marries and assumes the *dominant* role, then the marriage does not portend well. Unless the real life circumstances provide an enormous amount of compensatory cement to the marriage, the master will torment the slave subtly or overtly and finally leave him. If he assumes the *subjugated* role, then for much of the time he will resemble the narcissistic N type, and his aggressive component will be strongly muted.

In *The Ant and the Grasshopper* Somerset Maugham gives us this characterization of an NA type:

> For twenty years Tom raced and gambled, philandered with the prettiest girls, danced, ate in the most expensive restaurants, and dressed beautifully. He always looked as if he had just stepped out of a bandbox. Though he was forty-six you would never have taken him for more than thirty-five. He was a most amusing companion and though you knew he was perfectly worthless you could not but enjoy his society. He had high spirits, an unfailing gaiety, and incredible charm. I never grudged the contributions he regularly levied on me for the necessities of his existence. I never lent him fifty pounds without feeling that I was in his debt. Tom Ramsay knew everyone and everyone knew Tom Ramsay. You could not approve of him, but you could not help liking him

The female of this character type, in modern Western society, may lead a particularly turbulent life. And given her total confusion with regard to the instinctual demons that are propelling her through life like a speedboat out of control, one cannot help but be sympathetic toward her. She feels that she is destined to do great things, but somehow feels that she is not "normal." She wants to work with others, but bitterly resents being asked to do anything for anyone.

She has pride in her beauty, but somehow feels that she is physically flawed. It is perplexing to others how this individual, often frankly attractive, has almost invariably a negative body image of herself, a flaw, whether it be an imperfect complexion, an imperfect nose, this feature or that. Somehow her narcissistic voice demands physical "perfection" of herself. And in response to the disparate feelings of ugliness on the one hand, and the need to be a ravishing beauty on the other, she places an enormous investment in efforts to enhance her physical appearance. There is an emphasis on fashionable, perfectly-matching clothes, with price being no object, on accessories of dress, on her hair style, on nail polish, cosmetics and perfumes, and the like. In some cases this emphasis on the superficial qualities of personal adornment is so striking that one has the impression that she is gaudily attempting to gild a lily.

With regard to her sexuality, in which there is so much pride invested, total confusion reigns. Being an aggressive type she instinctively scorns weakness, but all the while dreaming of a superman who will overpower her and possess her, she paradoxically seeks men whom she can dominate and subjugate. And in doing so, she almost invariably chooses among those who long for submission. And in dominating she wonders why she cannot submit to true love, why she cannot be tender, and why she is, despite all of her investment in sexuality, often sexually unresponsive. And it is in this lack of responsiveness and in her need to dominate all aspects of a sexual relationship — sometimes to the fringes of normality — that one senses her desperation. It is the desperation of a person who, despite the whirlwind of activity surrounding her, is essentially alone and forlorn in her life. It is the desperation of a voice crying in the wilderness.

There is little doubt that an individual of the NA type (or N type) has a predilection for an older mate, as has been noted by literary figures of the past. For example, the playwright Henrik Ibsen, at the age of sixty-one, became involved in an affair with a young predatory female, Emilie Bardach, who was to become the character Hilda Wangel in two of his plays.

We can see, in fact, in Ibsen's play *The Master Builder* the broad outlines of what must have been the basic dynamics of Ibsen's relationship with Emilie. The character Solness is clearly Ibsen himself, and although he is the object of Hilda's sexually predatory nature, this aspect is secondary. Solness is, to be sure, a sexual being, and even a father figure, but he is above all Hilda's ideal image of her unbridled ambition. We realize, now, why the NA female is drawn to an older man. It is the ulterior motive of the narcissistic behavioral complex that is telling her subliminally: "He is your ideal image. He is your promise of glory. You and he are one and the same. Lose yourself completely in him. Merge with him..." Unfortunately, her aggressive nature is telling her, "Dominate him every day and in every way," hence her sadistic trends may ultimately doom any hopes for a stable relationship.

In competitive society, if a female of this character type attempts to achieve success in a career, and at the same time dominate her erotic sphere, conflicts are bound to arise, as in the following example from a recent letter* to newspaper columnist:

DEAR ANN LANDERS:

I'm a 26-year-old career woman and lately I've been experiencing severe anxiety attacks — shortness of breath, pounding heart, sweaty palms, feelings of weakness and dizziness.   Sometimes my hands shake so I can't hold anything.

I can't understand what is causing these symptoms.   I'm doing well in my career and enjoy my work. My social life isn't great, but this doesn't bother me. I'd rather stay home than go with some of the creeps I see around town. Can you help me find some answers?

— Wassau, Wis.

But such a conflict is, of course, not one limited intrinsically to the female. As with other character types, we must ask ourselves whether *any* individual is, by his basic nature, suitable for a given life situation.

Finally, we reiterate the wide range of emotional behavior of the NA type: his hypomanic highs and his abject depressive lows. He may be prone to obesity as he alternately overeats in splurges and then abstains from food entirely. He is particularly prone to psychosomatic illness, for example to headaches, intestinal problems, palpitations, or attacks of "nervousness." When intimidated he may split his personality abruptly to the NA– subdued state, where he becomes a narcissistic type with not a trace of aggression to be seen! This dramatic personality split, when it occurs, is truly remarkable.

As an example of the emotional lability of the NA type, the reader is referred to Maugham's short story *Rain,* where we find a young lady of the NA type, Sadie Thompson, abruptly change from a hypomanic state to one of abject depression:

---

* Ann Landers, Gannett Newspapers, *The Daily Item,* August 12, 1980.

> ...They saw Miss Thompson standing at the threshold. But the change in her appearance was extraordinary. This was no longer the flaunting hussy who had jeered at them in the road, but a broken, frightened woman. Her hair, as a rule so elaborately arranged, was tumbling untidily over her neck. She wore bedroom slippers and a skirt and blouse. They were unfresh and bedraggled. She stood at the door with the tears streaming down her face and did not dare to enter.

Later, after she achieves a vindictive triumph, we read:

> ...They walked up the steps and entered the hall. Miss Thompson was standing at her door, chatting with a sailor. A sudden change had taken place in her. She was no longer the cowed drudge of the last days. She was dressed in all her finery, in her white dress, with the high shiny boots over which her fat legs bulged in their cotton stockings; her hair was elaborately arranged; and she wore that enormous hat covered with gaudy flowers. Her face was painted, her eyebrows were boldly black, and her lips were scarlet. She held herself erect. She was the flaunting queen that they had known at first. As they came in she broke into a loud, jeering laugh.

Finally, if we add to the NA individual's behavioral repertoire the occurrence of the narcissistic-aggressive rage, we see the extent to which this individual is a captive of his emotions.

Narcissistic-Perfectionistic-Aggressive type (NPA)

There is no question about it. He is an extrovert. His complexion definitely tends toward the sanguine, especially if he is at all agitated. His voice has an unrestrained quality. It may be forceful. In the female it may have a sharp, piercing quality. His voice may be outright loud, even if he consciously tries not to be overbearing, which often he is. His voice may be very, very LOUD. It may be stentorian, especially in the male, and he may sometimes be heard at the far end of a railway car. It is a voice that seems to long for diffuse dissemination, as he speaks at and through his partner in conversation. It is a voice of narcissism ("how grand I am") and of aggression ("all of you had better listen to what I am saying"). Often he can be identified immediately by his forceful voice alone.

This individual is not relaxed. He radiates a certain intensity and activity, and like the NA type, he too may have difficulty keeping still. In manner, he may be only moderately outgoing, especially if he is in circumstances where he is chronically dominated by stronger individuals. On the other hand, he may be intensely outgoing, affably ebullient or outright truculent. If he is only moderately intense, he has a certain amount of real charm, but sometimes he is so frankly overbearing that his charm is lost in his abrasiveness. He is never at a loss for words. At times he may talk, and talk... and talk. And in his garrulity his smile and laughter may take on a forced intensity.

He has a moderate gait and carries himself with relative confidence. This individual seems to be going somewhere. He usually makes intense eye contact with his partner in conversation, whether the latter be strong or weak, important or unimportant. His prominent eyes have a spirited look about them and may seem to sparkle or even protrude.

In the mature adult, he has a non-seductive maternalistic or paternalistic character, even if he is promiscuous. In the female, she has more of a "wholesome" than a "sexy" demeanor. If her voice is very forceful, her detractors will say that she is "not feminine," "brassy," or even "masculine."

He is reliable, dependable and responsible. He is faithful to his family and friends. If he has children he has affectionate pride in them and talks about them. He is a "solid citizen." He has a sense of duty to his profession, his colleagues, his business, perhaps his church, and his country. He tends to have a sense of citizenship, of attachments, and of devotion to some ideals. He is, shall we say, ...rather conventional.

He is expansive, something of a perfectionist and a "doer." His perfectionist tendencies are not as constraining as in the NP type, and he actually gets things done. He likes to get the present job finished and move on to the next task, and he may be a true "workaholic," with his time filled with real activities. Although he may be a procrastinator, he dislikes intensely any ambiguous situations, incomplete information, work half-finished, or any feeling of "loose strings hanging." He tends to be impatient and sometimes impetuous. He cannot bear to stand and watch someone doing something slowly or fumbling about, and will immediately say, "Here, let me do it for you."

We usually see in him a person who is cordial, but who has an air of self-importance. And it is in this feeling of self-importance that there may emerge a barely concealed attitude of callousness, for example in being late for appointments or in downgrading the wishes and aspirations of others.

His gestures are expansive. He may have friendly body contact with his colleagues and acquaintances but, in contrast to that of the NA type, it is of a non-seductive nature. His clothes are usually fitting for the occasion, but he is not one to devote himself to superficial fads or fashions. In the male, especially, his shoes will be polished and his hair combed, but his clothes may even be somewhat ill fitting. He is, thus, more properly dressed than truly fashionably or gaudily dressed. To the extent that narcissistic-perfectionist tendencies predominate, which is usually the case, he will make an effort to have clear, legible handwriting, even when he is in a hurry.

In social situations, he may be at his best. He is the classic after-dinner speaker. He may be a master of inverted modesty in speaking before an audience (e.g., *"He* is an absolutely fantastic person"). He indulges in "hail-fellow-well-met." At a social function, he may shout across the crowded room to welcome a prestigious individual when the latter arrives. He is sympathetic at the core, from his sense of duty, but does not tend to spout forth spontaneous sentimental affection.

If someone does not uphold his standards of perfection, then he will become indignant. He will spread his arms at his sides, with the palms up, in that characteristic stance of perfectionist incredulity. "How could you let that happen? Didn't you realize that...? What is going on here?" If criticized he will respond immediately and arrogantly in defense, with no prior reflection on the merits of the criticism, like a porcupine bristling its quills. If angered, he becomes sarcastic and ill mannered; he becomes rude and he shouts. He may become involved in shouting matches, or even fisticuffs, in public with strangers. And if his sense of pride or vanity is trampled upon, he may be incited to the narcissistic-perfectionistic-aggressive blind super-rage, which has been described earlier. The intensity of this red-faced rage may shock others seeing it for the first time. Many NPA individuals are intensely aware of this tendency and consciously attempt to keep a tight lid on their emotions, so that in fact the rage appears only infrequently.

We include below an account* of an American baseball game, in which the aggressive and narcissistic components of a rage are clearly seen:

> NEW YORK — In case you missed it, Lou Piniella went 2-for-4 Tuesday night at Yankee Stadium as the Yanks handed the Oakland A's their sixth straight loss, 6-1.

---

* Gannett Newspapers, *The Daily Item,* September 3, 1980.

Problem is, Piniella thought he should have been 3-for-4. In the bottom of the seventh, he hit a liner to right and Tony Armas, after a long run, dropped the ball.

"Error, right fielder," said official scorer Harold Rosenthal, a retired member of the Baseball Writer's Association of America and well-known author.

At second base, Piniella spread his arms wide apart, in total disbelief.

That was half the fun. After the game, it was vintage Piniella, howling, screaming and storming. "I'm going to strangle him," Piniella roared. "I'm going to his house and choke him"...

"If they don't change that scoring and give me a hit," Piniella continued, "I'm through playing baseball for the year. I quit this game."

As a character sketch of an NPA individual, we offer Captain Forestier, from Maugham's short story, *The Lion's Skin:*

The trouble was that Captain Forestier was almost too perfect a type of the English gentleman. He was at forty-five (he was two or three years younger than Eleanor) still a very handsome man, with his wavy, abundant grey hair and his handsome moustache; he had the weather-beaten, healthy, tanned skin of a man who is much in the open air. He was tall, lean, and broad-shouldered. He looked every inch a soldier. He had a bluff, hearty way with him and a loud, frank laugh. In his conversation, in his manner, in his dress he was so typical that you could hardly believe it. He was so much of a country gentleman that he made you think rather of an actor giving a marvelous performance of the part.

When you saw him walking along the Croisette, a pipe in his mouth, in plus-fours and just the sort of tweed coat he would have worn on the moors, he looked so like an English sportsman that it gave you quite a shock. And his conversation, the way he dogmatized, the platitudinous inanity of his statements, his amiable, well-bred stupidity, were all so characteristic of the retired officer that you could hardly help thinking he was putting it on.

Another masterful description of an NPA individual may be found in the character sketch of Alroy Kear in Maugham's novel *Cakes and Ale.*

If an NPA individual is subjugated in a long-term relationship with a companion or mate, and in particular by a "power behind the throne," then for much of his daily life his aggressive component will be strongly muted. He may thus appear as an individual somewhat placidly riding the merry-go-round of life, apparently little motivated and somehow "lacking in ambition."

Moving on to an NPA individual who is overtly power seeking, he is sometimes grandiloquent, and he may lose all sense of propriety. One senses that he is often making a conscious effort to be not overly overbearing or too dominating. He assumes leadership because he thinks — nay he knows — that he is the best one for the job. But despite his efforts to convince others and himself that he is not aggressive in a cutthroat manner, he has a pervasive inward feeling that no one should push him around and that he should attain heights of excellence... that he should succeed to the very limits of his ambition. As in other aggressive types, but in a form muted by his sense of duty, there emerges the deep conviction that he should be above the masses and above the competition. It is a conviction fed by narcissistic ambition acting synergistically with his innate aggressive drive, with both being forged and tempered by the behavioral complex of perfectionism.

If he aspires to be "the boss," then the boss he is — there is no question about it, and his sense of restraint may fade. He may reveal a barely camouflaged arrogance. He then requires that his instinct for perfectionism be fulfilled. He requires that others pay constant attention to the details that his sense of "doing things well" requires. He demands "spit and polish." He insists on punctuality. He requires the continual approbation of his colleagues but would be embarrassed by their overt adulation.

He will dominate the conversation in a group. He tends to speak compulsively at conferences. If a difference of opinion arises, he will become agitated and will feel that he must have the last word.

He has pride in his honesty. Overt prevarication is anathema to him. However, to maintain his position of dominance, which he somehow feels in his bones is a right granted to him directly from the heavens, he will have no compunction with regard to the withholding of information that his opponents might use against him.

If he is not so much a boss than an "upcoming achiever," for example in a hierarchal structure, then he becomes an "expert" in some field. Once his area of expertise is established it tends to expand in two ways. First, it expands in his own mind so that he becomes an expert not just in his narrow field but also in the broader areas of technology, science, education or philosophy. Second, it becomes, by his own word of mouth, disseminated in time and space: everyone should have the benefit of his expertise.

Despite his often-overbearing demeanor, he has pride in his manners, in decorum, and he rarely would insult an individual to his face. Vindictiveness toward others and sadistic behavior may be overt, it is true, but usually it is well camouflaged and only subtly present. It may take the form of apparently good-natured but persistent teasing of a subordinate or opponent, being condescending toward him, overpowering him in public under a mass of sarcastic verbiage, or in that passive vindictive triumph that has been classic since time immemorial, simply leaving without saying good-bye.

As a leader or as a head of the family, he becomes an autocrat and cannot tolerate insubordination. He is the classic martinet with panache. The slightest indication of disloyalty may be dealt with unbounded harshness, and no punishment, even corporal punishment, will be excessive. Family members will not be exempt. Although others may stand aghast at the intensity of his vengeance, to his own mind he is a misunderstood man of mercy whose hand is occasionally forced to mete out harsh justice. In such circumstances his only saving grace is that he tends to mellow with age.

If he comes to absolute power, then he may become the standard bearer of unlimited power and glory. He becomes Friedrich Wilhelm I and Ivan the Terrible. Yea, he becomes Barbarossa and Akbar of Hindustan!

As an aggressive type, the NPA individual certainly "plays the game" of dominance and submission, but the "game" is often muted by his sense of duty. His aggressive tendencies are moderated by the behavioral complex of perfectionism, but in the framework of narcissistic behavior, the resultant character structure is very different from that of the PA type. In marriage he will usually be devoted and faithful to his mate. But if the relationship goes awry, he will become a somewhat passive "situational sadist" and may finally attempt to extricate himself. He may be sexually promiscuous. In other words, his sense of duty has limits.

Finally, like his cousin the NP type, he too has made a "deal with life," though in muted form. If he takes care of the business of life, then the business of life should take care of him. And if it does not, then life's failures are taken very, very hard, subjecting this active, vigorous individual to the depths of an abject state. When he falls, it is all the more painful, because it is, in his own mind, the mighty who has fallen.

In the final analysis, this individual, too, is a prisoner of his character structure, and he is puzzled by the demons that seem to be driving him in three different directions. His life may be one of constant turbulence as he tries to streamline his narcissistic, perfectionistic and aggressive traits into a cohesive unity, all the while trying to keep a lid on his NPA super-rage. And it is only in looking into his character structure that he may begin to create for himself a life situation that is compatible with psychic survival in his human society.

SUBMISSIVE TYPES

Compliant types (*C*)

This individual's character structure, whatever its resultant complexity and whatever his real accomplishments in life, seems to be constructed around a nidus of a feeling of shame. He is an introvert. He has the so-called "inferiority complex."

He may be blandly passive or a true "shrinking violet." He has a low, restrained voice, and he is not articulate. His countenance suggests a trace of sadness. His movements are tentative and his gestures reserved. His eye contact with others is poor, as he displays the averted eyes of a slave before his masters, which in his case includes almost everyone with whom he comes in contact. He shrinks in the presence of kings, but also in the presence of shopkeepers.

Meeting people in a business or social situation is an ordeal for him. He forgets names as soon as the introductions are uttered. If he is required to make the introductions himself, his mind goes blank and he may enter a state of panic. He is very uncomfortable before a group of strangers. He lives in fear of being called upon to speak extemporaneously. If he must give a speech he will write it out word for word or commit it to memory, fearing that his mind will go blank when it comes time to deliver it. When he does deliver it, his nervousness is apparent, all the more so if his audience is hostile or in the least bit threatening. For the same reason this individual will not be seen on television unless he is accompanied by his master.

He will be uncomfortable and feel anxious if he must go to a social function where strangers will be present. What he fears most is a medium-sized group of six to twelve persons, where he might suddenly become the center of attention. Paradoxically, though, he yearns for the presence of others, and to stay home in loneliness is a state of shame that he tries to avoid or hide at all costs.

He has a poor self-image but does not try to bolster it. Even though he may possess the narcissistic trait, he tends not to adorn himself excessively (but the NA= type may, indeed, adorn himself flamboyantly). His dress is usually reserved and may be outright shabby. In fact, any state of ostentatiousness is alien to him. If he sees himself suddenly in a mirror he may startle himself, and he is not particularly enamored by what he sees. He may at times consciously wish that he were not he, but someone else.

Having renounced competitiveness, he has all of his pride invested in helping others and in trying to please them. He is in his self-effacing way overhelpful, overkind, overcaring and oversympathetic. His handwriting, written as it is for the benefit of others, is nicely legible. And in his pride he sees himself, not as a selfish person whose claim is to be cared for throughout his journey through life, but as a selfless saint who is indispensable to his boss or to his mate and family.

His taboo on competitiveness and on any aspirations for himself pervades his entire life, from his important decisions of how to gain his livelihood and whom to marry, to the less important ones regarding the minutiae of his daily existence. Feeling as he does as a stowaway on the ship of life, he fears that if he does anything implying independence from his protectors then he might suddenly find himself in a lifeboat, alone at sea and having to fend for himself in the struggle for survival in his hostile world.

In space and time, he is something of a lost soul. He may have a magic circle, of ten-mile radius, from which he dare not leave. He may have only the vaguest idea of the locations of nearby states, cities or townships. He may not have the slightest idea where the maze of highways near his home actually leads. If he goes on a trip, he enjoys himself as he is piloted about by his protector, but he will have only the slightest idea in geographical terms where he has been. He is totally incapable of reading maps or transportation schedules, and may ascribe this inability to some kind of learning disability. In fact, if he is not taken somewhere, he would not dream of going alone. Hence, in contrast with the resigned type whose pride is vested in complete independence, the compliant type has all of his pride invested in complete dependence on his protector, protectors or subjugator.

As an example of a compliant type, we offer this account* from a recent letter to a newspaper columnist:

> DEAR DR. BROTHERS: Whoever said life begins at 40 was a terrible liar. For me, it seems to be the end of the line. Everything has happened wrong for me in the past two years.

---

* Dr. Brothers, *New York Post,* September 10, 1980.

> I had a husband who took care of everything for me, and then he suddenly died. He took care of all the family business; he never even allowed me to drive an automobile because he preferred to do it for me. I'm ashamed to say I don't know how to balance a checkbook or do any of the things that seem to be a part of daily living for a lot of people. I'd like to find another man just like my husband, but now I'm 42 and it seems unlikely that such a man could ever come along. I'm very depressed most of the time and I feel that my life is over.
>
> — A.G.

Non-Compliant types (*n-C*)

NPA– type. We focus first on the NPA– type. This individual appears to be an introvert who periodically seems to come out of his shell. His demeanor, in fact, is highly dependent on the circumstances of the moment.

His baseline personality is essentially that of an affable compliant type, and his facial features often show that characteristic trace of sadness. But his car is parked outside, with the engine running, and it may be a sport model capable of flying him to the pinnacles of the glory of his imagination. Thus, this individual has a subtle, or sometimes not so subtle, agitated, fidgeting demeanor as if he wants to go somewhere. He nervously fidgets with his hair, his hands, his moustache, as if they were masturbatory equivalents suggesting repressed sexual yearnings. In stressful situations his eye contact with others is abysmal, and indeed he often has the wandering eyes of perpetual apprehension. His handwriting often reflects his agitation, and it is often barely legible, with many corrections, or else well formed but jerky.

If he does ascend to a position of leadership on the basis of his affability or his intellectual qualities, he does not inspire confidence before his colleagues and subordinates. He is frequently very self-conscious. If at all intimidated, he becomes tongue-tied; his voice loses its forcefulness, and he tends to fidget and to stammer or mumble. Even if he is an expert in his field he cannot mount the podium with any confidence to speak extemporaneously. He exasperates others with his indecisiveness, as he tries to please everyone. He is vulnerable to the views of any strong personality, and as he is at the mercy of the last person with whom he speaks, he may be constantly changing his views. Thus, others will accuse him of being "good natured" but "weak minded." Often they will misinterpret his apprehensive reticence for cynical indifference. His basic attitude toward life is seen in his usual photograph: he is not comfortable looking at the camera, and often he is not smiling.

In love relations he may assume any role in the "game," and he may lead a rather quiet life, playing either the dominant or submissive role of subjugation. If he is not too shy and is successful in competitive society, he is exceedingly vulnerable to being subjugated by a "power behind the throne." Alternatively, he too may assume the role of a quiet hunter or "bird of prey." In fact, he and the NA type are often both out on the prowl, looking for each other, and when they meet sparks are bound to fly. More often than not, an affair of the "lightning and thunder" variety begins, and exciting and exhilarating as it may appear, it is usually destined to achieve no lasting stability. As has been described earlier, sadomasochistic elements inevitably seep into the relationship, the two partners never really gain an understanding of their own or each other's needs and motivations, and the relationship comes to an end when the NA partner moves on to a new companion to continue the chain reaction.

Finally, we see the frustration that the non-compliant type faces throughout his active life. He has a strong sense of narcissistic ambition but a stunted component of aggression. And during his jerky ride through life he finds that others do not particularly admire his jerkiness, are ill at ease with him, and he often finds himself isolated. Sometimes he, too, seems to be a speedboat out of control in a fogbound harbor, and it is only in looking at his own basic character structure that he can begin to lift the fog and bring himself under control.

Insight into the basic character of non-compliant types may be acquired in the study of the lives of historical figures such as Edward VIII, Chateaubriand, Joseph Conrad, Henrik Ibsen and Somerset Maugham (in particular the self-revealing character of Philip in *Of Human Bondage).*

As an example of a non-compliant type, we give below an excerpt* from a recent newspaper column:

> Far from strident, he was totally silent, never speaking in class discussions, and I was reluctant to call on him. Since he had a Spanish name, I wondered whether he might have trouble with English. Bureaucratic chaos was the order of the day, with the City University enacting in microcosm the confusion in the nation at large; it was not unusual for barely literate or barely English-speaking students to wind up in an Introduction to Literature class. His silence and his blank arrogant look could simply mean bewilderment. I ought to find out, but I waited.

---

* *Hers*, L.S. Schwartz, *New York Times*, October 10, 1980.

His first paper was a shocker. I was surprised to receive it at all — I had him pegged as the sullen type who would give up at the first difficult assignment, then complain that college was irrelevant. On the contrary, the paper, formidably intelligent, jarred my view of the fitness of things. It didn't seem possible — no, it didn't seem *right* — that a person so sullen and mute should be so eloquent. Someone must have helped him. The truth would come out in impromptu class papers, and then I would confront him. I bided my time.

After the first exam he tossed his blue book onto my desk, not meeting my eyes, and, wary and feline, glided away, withdrawing into his body as if attempting a disappearing act. The topic he had chosen was the meaning of "the horror" in Joseph Conrad's "Heart of Darkness," the novella we had spent the first few sessions on.

He compared it to Faulkner's "Intruder in the Dust." He wrote at length about racial hatred and war and their connection in the dark, unspeakable places in the soul from which both spring, without sentimentality but with a sort of matter-of-fact, old knowledge.

He knew Faulkner better than I did; I had to go back and skim "Intruder in the Dust" to understand his exam. I do know that I had never before sat transfixed in disbelief over a student paper.

The next day I called him over after class and asked if he was aware that he had an extraordinary mind. He said, yes, he was. Close up, there was nothing arrogant about him. A bit awkward and shy, yet gracious, with something antique and courtly in his manner.

Why did he never speak in class, I asked.

He didn't like to speak in front of people. His voice and his eye turned evasive, like an adolescent's, as he told me this. Couldn't in fact. Couldn't speak.

What do you mean, I said. You're not a kid. You have a lot to say. You write like this and you sit in class like a statue? What's it all about?...

NA– type. We now turn to the NA– type, keeping in mind the three basic attributes that rule his life: unbridled narcissism, non-perfectionism, and submission. In fact, the two traits of unbridled narcissism and submission seem to be continually acting in antagonism in this personage, so that sometimes he appears as an affable extrovert, and sometimes as a submissive introvert. In addition, lack of the behavioral complex of perfectionism often gives his character structure a labile, fragile quality.

Once the character structure of this personage is recognized, one can predict his general behavior, almost as if it were a foregone conclusion. In real life he appears as an affable, sanguine complexioned, somewhat nervous individual who, despite his generally submissive nature, is ever ready to respond to the call of the limelight. If his life situation supports him so that his unbridled narcissism comes to the fore, then he may be somewhat expansive or even charismatic. He tends to adorn himself or "dress up" — sometimes strikingly — to a much greater extent than the NPA– type, and indeed this is often the most obvious superficial difference between these two types. He may, in short, display any of the attributes that we have come to associate with the charming, but often flighty, N character type.

As a submissive type, he is exactly that — however much he would like to ignore or deny this facet of his personality. His tendency to be sympathetic and agreeable pervades his daily life, and his acquaintances will invariably consider him to be a "nice person." However, having a remnant of the trait of aggression, he reveals himself to be a "player of the game" in stressful situations, being susceptible not only to red-faced narcissistic rages but also — albeit rarely — to the aggressive-vindictive rage. In competitive society his submissive trait sooner or later becomes evident. When challenged, he is easily intimidated, he stammers, and others will say that "he can't seem to get it all together," or will accuse him of "bumbling" or of "histrionics." He frequently has a marked hesitancy in his speech. If he comes to power on the basis of his narcissistic charisma, at the highest level of society, then he requires much support from his entourage in his attempts to achieve his grandiose visions of unbridled glory. Hence, he is easily subjugated, and indeed, hovering in the wings there may be seen a protective, perfectionist "power behind the throne," always ensuring that everything is progressing with the precision of a well-oiled machine.

In adolescence the NA– type is an active but reserved individual who will often describe himself as being "on the shy side." Nevertheless, his unbridled narcissism is continually seeking to overcome his submissive shyness, and he may be attracted to expansive, soaring projects, again reminiscent of the activities of the self flaunting N character type. He may, in fact, be attracted to public speaking, to a career in modeling, to the stage, or to the dance.

If he aspires to be an actor, he may be highly successful in his portrayal of the affable, kind-hearted — but perennially abused — hero or heroine. Often, however, he is only moderately successful, since his generally submissive nature continually intrudes into his desire to be the forceful personality of his dreams. Although his likable qualities usually carry the day, onlookers will often titter behind his back for his general lack of talent, for his lack of "substance," and for his penchant for appearing as a second-rate actor in a grade-B production.

Like his cousin the NPA– type, the NA– personage may lead a successful, relatively calm existence if his life situation is well supportive. If it is not, then this labile personality is prone to transient masochistic love affairs, hypochondria, phobias, fugues, agitated reactive depressions, as well as episodes of euphoria and hysteria. Nevertheless — contrary to what might be expected — like the NPA– type, this somewhat fragile personage has a relatively low risk of succumbing to a state of schizophrenia (see Chapter 11).

PA– type. Non-compliant submissive individuals of the PA– type tend to be more profoundly schizoid than those of the NPA and NA types, and their relations with the opposite sex may be more disturbed. An historical example of the PA– type is King Ludwig II of Bavaria (Plate 46), who succumbed to a psychosis.

Finally, we note that submissive individuals who have the N trait tend to have, even more than their dominant cousins, a sanguine or flushed complexion. Fair individuals of the A– and PA– types tend to be more pallid.

RESIGNED TYPES

The resigned types NP-A and N-A are much like NPA and NA types in totally unstressed circumstances. The voice is often somewhat subdued, yet it is usually more forceful, and has greater range, than that of the NP type. The NP-A type preserves the intimate eye contact of his NPA cousin. With regard to countenance, the NP-A and N-A types tend toward a sanguine complexion, while the non-sanguine –A and P-A types (schizophrenic individuals) tend toward pallor.

The resigned type often reveals himself not so much by his mannerisms, the tone of his voice or his smile as by the detached life that he leads. In his serenity and in his non-involvement in close relationships he puzzles his colleagues and acquaintances, who may even suspect that he is homosexual, which he would bitterly resent. The dynamics of his lifestyle become understandable once his basic character vector is apparent. These dynamics were presented in Chapter 5 and will not be repeated here.

Somerset Maugham was obviously fascinated by resigned persons. The protagonist, Larry, of his biographical novel *The Razor's Edge* is a resigned personage. The reader could do no better than to read two other works of Maugham, to which we refer briefly below.

The first description is that of the character, Dr. Saunders, taken from Maugham's novel *The Narrow Corner:*

...Dr. Saunders lacked this sensitiveness. Unpleasant table manners affected him as little as a purulent ulcer. Right and wrong were no more to him than good weather and bad weather. He took them as they came. He judged but he did not condemn. He laughed. He was very easy to get on with. He was much liked. But he had no friends. He was an agreeable companion, but neither sought intimacy nor gave it. There was no one in the world to whom he was not at heart indifferent. He was self-sufficient.

His happiness depended not on persons but on himself. He was selfish, but since he was at the same time shrewd and disinterested, few knew it and none was inconvenienced by it. Because he wanted nothing, he was never in anybody's way. Money meant little to him, and he never much minded whether patients paid him or not. They thought him philanthropic...

He knew exactly what to say to alleviate the terror or pain of the moment, and he left no one but fortified, consoled and encouraged. It was a game that he played, and it gave him satisfaction to play it well. He had great natural kindliness, but it was a kindliness of instinct, which betokened no interest in the recipient: he would come to the rescue if you were in a fix, but if there was no getting you out of it would not bother about you further. He did not like to kill living things, and he would neither shoot nor fish. He went so far, for no reason other than he felt that every creature had a right to life, that he preferred to brush away a mosquito or a fly than to swat it.

Perhaps he was an intensely logical man. It could not be denied that he led a good life (if at least you did not confine goodness to conformity with your own sensual inclinations), for he was charitable and kindly, and he devoted his energies to the alleviation of pain, but if motive counts for righteousness, then he deserved no praise; for he was influenced in his actions neither by love, pity, nor charity.

The second excerpt is taken from *The Gentleman in the Parlour,* an account of Maugham's travels in the Far East:

> We are gregarious, most of us, and we resent the man who does not seek the society of his fellows. We do not content ourselves with saying that he is odd, but we ascribe to him unworthy motives. Our pride is wounded that he should have no use for us, and we nod to one another and wink and say that if he lives in this strange way it must be to practise some secret vice, and if he does not inhabit his own country it can only be because his own country is too hot to hold him. But there are people who do not feel at home in the world, the companionship of others is not necessary to them, and they are ill at ease amid the exuberance of their fellows. They have an invincible shyness. Shared emotions abash them. The thought of community singing, even though it be but *God Save the King,* fills them with embarrassment, and if they sing, it is plaintively in their baths. They are self-sufficient, and they shrug a resigned and sometimes, it must be admitted, a scornful shoulder because the world uses that adjective in a depreciatory sense.

> Wherever they are they feel themselves "out of it." They are to be found all over the service of this earth, members of a great monastic order bound by no vows and cloistered though not by walls of stone. If you wander up and down the world you will meet them in all sorts of unexpected places. You are not surprised when you hear that an elderly English lady is living in a villa on a hill outside a small Italian town that you have happened on by an accident to the car in which you were driving, for Italy has always been the preferred refuge of these staid nuns...

> They may not be very useful members of society, but their lives are harmless and innocent. If the world despises them, they on their side despise the world. The thought of returning to its turmoil is a nightmare to them. They ask nothing but to be left in peace. Their satisfaction with their lot is sometimes a trifle irritating. It needs a good deal of philosophy not to be mortified by the thought of persons who have voluntarily abandoned everything that for the most of us makes life worth living and are devoid of envy of what they have missed. I have never made up my mind whether they are fools or wise men. They have given up everything for a dream, a dream of peace or happiness or freedom, and their dream is so intense that they make it true.

NON-AGGRESSIVE WITHDRAWN TYPES

With reference to Figure 11 (Chapter 5), the non-aggressive withdrawn types are those of phenotypes N-, N=, N-P, N=P, -N and -NP. The type N- is relatively uncommon, while the types N=, -N, N=P and -NP are severely withdrawn or schizophrenic. We confine our discussion here to the N-P type.

We shall not dwell on the description of the N-P type, for he resembles an NP type who is in a prolonged abject state of melancholic depression. He may have a sanguine complexion, but because of his low energy level, his radiant smile and his narcissistic rage are nowhere to be seen. He tends not to be robust physically and may have many hypochondriacal complaints. He may seem to be somewhat "nervous." In fact, he may teeter through life on the brink of a schizophrenic break.

This individual is almost helpless in life. Like some of his dominant cousins of the NP type, he is cautious, persistent and ruminating. But not only is he unaggressive, he is unmotivated. Competitive society is alien to him, and he is aloof from close interpersonal relations of any kind. His sex drive could hardly be lower. He may indulge in endless melancholic ruminations on his unfortunate state, and may openly declare his desire to leave this world for a better one.

Given his precarious status in life, he may seek solace in art, poetry, religion or nature, where he is a threat to no one and no one is a threat to him. Thus, he may be *asthenic* and devoted to *aesthetics*. In fact, he may be the asthenic or neurasthenic personality of the psychiatric literature.

The N-P type is a figure of pathos, whether he be Rudolf of Mayerling or Vincent van Gogh. A suitable epitaph for the N-P type may be found in the opening lines of Basch's biography of Robert Schumann (who succumbed to psychosis), subtitled *A Life of Suffering:*

> By what comparisons may the life of Robert Schumann best be symbolized? It may be described as an idyll, with tragic elements latent in it from the first, and ending in a tragedy... Here is a human creature upon whom beneficent fairies had showered in profusion their rarest gifts: beauty, candour, a quivering sensibility, enthusiasm for all that is great and noble, communion with the marvels of nature and the mystery of men's souls, and genius to express them in language hitherto unheard; but whom a single jealous fairy crushed by her curse, which, at first comparatively ineffectual, ended by annulling all the virtues of this privileged being, and changing him into a lamentable and pitiful spectacle.

Q.   The character sketches, as I see it, describe phenotypically distinct individuals. But are there not large variations of behavior within each character type?

A.   Unquestionably so. For example, among N types there exists an unctuous or angelic subtype, and an obtrusive conceited one. Among NP types there exist phlegmatic, melancholic, elfish, negativistic, and "nervous" subtypes. In NA types, the baseline temperament can vary widely on the spectrum of hypomanic-depressive behavior. In NPA– types, the degree of motivation, shyness, activity and lability may be quite variable. And so on.

There are without doubt many genes that influence personality, not just the three loci that determine NPA character type. In particular, human temperament, or an individual's general level of activity or excitability in the Pavlovian sense, very likely will be found to have its own separate genetic bases.

We believe that the genetic basis of some subtypes within the major character types will, in time, be worked out. That the various subtypes do indeed have a genetic basis is indicated from studies of twins, especially those reared apart. It has become common knowledge that there exist amazing concordances, even down to fine details, in behavioral mannerisms of identical twins. These concordances have been found to exist even in those twins who have been raised apart in very different environments.

In a commentary reviewing recent studies of identical twins reared apart (conducted at the University of Minnesota by Dr. Thomas Bouchard and co-workers) Malcolm Browne (1980) stated:

> Dr. Bouchard and his team have been scrupulously careful to avoid any social or political inferences that might be drawn from their work, but nonetheless their findings have been provocative. The similarities between twins astonished the researchers. Physical similarities between newly reunited twins were often striking, of course, but so were I.Q. test scores, brain waves, mannerisms, moods, habits, vocational abilities, marital relations, fondness for sports and hobbies, attitudes toward society, and preferences for kinds and amounts of jewelry... Dr. Bouchard's group had difficulty finding financial backing. Many people believe there is a dangerous risk that scientific findings will be used out of scientific context to buttress the views of racists and other bigots. But however the chips may eventually fall, genetic research may tell us something important about ourselves and the limits of human capacity, unless we choose to suppress it.

Q.   Why do you say that the N, P, and A traits are discrete entities? Does not everyone have *some* narcissistic, perfectionistic and aggressive tendencies?

A.   No, not in the sense that we are using the terms. We are proposing that the traits N, P, and A are genetically determined behavioral complexes having their own particular neural pathways at the level of the central nervous system. If an individual does not carry the required genes **n**, **P** or **a**, then his character structure does not include the corresponding trait.

Perfectionism. Considering first the entity *perfectionism,* character types that lack the trait P (e.g., N, A or NA types) can nevertheless deal with various necessities in their lives requiring repetitive activity and the maintenance of order. This must be so, or else the individual would not survive in his society. The very act of eating, for example, requires a certain amount of ordered, repetitive activity. However, when it comes to the individual's drives of ambition, such an individual is simply not drawn to activities of the sort that we have associated with the behavioral complex of perfectionism. The latter is a powerful behavioral trait that is not easily compensated for by other modes of behavior. In fact, if such an individual is forced into a long-term, real life situation requiring perfectionistic activity, he will become deeply frustrated. He will then develop profound anxiety, and may decompensate by flying into a rage or withdrawing into an abject state.

Aggression. Considering next the trait of *aggression,* character types that lack the trait A (e.g., N or NP types) must nevertheless deal with various necessities in their lives in the realm of aggression and defense. Again, this must be so in order for the individual to survive, especially in a society in which many other individuals have the aggressive trait. However, the ways in which such an individual responds to the challenge, interesting and complex as they may be, are based on behavioral modalities other than the aggressive trait. This is best seen from the fact that such an individual is *incapable of mounting an aggressive-vindictive rage.* This is so, but not because he is being a "nice guy" about it. Rather, it is so because this behavioral modality is outside of the realm of his character structure and beyond the capability of his central nervous system. Hence, the compensatory behavioral modalities of an individual lacking the A trait, whatever they may be, we call *pseudoaggression.*

Narcissism. Finally, considering the entity *narcissism,* character types lacking the trait N (e.g., A or PA types) must nevertheless have a certain amount of pride invested in their personal appearance and in their ability to succeed without perpetual conflicts with others. In particular, such an individual may indulge in something of a "narcissistic display" in his attempts to attract a sexual partner. However, whatever compensatory behaviorisms he adopts, they will be based on behavioral modalities other than the narcissistic behavioral complex. Again, this is seen from the fact that under stress such an individual is *incapable of mounting a narcissistic rage.* Such an individual's compensatory behaviorisms we shall call *pseudonarcissism.*

If beyond fairly superficial pseudoaggressive and pseudonarcissistic behavior an individual is forced into a long-term life situation where, against his natural inclinations, he must be overtly aggressive or narcissistic, he will become highly frustrated. And, predictably, he will burst forth in a rage (his own real one!) or retire from the scene in an abject state.

Q.   You have presented pseudoaggressive and pseudonarcissistic behaviorisms in the context of compensatory behavior that the individual adopts selectively in order to survive in his environment. Is there any other context in which the concepts of pseudoaggression and pseudonarcissism are relevant?

A.   Yes. If the individual is severely stressed in his early developmental period, then we propose that he may deeply invert his aggressive or non-aggressive tendency, again as a desperate compensatory mechanism in order to ensure his psychic survival. Such profound inversions of the individual's original basic character structure may lead to pseudoaggressive or pseudonarcissistic behavior that may emerge as a predominant facet of his life-style at maturity.

Thus, according to the present model, an inversion of the non-aggressive trait of the N or NP individual may produce a rebellious *pseudoaggressive sociopathic personality,* examples of which would be the "con artist" and the compulsive bank robber.

Similarly, an inversion of the aggressive trait in an individual not possessing the N trait may produce an *exhibitionistic pseudonarcissistic individual.* Such an individual might be drawn to extreme measures of self-adornment, to the dance, and to other activities that we would associate with the trait of narcissism. On the scales of temperament and emotional tone he could be either an introvert or extrovert.

Q.   You are regarding pseudoaggressive and pseudonarcissistic inversions to be compensatory defense mechanisms. What exactly are these inversions defending against?

A.   They can be, as a last resort, psychic defense mechanisms against a suppression of ambition. Thus, they can be mechanisms of defense against the loss of psychic identity, or schizophrenia.

We mentioned earlier our concept that schizophrenia is a withdrawn state that can develop if an individual's aggressive or narcissistic drives of ambition are suppressed by environmental factors. For example, considering an NP individual, if his narcissistic ambition is crushed by unfavorable factors during maturation, then he is pushed to the brink of a schizophrenic withdrawal. By adopting a pseudoaggressive sociopathic personality he maintains a semblance of a drive of ambition, but at the expense of a life-style that is at variance with that of the majority of other individuals of his society.

Considering an individual lacking the N trait (for example, a PA type), if his aggressive ambition is suppressed by unfavorable factors during maturation, then once more he is pushed to the brink of a schizophrenic withdrawal. By adopting a pseudonarcissistic personality he compensates for his blunted aggressive drive, but again at the expense of a life-style that may be at variance with that of the majority of the individuals of his society.

Finally, we note that the character vector of submission may be considered to be a dramatic inversion of the drive of aggression. However, individuals of the NA= and NPA= types have a full measure of the trait N, hence the concept of pseudonarcissism is not relevant to these types.

For a discussion of the possibility that schizophrenia, as well as the traits of pseudoaggression and pseudonarcissism, may be determined primarily by genetic factors, see Chapter 11.

Q.   I know of some individuals whom I cannot clearly delineate into one of the character types of the model. What is wrong?

A.   You probably do not know the individual well enough. In unstressed circumstances all of the personages may be rather bland, good-natured individuals. In fact, from a behavioral standpoint three of the most useful questions to answer initially are:

- What kind of rage does the individual display, i.e., N or A-rage, or both?
- How does the individual react in the limelight?
- How does he interact in complex social situations?

One can best answer these questions by being well acquainted with the individual and by observing his behavior in a stressful competitive society.

Q.   What do we mean when we say that an individual is "not ambitious"?

A.   According to our model, all individuals have ambition, based on the narcissistic and aggressive behavioral complexes. Indeed, otherwise an individual would be schizophrenic and, except to satisfy primitive physical needs, would not be motivated to get out of bed in the morning.

But, it is true, an individual's ambition can be blunted.

An individual's aggressive drive can be inverted, and he may adopt the character vectors of submission or resignation. His narcissistic drive can be blunted, and he may become a borderline autistic individual. If he has only one component of ambition and is crushed during maturation, then his psyche may take refuge in pseudoaggression or pseudonarcissism, and he may be branded as "unwilling to compete on an equal footing with others" in his society.

If an individual is subjugated, or simply dominated, by his companion or mate, and in particular by a "power behind the throne," then he may seem to be "lacking in ambition." Or, his life situation may be so lacking in competition and stress, with everything being handed to him on a silver platter, that he will simply not be motivated to be "ambitious." Or, at the other end of the spectrum, if he is deeply frustrated and is depressed into a prolonged abject state, he will certainly be relatively "unambitious."

Finally, an individual may have an organic illness. If he is afflicted with a fundamental metabolic disorder, whether inherited or not, whether of intrinsic or extrinsic origin, and whether self-inflicted or iatrogenic, then the individual's mental state can be affected. And if his mental state is altered, then his functioning level of "ambition" may be altered, insidiously or dramatically, as well.

Thus, the apparently simple concept of "being ambitious" has many nuances, and touches virtually all of the facets of our model.

Q.   There is something bothering me with regard to the submissive individuals. In the animal model you stated that these individuals were "trained" and "flogged" into submission. Do you wish to apply such terminology to the human model?

A.   No. We believe that it is not well understood exactly why, or at what point in the human individual's development, he adopts a submissive basic character vector. Perhaps, rather than some adverse condition or negative influence having been inflicted on the individual, there may have existed a lack of some positive factor in the realm of his interaction with his parents or caretakers.

This may have occurred during the first year of the individual's life, perhaps even during the first few months or weeks. This possibility must not be ignored in view of what we know about the phenomenon of *imprinting* in lower animal forms.

Q.   What is "imprinting"?

A.   Imprinting is a process by which an animal adopts a pattern of behavior after having been exposed to certain environmental conditions during a critical period early in the animal's life. If the necessary environmental conditions do not occur, then the normal pattern of behavior is stunted forever.

For example, a newly hatched gosling must learn to identify its mother and learn to follow her within the first 36 hours of hatching. If it does not, then it will never learn to follow her. Or, if during that period the gosling is exposed to a dog, or even to a toy train, then it will forever be "imprinted" on this surrogate mother. Similarly, some birds must learn their song by imitation during a critical period of their early lives, and if they do not, then their ability to do so is lost forever.

There is some evidence that such environmental factors should be considered in the formation of the submissive basic character vector, at least in some cases. For example, we might speculate that early in its life the infant must receive some stimulus from the mothering individual in order to be able to fully develop the traits of aggression or narcissism.

Q.   Well, I must repeat, those characters make up quite a rogues gallery! I must confess, though, that I found myself among the personages...

A.   We knew that you would.

Q.   ...but I did not agree with all of your descriptions.

A.   Feel free to disagree. It is quite possible that some of the behaviorisms that we described were more in the realm of local cultural adaptations than basic qualities of the character structures. In addition, besides being caricaturized, the various personages were described in a biased way, having been seen through the jaundiced eye of an individual of one particular character type. Finally, it is simply not possible to distill the variability in temperament found in the various character subtypes into unique character sketches that would do justice to all of them.

Q.   Could you summarize, in a concise list, the various character types?

A.   Here they are:

CHARACTER TYPES OF NPA MODEL

I. DOMINANT TYPES

N	Narcissistic. Sanguine personality. "The non-aggressive non-perfectionist." "The self-anointed glory seeker."
P	Perfectionistic. Stillborn child. "Failure to thrive" infant. Endogenous schizophrenic.
A	Aggressive. Choleric personality. "The arrogant dynamo." Non-sanguine autocrat.
NP	Narcissistic Perfectionist. Obsessive-compulsive personality. Phlegmatic-melancholic personality. Bovine personality. "Nervous bird" personality. "The quiet achiever."
NA	Narcissistic Aggressive. Cyclothymic, hysterical or hypomanic-depressive personality. "The ambitious predator."
PA	Perfectionistic Aggressive. Passive-aggressive personality. Austere melancholic personality. Paranoid personality. Pseudonarcissistic extrovert. "The chronic complainer or criticizer." "The sardonic wit." "The suspicious manipulator." "The Power behind the throne." "The brooding non-sanguine autocrat."
NPA	Narcissistic-Perfectionistic Aggressive. Explosive personality. "The overbearing achiever." "The sanguine autocratic tyrant."

II. SUBMISSIVE TYPES

Non-Compliant types (n-C)

A–	Schizoid personality. Non-perfectionistic, non-narcissistic agitated achiever submissive.
NA–	Narcissistic non-compliant. Active non-perfectionistic submissive. Histrionic introvert.
NPA–	NP-like non-compliant. Agitated achiever submissive. Reactive depressive personality.
PA–	Schizoid-paranoid personality. Non-narcissistic submissive. Perfectionistic introvert. Pseudonarcissistic introvert.

Compliant types (C)

A=	Juvenile or maturity onset paranoid schizophrenic.
NA=	Narcissistic compliant. Quiet non-perfectionistic submissive. Narcissistic masochist.
NPA=	NP-like compliant. Quiet achiever submissive. Depressive, masochistic personality. "The shrinking violet."
PA=	Juvenile or maturity onset paranoid schizophrenic.

III. RESIGNED TYPES

–A	Maturity onset schizophrenic.
N–A	Narcissistic resigned. Non-perfectionistic resigned.
NP–A	NP-like resigned. Quiet achiever resigned.
P–A	Maturity onset paranoid schizophrenic.

In addition, we summarize below separately the various character types that lack any aggressive component. We confine ourselves to two levels of depression of narcissistic ambition before maturity (e.g., N–P and N=P) and one level after maturity (e.g., –NP):

NON-AGGRESSIVE TYPES

I. DOMINANT TYPES

N	Narcissistic (see above).
NP	Narcissistic Perfectionist (see above).

II. WITHDRAWN TYPES: *Juvenile and Maturity Onset*

Juvenile Onset

N–	Delayed speech in childhood. Schizoid personality. Non-perfectionistic introvert.
N–P	Delayed speech in childhood. Perfectionistic introvert. Borderline or "successful" autistic personality. Pseudoaggressive sociopathic personality. Asthenic personality. "The asthenic aesthete."

Juvenile or Maturity Onset

N=	Narcissistic schizophrenic.
N=P	Autistic child. Maturity onset schizophrenic.

III. WITHDRAWN TYPES: *Maturity Onset*

–N	Narcissistic schizophrenic.
–NP	Schizophrenic break at maturity (hebephrenic, catatonic and acute onset paranoid forms).

Q.   You mentioned sporadically in your descriptions references to these characters' *complexions, smiles, gestures, photographs* **and** *handwriting.* **Could you summarize what you found?**

A.   There are no inviolate rules in these categories. Nevertheless, we found the following trends:

COMPLEXIONS

It goes without saying that an individual's facial complexion must be interpreted in light of racial, regional and environmental factors, and in light of the state of the individual's health:

Tending toward sanguine or flushed: N, NP, NA, NPA. Also, submissive, resigned and withdrawn types having the N trait, including the schizophrenics N=P and –NP.

Tending toward pallid or sallow (non-sanguine): A, PA. Also submissive and resigned types lacking the N trait, including the schizophrenics PA= and P–A.

SMILES

Dominant types

N	Radiant, gingival smile.
A	"Pleased-with-self" grin.
NP	Sudden warm, radiant sheepish smile, like Cheshire cat appearing in the mist. Often gingival smile.
NPA	Warm maternalistic or paternalistic smile.
NA	Flashy, glamorous, toothy smile of movie star.
PA	Non-symmetric grin or grimace. Mona Lisa smile. Sardonic smirk. Frozen, toothy but non-gingival grin. Half-open mouthed grin. Grin with a short repetitive laugh.

SMILES (ctd.)

Submissive and Withdrawn types

NA– NPA– NA= NPA=	Warm smile when at ease. Nervous smile otherwise.
PA– A–	See PA.
N–P N–	Very rare sheepish or radiant smile.

Resigned types

N–A NP–A	Smiles easily. Similar to subdued NA and NPA.

PHOTOGRAPHS

Dominant types

N	Looks at camera. Broad charismatic gingival smile. Starry-eyed smile.
A	Looks at camera. Relaxed with trace of "smile."
NP	Looks at camera. Relaxed face; sheepish or radiant "limelight" smile.
NPA	Looks at camera. Warm maternalistic or paternalistic smile.
NA	Almost invariably looks at camera. Flashy "movie star" smile.
PA	Usually looks at camera. May pompously look away from camera. Not relaxed. No smile, tight lipped sardonic smile, non-symmetric grin or grimace. Non-gingival "frozen smile" showing teeth. Laughs, or tries to laugh, showing expressive extroverted countenance.

PHOTOGRAPHS (ctd.)

Submissive and Withdrawn types

NA– NPA–	Often avoids looking at camera. Not at ease and not smiling. Tries to laugh, uses props, mugs for camera. Sometimes flamboyant "narcissistic arms" gesture without smile.
PA– A–	Often avoids looking at camera. No smile.
NA= NPA=	Looks at camera with warm smile when with protectors. Otherwise ill at ease before camera.
N–P N–	Looks at camera. Melancholic countenance.

Resigned types

N–A NP–A	Looks at camera. Smiles easily. Similar to subdued NA and NPA.

Q.   **You indicated that individuals of some of the character types, namely those lacking the narcissistic trait, may in certain social situations substitute laughter for the smile of recognition. What is the difference between the two?**

A.   An individual's natural smile is his response to his recognition in the eyes of other individuals. Laughter is of much more complex origins, but it is essentially a reaction of pleasure in a real or vicarious situation which is in harmony with the individual's basic character structure.

For example, an individual having sadistic trends may laugh when someone is injured in ironical circumstances. A submissive individual may laugh in vicarious amusement when he views an animated cartoon of the fable depicting the tortoise overtaking the hare. A resigned individual may find detached amusement in observing others frantically "playing the game," and so on. Conversely, an individual certainly would not laugh in a situation that is at variance with his character structure, and in fact he may be very ill at ease in such a situation. We see, therefore, that the beginning and "tail end" of a response of laughter may produce a facial expression that can mimic the warm smile of recognition, but the two responses obviously have very different roots.

GESTURES

The subject of facial expressions and arm gestures is too complex to be considered here in detail. We shall mention only a few.

The gestures and poses associated with the trait of *aggression* are well known to us: the haughtily cocked jaw, the furrowed brow, the strained tight-lipped mouth, the aggressive finger point, the upraised clenched fist, and the intimidating glare.

The gestures associated with the trait of *narcissism* however, have not been generally recognized. We should like to emphasize one in particular, namely the *narcissistic arms gesture* in which the arms are extended in front of the individual, with the palms up and the fingers somewhat spread apart. It is a pose often assumed by singers and by religious leaders when praising their gods.

Our interpretation of this gesture is as follows: it is a "limelight" gesture of recognition, saying essentially, "Hey, I'm the greatest!" Thus, it is an arm gesture that complements or takes the place of a smile. In the singer it replaces the smile because the individual's lips are occupied with singing. In a public speaker, or in the religious leader, it replaces the smile because the individual is praising himself in inverted terms, and a true smile would be inappropriate. Finally, this gesture sometimes appears in informal photographs of some $n\text{-}C$ submissive types as the equivalent of a smile, since these individuals have a strong narcissistic component in their character structures, but in their self-consciousness simply cannot smile at will for a photograph.

The reader can rapidly verify for himself that the "narcissistic arms" gesture and its variations are widely used in various contexts. We do not believe that it is a frivolous gesture. Rather, we believe that, like the smile of recognition, it has powerful instinctual origins.

As an example of the gesture, we may quote Herndon's description of Abraham Lincoln on the podium:

> Sometimes, to express joy or pleasure, he would raise both
> hands at an angle of about fifty degrees, the palms upward,
> as if desirous of embracing the spirit of that which he
> loved.

Examples of the "narcissistic arms" pose may be seen in in an informal photograph of Somerset Maugham (Plate 15) and in Bellini's painting of "St. Francis of Assisi in Ecstasy" (Plate 16).

Other narcissistic mannerisms are the *Napoleonic stance* of supreme vanity (Plate 40) and what we call the *Joan of Arc pose,* in which the individual's eyes are directed toward the heavens when accepting recognition in the limelight. Finally, the *deep bow,* often accompanied by a sweeping arm, is another narcissistic gesture of recognition. This gesture is sometimes mentioned in accounts of megalomanic schizophrenics (see Chapter 11).

HANDWRITING

Dominant types

N	Very variable. May be beautifully well-formed, with embellishments, but non-perfectionistic. Scribbles to himself.
NP	Almost invariably well-formed letters. Sometimes striking calligraphic quality.
NPA	Highly variable among individuals. Usually well-formed letters but may be execrable. Signature often becomes illegible as dominant position in society is attained.
NA	Usually good in female. Often slurred in male.
A PA	Usually good in female. Often slurred or bold illegible scrawl in males. Sometimes messy corrections. Sometimes bold flourishes.

Submissive and Withdrawn types

n-C	Usually legible but "jerky" quality with large variations in size of letters. Sometimes slurred. Sometimes messy corrections. May be highly perfectionistic in PA type.
C	Usually neat and legible.
N–P	Usually scrupulously neat. May be calligraphic.

Resigned types

N–A NP–A	Usually neat and legible.

Q.   Where do we go from here?

A.   We are going to look at the character structures from a subjective point of view, that is, from the point of view of the individual himself. How does he see things? And we shall see that he views life very differently from the manner in which we have portrayed him. But this is not surprising since, not knowing his basic character structure, he goes through his daily life like an automaton blundering his way through a thick fog.

CHAPTER 7

JUSTIFICATION OF ONE'S EXISTENCE

I am the master of my fate: I am the captain of my soul.

W.E. Henley (ca.1890)

Everybody acts not only under external compulsion but also in accordance with inner necessity. Schopenhauer's saying, that "a man can do as he will, but not will as he will," has been an inspiration to me since my youth up, and a continual consolation and unfailing well-spring of patience in the face of the hardships of life, my own and others'.

Albert Einstein (1949)

OUTLINE OF CHAPTER SEVEN

POWER & GLORY

RELIGIOUS ECSTASY

ANXIETY & HAPPINESS
    Self-esteem, or pride
    Expectations in life

TRIUMPHS
    Aggressive-vindictive
    Narcissistic
    Submissive
    Resigned

DEFENSE MECHANISMS
    General
        Rationalization
        Intellectualization
        Denial
        Externalization

    Specific

HEDONISM

SOURCES OF INTRA-PSYCHIC STRESS
    Insufficient self-esteem relative to expectations
    Streamlining the behavioral complexes
    "Playing the game"

SELF-HATE & SELF-DESTRUCTION

Q.   What do you mean by the justification of one's existence?

A.   Each individual is being continually subject to severe intra-psychic stresses, whose bases lie in the inborn character structure that remains for the most part unknown to him. Whatever his station in life, he desperately attempts to maintain a semblance of psychic equilibrium. He attempts to streamline himself and his surrounding environment into some sort of sense of cohesion. If he did not do this, then his psychic unity would collapse like a house of cards. He would not be able to justify his existence, and he would follow a path to self-destruction.

Q.   How is it that the basic nature of the human character structure has gone unknown for so long? How can it be that man's desperate attempts to maintain a sense of psychic unity have gone unnoticed?

A.   These attempts have not gone unnoticed. As we have noted, each individual goes through life in a continual state of anxiety, not understanding the real reasons for his anxiety or why he assumes, sometimes rather illogically, states of rage, exhilaration or abjectness. He will on occasion state aloud that he does not understand himself. He will for fleeting seconds ponder the unthinkable — whether he is abnormal, or whether he is frankly insane. Accidents or self-destructive tendencies in others may horrify him, since they remind him of his own vulnerability. And silence and darkness make him uneasy because they give him a glimpse into the void in his life that he somehow just cannot fathom. And it is this void and the profound intra-psychic conflicts that surround it that lead him to the search for solace in the psychiatrist. And, of course, the psychiatrist can help him but little, since he too does not understand himself.

To continue further, philosophers throughout the ages and the great literary figures of the past have in dramatic terms described man's search for his inner soul, his search for an understanding of the powerful forces that drive him to ambition and to destruction.

Henrik Ibsen in his remarkable play *The Master Builder* vividly portrays his understanding of man's innate drives for power and glory. At the same time, Ibsen demonstrates that man understands little of the true nature of his drives. In his character Hilda we find a young lady of the NA "bird of prey" type. She enters the life of Solness, a great architect and builder, who represents to her an idol, a father figure and the ideal image of her own internalized passions for ambition. She is sexually predatory, as is shown by the reference to her as a "flirt," and by her inverted denial of desiring an escapade with Solness, supposedly because she would not want to hurt his wife ("I just *can't* hurt somebody I *know!*"). She exhibits sadistic behaviorisms and is drawn to use the word "thrilling" repeatedly. When Solness begins to relate an incident in which his house burned down, Hilda cries out excitedly, "What happened? Go on! Was anyone burned?"

The other aspect of Hilda that emerges clearly is her unbridled ambition. She wants a kingdom with a castle: "My castle must stand up — very high up — and free on every side. So I can see far — far out." Later, she wants "castles in the sky." Thus, she shows her narcissistic longings in typical human terms: she wishes to achieve, literally, great heights, but she focuses exclusively on the ends and not on the means. She wants her castle to be *given* to her. The fact that her castle may turn out to be only a "castle in the sky" signifies both that her ambition is out of proportion to reasonable human expectations and that it is destined not to be fulfilled. But Hilda, with Solness, dreams of flying with her legs tucked in under her like a bird to the dizzying heights of undirected ambition. Finally, despite her deep desire to dominate others, Hilda longs for a superman — a Viking — to capture her, carry her off and dominate her. We see in Hilda, therefore, the classic intra-psychic conflicts of the narcissistic-aggressive personality.

Ibsen makes a more general statement of man's craving for ambition in the character of Solness, who represents Hilda's idol and ideal image. Solness carouses of the liquors of life like a vortex out of control creating ever-expansive buildings and ever-taller towers. However, he repeatedly questions himself as to whether he might be insane. At last, in the final scene, spurred on by Hilda, he drives himself to climb the highest of high towers. He is seen as a tiny, struggling figure, high up on a scaffolding, and at those dizzying heights he is seen to be conversing with someone... to be conversing with God. But no, he is not only conversing with God, he is shouting his defiance to God! And the futility of the immensely powerful, but completely misunderstood, drive for ambition is shown when abruptly Solness' body is seen to come crashing down to earth.

A similar theme is presented in Spielberg's film *Close Encounters of the Third Kind.* In this film the protagonist, Roy Neary, is a young worker who finds himself leading a rather aimless existence. He obtains no special satisfaction from his job, and his relationship with his wife and children is a distant one. He overtly shows sadistic trends in mocking his children's faults, and in a striking scene he deliberately causes a wreck of their toy trains.

One night in a dream-like state he sees, in his mind's eye, what are apparently airships from outer space. He becomes overwhelmed with obsessive thoughts, and he leaves his home to follow compulsively a path leading to the heights of a rocky peak. Along the way he has met a young woman who was leading an equally aimless life and is equally confused. Even when he is about to climb to the summit of the peak, he does not understand his motivations. He is interviewed by a man who does not even speak his language. He cries out in desperation, as if to say, "Will someone tell me what the hell is going on here?" The film ends when our hero climbs to the summit, despite all odds against him, and is finally, apparently, assimilated into the bosom of a ship from outer space. There he stands, in an aura of colored lights and the most glorious music imaginable, dazzled, serene, and for the first time in his life at peace with himself. He lifts his eyes toward the iridescent paradise that is enveloping him, and there is no doubt in our minds that he is undergoing a profoundly moving religious experience.

What we finally realize — and the realization comes so forcefully that there is no escaping from it — is that man's drive for ambition, for accomplishment, and his drive to build, to climb and to fly, and his desire for power and glory, are in their acute intensity ultimately registered as a religious experience. In the writings of Christianity we find:

> "Our Lord and God! You are worthy to receive glory,
> honor, and power. For you created all things, and by
> your will they were given existence and life."

Revelation 4:11

We now realize that in honoring the power and glory of God, man is in inverted terms honoring the power and glory of himself! He is glorifying his own ambition, which is rooted in his behavioral complexes of narcissism and aggression.

Once we have come to this realization, we begin to understand the true nature of religion in its human context. On the one hand, we realize why composers of music, mountain climbers, aviators and astronauts are sometimes inclined to be enveloped in a religious-like aura of awe, ecstasy or humility. On the other hand, we begin to understand the powerful need of man — in virtually all cultures — to create religions, so that he may render homage to his own ambition. This need has proven throughout the ages to be so powerful, so incapable of being stifled, that it is indeed fitting, proper, and not at all surprising, to realize that it is written in the genetic code of our narcissistic and aggressive behavioral complexes. And it is man's perfectionist nature — not, as Freud would say, his repressed sexual urges — that allows him to actually build those cathedrals and statues, those steeples and those mountaintop pagodas of his religious fervor.

Finally, man's narcissistic nature promises him, to himself, accomplishments that are not realistic. They are often so outlandish as to be absolutely incredible. And they are often, like an object of an ulterior motive, expected to come to him effortlessly, to be dropped into his lap so to speak, with no effort having been made on his part. And, understanding man's narcissistic nature, we must ask ourselves whether man is realistic when he promises himself immortality, whether it be a reincarnated existence in his own world, whether it be life everlasting in a self-created paradise, or whether it be an immortal soul of man fulfilling a preordained human destiny.

> "To him who sits on the throne
> and to the Lamb,
> be praise and honor,
> glory and power,
> *forever and ever!*"
> "Amen!"
>
> *Revelation* 5:13

Q.   Could you kindly come back down to earth and return now to our individual. How does he view himself in relation to others?

A.   Only on rare occasions will he admit that within himself reigns near-total confusion, and that this is at the same time the source of his inability to really understand and to sympathize with the plights of other individuals with whom he interacts. The perennial story of the devil's pact symbolizes this confusion. Man, in quest of an illusory fulfillment of a poorly understood ambition for power, glory and material wealth, will at the slightest nod from the devil sell his soul, by dominating others by misguided narcissism or aggression. And in so doing, he renounces his ability to give to others his sympathy, compassion, empathy and love. And in losing his ability to be close to others, he loses at the same time, without realizing it, a portion of his free will.

Somerset Maugham in the following passage from his *Writer's Notebook* stated succinctly this nagging inability of the individual to come to terms with himself and with others:

Sometimes one feels rage and despair that one should know so little the people one loves. One is heart-broken at the impossibility of understanding them, of getting right down into their heart of hearts. Sometimes, accidentally or under the influence of some emotion, one gets a glimpse of those inner selves of theirs, and one despairs on seeing how ignorant one is of that inner self and how far away from one it is.

Q.   Let us go back to the question of how the individual perceives his own life. How does he see things?

A.   Although he is in a fog, he desperately seeks security in a feeling of unity. He must have a reassuring feeling of an integrated self, by which he may justify his very existence. If he did not have a feeling of pride in himself then he would not be able to get up in the morning and manage to live through his daily activities.

Q.   What is pride?

A.   Pride is self-esteem, in a non-pejorative sense. That is, one may have pride in one's family, one's intelligence, one's abilities to work well with others, and so on.

Q.   What is one's pride based on?

A.   It stems from two sources. First, it is based on *real qualities* or merits that the individual possesses. These merits are themselves determined both by genetic and environmental factors. For example, among the real merits that an individual possesses we might list an intact healthy body, good physical coordination, intelligence, articulateness, artistic qualities, and so on. A good many of these qualities would be highly dependent on the individual's family life and educational experience during his developmental period.

Second, whatever real merits an individual may have developed, these are overlaid on his *basic character structure,* which, as we have discussed, is also both genetically and environmentally determined. Now, the individual's character structure is the basis of immensely powerful intra-psychic forces. *These forces are essentially inviolable.* And it is in their inviolability that the individual is compelled — whatever the logic of his life situation — to think obsessive thoughts and act out his often bizarre compulsive actions.

In order that the individual be able, literally, to live with himself, he attempts to integrate the requirements of his basic character structure with the real merits that he has acquired in life. However, he simply does not see the two separate sources of his behavioral structure. The two sources are inextricably combined in his mind. If you mention to him his obsessive-compulsive qualities, he will not, for the most part, have the slightest notion of what you are speaking. In the final analysis, then, the individual develops an integrated sense of pride, of self-esteem, based not only on his innate or acquired real merits, but also on the obsessive-compulsive demands of his basic character structure.

For example, an *aggressive* person will not only have pride in his ability as a leader or public speaker, but also in his ability, compulsively, to dominate all other individuals whatever the circumstances. He will always feel self-justified in his actions. In his own mind, things could simply not be any other way.

Similarly, a *narcissistic* person will have immense pride in his creativity and in his future abilities. He becomes in his own mind an anointed visionary, a prophet. A *narcissistic-perfectionist* person will not see himself as a plodder. Instead his pride rests in his sense of duty and in his sense of doing things meticulously well.

A *submissive* person will have pride in his helpfulness, his considerateness, his lovability, and his saintliness. A *resigned* person will have quiet pride in his lack of imposition on the rights of others.

Q.   Are these individuals happy? In fact, what is happiness?

A.   All these individuals have their psychic integrity vested in their feelings of self-esteem or pride. Happiness is a subjective measure of the individual's accomplishments and feelings of pride in relation to his expectations in life.

To summarize, an individual's pride rests on his feelings of self-worth or self-esteem. It is based on his basic character structure, as well as on the real merits that he possesses in his particular life situation. Happiness is a measure of the adequacy of the individual's feelings of pride, in what he is and what he has accomplished, in relation to his expectations in life.

Q.   What determines an individual's expectations in life?

A.   An individual's expectations in life stem from both intrinsic and extrinsic sources. The main intrinsic source is the individual's bask character structure itself.

For example, *dominant* individuals are motivated by the components of ambition, that is, by the requirements of narcissism, aggression or both. Such individuals will tend to be competitive, whether quietly or overtly, and will tend to have high expectations of themselves. In fact, their expectations of the attainment of power and glory may be frankly outlandish.

Submissive individuals may have a taboo on ambitious ventures based on competition with others. The expectations in life of such individuals lie chiefly in being cared for and protected by others. However, submissive individuals may possess ambition in camouflaged forms. The defensive thinking and opportunistic means by which a submissive person may satisfy his needs for ambition will be discussed later in this chapter.

Finally, the expectations in life of a *resigned* person are muted, as they are in the case of submissive individuals. These expectations lie in the realm of various deeds that the resigned person accomplishes in the splendid isolation of his ivory tower.

The second main source of an individual's expectations in life lies in factors originally *extrinsic* to the individual. These lie in the realm of the expectations of the individual's society, his family, his boss, his mate, his children, and so on. These expectations of others, to the extent that they become internalized, become assimilated by the individual, hence they become expectations that the individual demands of himself. In some *dominant* individuals, the expectations of others may be readily assimilated, perhaps even wildly surpassed. In *submissive* and *resigned* individuals, the expectations of others may be tenaciously resisted or only grudgingly accepted.

In the final analysis, the sum total of an individual's expectations in life is a complex entity. It stems from both intrinsic and extrinsic sources. Some aspects of the individual's expectations in life lie in the realm of aspirations, that is, his actively seeking power and glory. Others lie in the realm of defense, of his building fortifications against the intrusions of life, and of consolidating his gains.

Finally only a fraction of the individual's expectations in life are consciously perceived for what they are. To be specific, those aspects of an individual's expectations in life that are based on his basic character structure are either totally unknown to him, or at best are badly misinterpreted by him. It is as if each individual possessed a secret portrait of his idealized self, representing his expectations in life. Unfortunately, this idealized portrait is grossly distorted because of the inability of the individual to perceive his true motivations, stemming from his basic character structure.

We come to realize why the individual is in such a state of free-floating confusion. His happiness is dependent on how well his pride structure matches his expectations in life. But both his pride structure and his expectations in life are dependent in good part on his motivations, stemming from his basic character structure, which he consciously does not understand one whit. He does not realize the fantastic claims on life that his character structure engenders. He does not realize the self-appointed status of megalomania to which he elevates himself.

Hence, when the individual's expectations exceed the more modest accomplishments that his life situation can generate, he is not only in a subjective state of unhappiness, but he is also profoundly confused. And it is this state of confused unhappiness that characterizes most of the lives of most individuals. It is what the psychiatrists would call *anxiety.* And to the extent that the true causes of the confusion remain a mystery to the individual, the psychiatrists would call it a *free-floating anxiety.*

Q.   As the individual becomes anxious, what does he do?

A.   He attempts to *compensate* for his inability to match his life situation to the requirements of his character structure (or his inability to match his accomplishments to his expectations). His first instinct tells him not to abdicate. Rather, he will seek victory. He will seek to triumph over his particular life situation.

Q.   How does he seek to triumph over his life situation?

A.   Each character type has, so to speak, his own way of doing things. We will summarize how he seeks to triumph in his life, with the understanding that each personage has little awareness of the particular forces that drive him to seek his particular mode of triumph.

Aggressive personages — In the aggressive approach to life, the individual elevates himself to a self-appointed status of omnipotence. He requires power over all others and at all times, whether he merits it or not. If all other individuals yield to him at all times, then this is a successful solution to life for him. He will accept their abdication and will usually accept their recognition of him. He may even "smile."

If others do not yield, however, then he seeks victory by intimidation in the form of the *vindictive triumph.* This is usually done in the context of "playing the game," but he may sometimes seek diffuse vindictive triumphs over unseen persons or over large groups of persons.

If his efforts toward a vindictive triumph are not successful, he may be incited to the *vindictive rage,* which may finally produce the desired result.

Narcissistic personages — In the narcissistic approach to life, the individual elevates himself to a self-appointed status of a glorified being. He requires that others flatter him, glorify him and support his self-anointed status. If other individuals accept his status and cooperate with him, then this too becomes a successful solution to life for him. He will accept the honor and recognition that others give him. He will, most willingly, accept his position in the limelight, and the smile of smiles that he displays could not possibly be more sincere.

If others do not yield, then he will seek victory by intimidation in the form of a *narcissistic triumph.* He will caution others that he is the only one who can solve the difficulties to be conquered, and he will warn them that if they do not cooperate, then he may be forced to go elsewhere.

If his efforts towards a narcissistic triumph are not successful, then he may be incited to the *narcissistic rage.* If after he leaves, others run down the corridor and beg him to return, then the rage will have produced the desired result.

Submissive personages — In the submissive approach to life, the individual elevates himself to a self-appointed status of lovableness, helpfulness and saintliness. If others accept his presumptuous status, then his solution to life is a successful one. He will accept his position of subjugation so long as others care for him, protect him and totally provide for him.

If others do not yield to his demands for care, then he will seek victory by intimidation in the form of a *submissive triumph.* He will demand retribution in compensation for his helpfulness and fidelity. He will weep. He will threaten to give up all active life. He will threaten to "go to pieces." He will threaten self-destruction.

If his efforts towards a submissive triumph by intimidation are not successful, then he may go into an abject withdrawn state of *masochistic suffering.* In this state he will make it plain to his master or protectors that he must be rescued and once more taken into protective custody, otherwise he may simply shrivel up and die.

If his efforts to intimidate his protectors by the mechanism of masochistic suffering are, they too, not successful, then he may, at long last, finally become vindictive. He may, in fact, fly into a rare vindictive rage. Although the rage is unlikely to accomplish the desired objective, it will certainly surprise his protectors and may leave the individual himself aghast at the sight of the aggressive-vindictive demons that lurk in his soul.

Resigned personages — In the approach to life based on resignation, the individual elevates himself to the self-appointed status of splendid independence. If others accept his status and he manages to achieve a life situation in which his meager needs are satisfied, then his solution to life is a successful one. He will be happy to do what he can for others, so long as they do not coerce him with demands, requests or expectations.

If others do not respect his magic circle of detachment, then he will seek victory by intimidation in the form of a *triumph of detachment.* He will become recalcitrant and brooding. He will make others ill at ease and make them feel that they are imposing on him with their demands. He will threaten to leave.

If his efforts towards a triumph are not successful, he may be incited into a *rebellious rage,* essentially narcissistic in content, which may finally produce the desired result.

Q.   If all other individuals cooperate with our personages, then this registers subjectively as happiness. Is this correct?

A.   Yes. But judging from experience, this is a rare occurrence. More often than not, in a competitive society, the individual is forced to seek compensation in the form of triumphs in an overlay of anxiety.

In the case that the individual's solution to life works exceedingly well, he may enter an *exhilarated state.* This may occur in acute situations, for short periods of time, and would be characterized by euphoria, laughter and, for a dominant individual, an accentuation of his ambitious desires. In rare cases, an individual can remain, for long periods of time, in a chronic exhilarated state if his mate or life situation supports, in all of its aspects, the requirements of his character structure. In the narcissistic individual such a state may manifest itself as a chronic exhilarated state of evangelical fervor.

Q.   If the individual's life situation does not perfectly match the needs of his character structure, how does he retain his psychic integrity?

A.   He does this in many subtle ways, each one more ingenious than the other. The important point to recognize is that his objective is to convince himself, if not others, that the needs of his character structure fit perfectly any and all life situations that he encounters.

He uses all kinds of alibis for the *self-justification* of his actions. In victory, but especially in defeat, he is the master of *rationalization;* he didn't really care what happened or happens; he wasn't really trying; someone else's victory was just luck, and so on. There is, in short, no failure of his ability to meet the requirements of his character structure that cannot, somehow, be rationalized so that he emerges unscathed, like the red, red rose.

He may use the mechanism of *intellectualization* to raise any of his failures to a higher plane, and then convince himself that since he understands his failure in all of its aspects, then it simply no longer exists. Hence, he cultivates *blind spots* and uses the defense mechanism of *denial.*

In an effort to maintain his pride structure, he will use the mechanism of *externalization.* That is, he will attribute any or all of his failures to the failures of others. And again, there is virtually no problem in his life that he cannot, somehow, lay at the doorstep of someone else. His reasoning in this regard is unconscious and instinctive. His actions are reflexive and compulsive. When he states that others are the cause of his problems, there is no hope of using logic to dissuade him. He believes that he is right because his character vector is unfailingly pointing the way. He believes that he is right, down to the very marrow of his bones.

Finally, if these mechanisms of defensive thinking are not sufficient, the individual may find himself leading a fragmented life. By compartmentalizing his daily life into sequences of unrelated activities he blunders through life like a blind automaton, but somehow manages to convince himself that his life has an integrated, unified structure.

Q.   I see that the above thought processes are general mechanisms that the individual uses to defend his psychic unity. How, specifically, does he approach life?

A.   As the individual grows to maturity he develops a feel, instinctively, for situations that are in harmony with his basic character structure. As he begins to realize that a potentially hostile world awaits him outside, he will by necessity perceive these situations as ones that promise him safety. Conversely, the individual develops, instinctively, a feeling for situations that are in discord with his basic character structure. He perceives these latter situations as ones of danger to his psychic integrity.

In light of the above, we see that by necessity the individual is severely limited to a narrow path in life where he may safely tread. If he is a motivated dominant individual, he may scorn the idea that he is in any way limited in his choice of actions. However, if we examine his character structure and the actions he actually takes, then we will see that he, too, treads a narrow path of compulsive activity, as a prisoner of himself.

Given the fact that the individual growing to maturity senses subjective feelings of "safety" and "danger," he will avidly, and perhaps desperately, seek to exploit any particular life situation to his advantage. He may exploit his family situation, the prestige of a group, or he may enlist in a cause. He may gravitate to an occupation that presents the least threat to the demands of his character structure. He may permanently enter military or religious service. He may find an area of expertise and use it as an island of safety. If he is handsome or beautiful, then he will by necessity use these qualities synergistically with his sexuality to find safety in life.

Even apparent disadvantages may be turned into opportunistic solutions to life in the individual's hostile world. If he is a hyperactive NA type or a high-strung PA type, he may exploit his zaniness and extroversion to become a successful comedian. If he is a schizoid loner of the PA type, he may become a poker-faced actor, or he may exploit his intellectual qualities to become the beloved sardonic wit of television fame.

Even if he has a physical disability, he may turn this into an advantage. For a submissive type this could be used as a means of ensuring that he will be cared for and protected for the rest of his life. A poorly adjusted dominant person may use a handicap, a marginal physical disability, or simply his knowledge of medicine, to gain repeated admissions to scores of hospitals around the country, this becoming his tenuous solution to life (in medical circles this is known as the *Munchausen syndrome,* after Raspe's Baron von Munchhausen, the legendary teller of tall tales). Even the abject state of suffering, in the submissive person, may be turned into an advantage in the form of a claim on others to come to his rescue. In short, there is virtually no situation in life that is not amenable to be taken advantage of, in some way, by some of the character types.

Finally, we note that the individual, as he seeks safety in his hostile world, attempts to find a sense of equilibrium between himself and his life situation. As he does this, he is, as we have seen, intensely selfish and arrogant. But he is even more than this. Each personage attempts, in his own way, to find pleasure in his life. Each personage is, in his own way, hedonistic.

Q.   Hedonistic? How about the selfless individual who devotes himself to others or to a religious order?

A.   Permit us to quote Somerset Maugham, from his *Writer's Notebook:*

> It is obvious that the hedonic element is very present to the
> mind of the religious man, and influences his action as
> profoundly as it influences that of the hedonist pure and

simple — only he puts a future happiness as the reward of his deed rather than an immediate one...

But by a curious refinement of emotion some deeply religious persons persuade themselves that they act with no hope of reward, but merely for the love of God. Yet here too, if the feeling is analysed, a hedonic element will be discovered; the reward is in the intimate self-satisfaction of virtuous action, in the pleasant consciousness of having done right; and this for emotional natures can be more satisfying than any grosser, more obvious benefits...

Q.   You stated that if an individual's life situation did not fulfill the needs of his basic character structure, then anxiety would result. Could you summarize the sources of intra-psychic stress that produce this anxiety?

A.   In our competitive human society intra-psychic stress, producing a state of anxiety, stems from, three main sources:

First, if the *individual's expectations in life exceed his accomplishments,* then he will be in a state of stress. Here, the individual lacks sufficient self-esteem, or pride, in relation to the demands that he places on himself.

Second, if the *individual's character structure is based on more than one behavioral complex,* then in some life situations he may seem to be torn in conflicting directions. He must, somehow, streamline the demands made by two, or even all three behavioral complexes, into a single integrated personality structure.

Third, in individuals having the behavioral complex of aggression, *"playing the game"* imposes frequent and sometimes highly conflicting demands on the individual to split his personality.* In addition, this is the source of symbiotic relationships, for example the "morbid dependency" and the "power behind the throne," which may not only subject the individual to high degrees of stress, but also severely limit his activities in life.

To summarize, the main sources of intra-psychic stress are those of:

- insufficient self-esteem or pride,
- streamlining the behavioral complexes, and
- "playing the game" of dominance and submission

------

* However, individuals not having a measure of the trait of aggression (i.e., N and NP types) may also be subject to stresses from social interaction.

All three sources of stress come about from the individual's inability to match his real life situation to the demands of his character structure. However, the three sources of stress are not completely separate entities. In fact, in the individual's mind they are inexorably jumbled and blended. They certainly are not in the least understood.

All individuals are subject to stress from the first source, that of insufficient self-esteem. In fact, in an individual who is subject to one or both of the other two sources of stress, his self-esteem may depend to a large extent on his ability to cope with the stresses generated by the other two sources, that is, on his attempts to streamline his behavioral complexes and on his attempts at "playing the game."

The stresses imposed on the individual of "playing the game" have been mentioned earlier and will not be repeated here. Details of the dynamics of subjugation (the "morbid dependency") and of the "power behind the throne" type of symbiotic dependency will be discussed in Chapter 9.

With regard to the types of conflicts that may arise in individuals who are forced to streamline two or more of the behavioral complexes, one need only look at the turbulent lives of these caricaturized personages. The NPA type, in particular, is vulnerable to all the possible sources of stress. His pride must live up to his ideal image of himself. He plays a stressful "game" with others. He has within his soul the full double dose of ambition in the form of the narcissistic and aggressive behavioral complexes. And finally, he desperately tries to maintain the lid closed on his Pandora's box by the means of the dutiful "deal with life" that he has made on the basis of his perfectionist behavioral complex. It is no small wonder, then, that this individual often consciously, and almost literally, attempts to keep a lid on his emotions. And it is no surprise to us that when the lid blows off, he becomes the volcanic individual having an "explosive personality disorder."

Q.   You have described an individual in a competitive society, struggling to maintain his psychic integrity in a general state of anxiety. You have described how he attempts to cope and how he attempts to triumph. What happens if he is frustrated in his attempts?

A.   As the individual attempts to cope with life he achieves transient states of triumph, but he also becomes periodically wounded in miserable, embarrassing defeats. He encounters danger repeatedly, and intermittently leaps to islands of safety in the shelter of his real life situation or in his psychic structure. He is alternately subject to euphoric feelings of optimism and hopefulness on the one hand, and to abject feelings of pessimism and hopelessness on the other hand. If feelings of hopefulness predominate, he may enter the state of exhilaration. If, on the contrary, feelings of hopelessness prevail, then he may enter the abject state of depression.

Q.   At what point may an individual become self-destructive?

A.   An act of self-destruction is the endpoint of a process that may be looked upon as one of hatred of oneself.

If the individual cannot measure up to the demands that he places on himself, to fulfill the requirements of his basic character structure, then he begins to despise himself. It is as if he looked up to the portrait representing his ideal image and found it leering contemptuously down at him, proclaiming his inadequacy: "You are failing to live up to your expectations, you hopeless, hapless, helpless, inadequate fool!"

The process of self-hate, however, may show itself, in its early stages, not in any overt manner, but in actions and subtle mannerisms that may totally elude the consciousness of the individual, as well as escaping the notice of others. For example, the individual may become less efficient and listless. He may procrastinate until it is too late, thereby forcing him to absorb the punishment of the consequences that ensue. He may unemotionally watch the approach of coming catastrophes casting their shadows before them, and take only weak actions in a barely mustered effort to prevent their occurrence. He may become accident-prone. He may somehow botch important tasks and find himself fired from his job. All these actions may occur without his recognizing that he is an individual in psychic disequilibrium, disenchanted with what he has, with what he does, and with what he is. He has become an individual profoundly dissatisfied with his own self.

The individual's self-hate may become more overt. He may call himself a "stupid idiot" aloud. He may exhibit what we call the "perfectionist wince": he may be sitting quietly by himself, and as he ponders some aspect of his inadequacy, he may visibly wince. He may begin to mutilate his own body by striking himself in frustration, pulling on his hair, hitting his fist against a wall in anger, biting his lips or his nails, or subjecting his body to excesses of tobacco, alcohol or drugs. He may take open risks to his well being. In a hospital setting he may flagrantly expose himself to X-rays. He may take risks while driving an automobile, or he may involve himself in sports such as mountain climbing or skydiving.

Finally, at the very end of the road, he may find himself pointing a gun to his own head. But even here, in the final stages, he may not realize that it is an internal process of psychic disequilibrium, or self-hate, that is driving him to destruction.

He may, in fact, openly accuse others of being responsible for his suicide, as the ultimate vindictive triumph, his final act of retribution with respect to his tormentors in life. What occurs, though, is an ultimate act of irony: the individual is born into his world by no will of his own; he has lived an erratic life with little free will, and he dies by his own hand not so much with a will to die as a desire to escape a confused existence into which he has had little insight.

Q.   Can an individual come to terms with his confused state of mind?

A.   We believe that, in fact, the individual can come to understand his character structure and come to terms with the sources of the intra-psychic conflicts that it inexorably generates. And in understanding his conflicts, he can come to recognize his tendencies to frustration and anxiety, and finally to self-hate and self-destruction.

In blunt terms, we cautiously suggest that this approach to finding one's real self, one's real motivations behind one's actions, may in the final analysis prevent countless acts of needless suicide.

CHAPTER 8

LOVE AND EVIL:
THEIR COUNTERPARTS
AND COUNTERFEITS

A factory worker who sees a fly on his machine will hit and kill it if he has not become aware of the wonder of life. But once he knows, he will not kill it thoughtlessly, but help to free it if he sees it caught in a room. He will open the window and let it go.

Albert Schweitzer (1959)

In Warsaw conditions are chaotic not only in the Ghetto, but also among sections of the Polish population. It would be a good thing if the post of governor general were filled by a new man — a person who could put an end to these difficulties.

Worry after worry, both in my official and personal life. But that's war for you! Nevertheless we must never forget that, once these anxieties are a thing of the past, they will later constitute our most beautiful memories.

Joseph Goebbels (ca. 1943)

OUTLINE OF CHAPTER EIGHT

LOVE

      "True" love
         Narcissistic
         Aggressive: the "game" of dominance &
            submission
         Perfectionistic
         Submissive
         Resigned

      Counterfeit love

EVIL

      Coercion
         Narcissistic
         Aggressive: sadism
         Submissive
         Resigned

      Sadism: manifestations
         Enslavement
         Punishment
         Exploitation
         Frustration
         Control of emotions
         Violence
         Emotional barrenness

      Sadism in various character types: schematic diagram

      Sadomasochism in medicine

Q.   What is this chapter about?

A.   In the previous chapter we found that the concept of human happiness arose quite naturally from our consideration of the needs of an individual's basic character structure. Here, we find that, with little effort, we can focus on the concepts of human love and human evil.

Q.   Let us start with love. What is exactly the love that two individuals feel for each other?

A.   If we base our answer on Aristotle's concept of true friendship, then "true love" is the act of one's showing sympathy and compassion toward another individual, in the context of desiring to share one's life with him, *without there existing any ulterior motive in one's mind.*

Q.   Does such "true love" exist?

A.   If we consider the dynamics that occur between individuals, who in their confusion attempt to fulfill the powerful needs of their character structures, then we are drawn to the conclusion that such "true love" just does not exist.

For example, the ulterior motive of the *narcissistic* individual is by now well known to us. The attraction that he has for another individual is rooted in his desire to merge with someone who will be his ideal image in his search for glory. And here the love of this ideal image may become inextricably entwined with the love of his god. The love that he professes for another individual is, at best, an inversion of the love that he has for himself and of the love that he expects to receive from others because of his glorious qualities. In the final analysis, the narcissistic individual's love is the response he gives to another individual in acknowledgement of the recognition that he feels he so richly deserves.

In *aggressive* individuals the needs of love are satisfied in the ulterior motives of "playing the game" of dominance and submission. Unfortunately, this is a "game" involving master and slave, and as Philip states in *Of Human Bondage,* "There's always one who loves and one who lets himself be loved." Such a "game" is never played on an equal footing between the two individuals, and since the rules are completely beyond the ken of the players, the sadomasochistic elements that gradually creep into the relationship often ensure its demise.

In the *narcissistic-perfectionist* individual the ulterior motive that must be satisfied is the individual's powerful need to be dutiful to his own sense of perfection. Hence, by extension, he loves another because of the voice within him, reminding him of his devotion to himself, telling him that he *should* love another, that he *must* love another. And if he does not find the true love that he expects of himself, then it is a matter of unfinished business that he will not allow to be forgotten. It is the raven of unfinished business, ever sitting above his chamber door, reminding him of his duty to himself.

In the *submissive* individual the ulterior motive that must be satisfied is, usually in the context of "playing the game," the requirement of care and protection that he so desperately needs. He will worship on the altar of tender love, total love, masochistic love, love in which he wishes to merge completely with another being. It is as if, in his self-effacing way, he wishes literally to erase himself from the face of the earth and to become assimilated, body and soul, by his master and protector. If he is a *non-compliant* type, he may find himself living the life of a dominant subjugator, but even here, he is ever vulnerable to the call of submissive love. Using the blunt terms of psychiatric jargon, we would say that he shows both sadistic and masochistic behavioral qualities, but that his predominant trend is one of masochism.

In the *resigned* individual we find a desire to love and to be loved in subtle forms. It may appear as a love of nature or of God. It may appear as a compulsive search for sexual partners, this being the individual's only bridge to others. Or it may appear in the form of pride in a basic integrity in his relations with other individuals, in the context of somewhat nebulous interactions with them. He may be a vagabond in search of a will-o-the-wisp, and although others may have little idea of where he goes or what he actually does, they assume that this person of kindness and integrity cannot be but a lovable being.

Q.   Does the individual realize what the dynamics of his love relations are?

A.   Certainly not. He may realize, at best, that powerful forces are working in bringing him, like a magnet, toward a relationship with another individual. He realizes, in fact, that the forces are so powerful that he yields to their demands; he surrenders. This is the surrender of love of which the individual has heard so much. Since his society demands of him that he be lovable, both in giving and receiving affection, he cannot help but consciously assimilate this expectation of his culture. Unfortunately, often his choice of partner is frankly inappropriate, and his lack of understanding of the dynamics of the relationship may make it a foreordained failure.

Q.   How does he attempt to deal with the demands of lovableness that his society places on him?

A.   He may decide, consciously or unconsciously, that "true love" is not for him. An individual may find that narcissistic love or "playing the game" of dominance and submission always leads to pain and disaster. Another may find that his perfectionistic nature will not allow him to choose himself a partner at all, or he may find himself to be henpecked to death by an aggressive one. He may look at his marriage and at those around him and see not love, but primarily Kafkaesque relationships that have gone far, far astray.

He may then go even farther astray to lose himself in vicarious love or counterfeit lovability. He may lose himself in the false pride of sexuality, or in promiscuity, and try to convince himself that he is, after all, a lovable being. He may devote himself to maintaining the attractiveness of his physical self. He may devote himself to a cause, to volunteer work, to religion, to faith, hope and charity, and thereby prove his lovability. He may devote himself to an individual, perhaps to an "exception that proves the rule," or to his children. He may devote himself to masculine or feminine rights. He may suspect that his love relations will be a disaster, but he may seek shelter in a marriage all the same. And like the professor who does not really understand the subject of his specialty, he may intellectualize about love or even write a book about it.

Q.   You sound cynical, I must say!

A.   No. On the contrary, we remain unflappably optimistic. An individual's approach to love relations should be made on a more firm foundation of his own self-knowledge and of his knowledge of a potential partner. And there is no doubt in our mind that he can avoid the numerous pitfalls that contribute little to human intercourse other than sending the individual on wild goose chases to pinnacles of anxiety and nadirs of suffering.

Q.   Well, if that was an optimistic view of love, I can't wait to hear about evil!

A.   As you probably have guessed, the roots of human evil lie in the selfishness and arrogance of each personage as he desperately attempts to mold his real life situation around the powerful requirements of his basic character structure. All of the personages, in the final analysis, have little respect for the truth, when their particular character structures are at stake.

The *narcissistic* individual coerces or bullies others into leading him to the limelight, whether he merits it or not. Others may have to put up with his narcissistic rages. As a false prophet he may lead others astray or even to their doom.

The *aggressive* individual coerces or bullies others by crushing them with the iron fist of his self-justified vindictiveness. But more than this, he has at his disposal a whole spectrum of sadistic activities that, unfortunately, may become a way of life for him.

The *submissive* individual may coerce others into providing for him by making them feel sorry for him or making them feel guilty.

The *resigned* individual seemingly is completely independent of others, but this is often a delusion. No man is truly an island entire of itself, and although he may coerce others into leaving him alone, the time will come when he will become dependent on others, in illness or in some other one of the misfortunes of life that await most of us.

Q.   I see that all of the personages may, in their own way, coerce others because of their special needs. Do the personages realize that they are coercing others?

A.   Not as a rule. As we have stated, each one has developed a magnificent assortment of rationalizations and internalized alibis, so that he can proceed with seeking his triumphs in confident self-righteousness.

Q.   But in their relations with others, are not some of the personages more benign in their coercion than others?

A.   Unquestionably so. And this brings us to the consideration of *malignant trends* of narcissism (narcissistic personality disorder, or NPD) and of aggression (sadism, or sadomasochism) in the human character structure.

Q.   What is the source of sadistic behavior and how is it manifested?

A.   As we have mentioned, the source of sadistic behavior in man is the innate behavioral complex of aggression. It is based on the inborn instinct of an individual to attempt to achieve dominance over others as his solution to survival in life. It is an inborn instinct to become master over slaves, of compulsively reigning over all others. And it is in achieving mastery over all others that the specter with the evil horns looms into view.

Sadistic trends may be manifested in various degrees of openness. They may be *overt, covert,* or *inverted.*

First of all, aggressive-sadistic trends may be manifested in an *overt* form. This is often the *modus operandi* of the A and NA types, and sometimes that of the PA and NPA types. This is the arrogant individual who is loud and openly vindictive as he goes through life bludgeoning others verbally and, if necessary, physically. His self-righteous vindictive rage cannot be missed. Since he obtains satisfaction from continually achieving triumphs over others, who in his mind are weaker than he, then we see that the pleasure that he derives is the very essence of sadism.

He is self-righteous, by necessity, to be sure. But if his aggressive arrogance is accepted by those with whom he deals, because of their impotence, their fear, or because of the special needs of their own character structures, then the road to purgatory becomes the road that is taken. For as others submit to the aggressive individual's iron fist, then his self-righteousness and self-justification are reinforced. And as his acts of aggressive domination become more callous and more overt, they become, paradoxically, less recognized for what they are, more rationalized, and more tolerated.

Secondly, sadistic trends may be manifested in a *covert* form as well, revealing their horns in an insidious, diabolical way. If overt forms of sadistic behavior are often not recognized for what they are, then this is all the more true for covert forms. The various covert forms of sadistic behavior may be found in all individuals having the aggressive trait, but it is the perfectionist-aggressive (PA and PA–) personages who are at the highest risk of adopting furtive vindictiveness and the insidious manipulation of others in an atmosphere of paranoia, as a general approach to life.

Finally, sadistic trends may show themselves only subliminally in an *inverted* form. Here the individual bends over backwards, deep in his subconscious mind, to deny to himself and to others the presence of the sadistic demons that are an inherent manifestation of his aggressive gene. But the demons lurk there all the same, and they may show their little horns in a wide variety of devilishly subtle activities and mannerisms. Inverted sadistic trends are to be found especially in the NPA dominant individual, and in submissive and resigned types.

Q.   What are the forces that operate to contain aggressive-sadistic behavior in an individual, or cause it to be inverted?

A.   First of all, the parental instinct to give sympathy and care to one's children may be, consciously or unconsciously, generalized and extended to others.

Secondly, the individual's instinct to obtain succor, comfort or sexual satisfaction may require him to modify his aggressive behavior towards others, at least to some extent. To quote Desmond Morris (1967):

> If a female monkey wants to approach an aggressive male in a non-sexual context, she may display sexually to him, not because she wants to copulate, but because by so doing she will arouse his sexual urges sufficiently to suppress his aggression. Such behavior patterns are referred to as remotivating activities. The female uses sexual stimulation to remotivate the male and thereby gain a non-sexual advantage. Similar devices are used by our own species.

Third, the pressure of the individual's society, culture or of religion may require a conscious dedication on his part to activities consistent with the general dignity of mankind.

Fourth, the behavior patterns generated by one or more of the behavioral complexes may repress or invert aggressive-sadistic trends. For example, the NA individual has a strong urge for recognition in the eyes of others and for the maintenance of a positive self-image. The NPA individual may have a strong sense of perfectionist duty that would suppress overt sadistic trends. A submissive individual lives in a continual free-floating anxiety of being abandoned by his master or protectors, hence he would not dare to show aggressive-sadistic behavior toward them — and by extension — to anyone. The resigned individual, unless he is coerced, is dedicated to non-involvement with others, hence his sadistic trends are of the most passive kind.

Fifth, the instinct for self-preservation, or the outright fear of retaliation by others, may cause the aggressive individual to modify or moderate his sadistic behavior.

Q.   How does sadistic behavior find its way into the life of an individual?

A.   We believe that the tendency of an individual to develop sadistic trends depends on his inherited character structure, his intelligence, his upbringing, his society, his general station in life, other subtle qualities of the human spirit, and finally, sheer luck.

Sadistic trends may mark an individual in several ways:

First, an individual may attempt to lead a normal life in his society, but may find himself drawn periodically to commit, consciously, overt acts of cruelty. Such acts are committed to satisfy powerful but confused cravings for satisfaction, and are oriented against selected vulnerable victims. A dramatic example of this type of behavior in a PA individual is presented in Appendix A.

Second, the individual may actively or subconsciously seek a life situation in which he may subjugate a companion, co-worker or mate in order to satisfy the internal demands of his character structure. Many relationships, some very stable and some wildly unstable, and many marriages, are entered on this basis. The specific dynamics of the "morbid dependency" and of the "power behind the throne" will be discussed in the next chapter.

Third, if an individual is, because of an inappropriate choice of companion or mate, or simply because of bad fortune, drawn into a situation from which he cannot extricate himself, then he may become a *situational sadist*. He may realize that he is behaving in a gloomy, sadistic manner and he may, eventually, begin to thrash about wildly in an effort to disentangle himself from the relationship.

If an individual has a personality structure in which sadistic trends come to a fore more readily, then he becomes an *opportunistic sadist*. If he finds himself conducting college hazing rites or directing maneuvers in time of war, then he may see himself, before his very eyes, devising unthinkable plans and carrying out unspeakable acts.

Finally, a person may adopt a brooding, diffusely cynical approach to life. He simply believes, as he looks out of his window, that there is just no "good" out there at all, and in his diffuse pessimism he manages to concoct a passive-aggressive, sadistic approach to life based on the vindictive triumph.

Q.   What are the general manifestations of sadistic trends?

A.   The individual with sadistic trends* acts in the following manner:

- He *enslaves* his victim. He is the unquestioned master. He has an attitude of invincibility toward his victim and can show no fear toward him. He can never be wrong. His motto is "I have all the rights. You have none."

---

* Horney, K. (1945, 1950)

- He has the right to *punish* his victim with psychic or physical pain. Since the slave, in his weakness, can never live up to the demands of the invincible master, the verbal abuse or corporal punishment that he metes out is always justified retribution. His motto is "You must suffer for the defects in what you are, what you have done, or what in your weakness you cannot do."

- He has the right to *exploit* his victim, whether for time, money or sexual pleasure. His motto is "Since I am the master, I have the right to all that you can give me."

- He has the right to *frustrate* his victim. Any requests of the master by the slave are automatically, either actively or passively, turned down. At best, such a request is complied with in the most grudging manner. The purpose of the request is then twisted by the master out of its original context so that the end result appears to be, at least in part, to the advantage of the master. The only way the master can accede to a request in good humor is if it is preceded by the preamble, "Would you do me a favor...?" His motto is "Since I am the master, you dare not threaten me with any requests for the purpose of your own well-being."

- He has a right to *control the emotions* of his victim. When the master becomes brooding and gloomy, the slave should desist from any cheerfulness — immediately. The master has a right to play on the emotions of the slave as if on a musical instrument. His motto is "Since I am master, not only your body, but also your emotions belong to me."

- He has a right to threaten, and to actually use, *violence* on his victim. If the slave shows the slightest tendency toward independent thought or action, or if he challenges the master on even the most picayune detail, then his vindictive actions may be transformed into his vindictive rage. His motto is "If you do not wish to obey the master voluntarily, then you will be *forced* to obey him."

- Finally, in his callous disregard for the slave, the master can show no sympathy or tenderness toward him. He may, intellectually, be able to empathize with the plight of the slave, to literally feel what his companion is feeling, but be unable to sympathize with him. The master has an *emotional barrenness*, a complete lack of ability to show any positive feelings toward his subjugated companion. His motto is "Since I am the master, you dare not threaten me with requests for submissive love, sympathy or remorse."

Q.   How does the master, in a given relationship, see things?

A.   Unbelievable as it sounds, all of these mottos are usually deeply repressed and rationalized.   The master simply does not see the sadistic trends that determine, almost in their entirety, his relationship with his companion. He may, in fact, rationalize that he is trying, despite all kinds of difficulties, to help his weaker companion stand on his own two feet. And, of course, it is very easy for the master to externalize his difficulties, to lay them at the doorstep of his subjugated companion. In short, the master will be deeply shocked when he realizes that his internal alibis hold no water at all, and that the horns of the dilemmas in his life stem from the horns growing from his forehead.

Q.   How about the subjugated slave. How does he see things?

A.   Again, unbelievably enough, he does not see himself as subjugated at all. He may, from time to time, feel that he is abused, but this is quickly rationalized and repressed. He is so unconsciously afraid, so anxious that his master will leave him that he is prepared to absorb ever greater and greater degrees of abuse. He will even be annoyed if his friends or colleagues point this out to him. In short, he is in his own mind so at home in this solution to life that, so long as his master does not leave him, he will be ever faithful to him.

Q.   So the subjugated partner lives a life of repressed fear that his partner will leave him, without this fact ever coming to his consciousness.

A.   Yes, even to the last days of a doomed relationship.

Q.   Where do we go from here?

A.   *Overt* manifestations of aggressive-sadistic behavior are well enough known to us that we need not describe them here further. We shall now consider in some detail covert and inverted modes of sadistic behavior.

The manifestations of *covert* sadistic behavior may be found in one form or another in the various personages exhibiting the aggressive trait, but they are the especial specialty of types having perfectionist-aggressive trends.

If one reviews the qualities of the perfectionist behavioral complex, one understands why this type, when acting alone, may choose his victims carefully, court them, stalk them, with an emphasis on repetitive actions and ritualistic deeds. He does his job slowly, teasingly, while playing hard to get, while playing cat and mouse, while playing the waiting game. He relives his thrills; he keeps notebooks with careful entries of dates, times, names and outcomes. He is as much involved with the means to his ends as the ends themselves. We recognize him in his schizoid loneliness, and we realize that his link to his society may be by the most tenuous of threads.

In the PA individual who is not isolated from his society we may find in a diffuse, muted, covert form, but always liable to burst forth into the open, all the manifestations of aggressive-sadistic behavior:

As master of all others, which is how he sees himself, he not so much enslaves others as maintains a *passive dominance* over them. He is committed to maintaining his position of strength, of superiority, so all others must be made impotent if they have the slightest potential of challenging his position of self-righteous dominance. He lives in an atmosphere of defensive suspicion. He has a mistrust of strangers and he looks for weaknesses, for the soft underbelly, in new arrivals. These are weaknesses, of course, that he may have to exploit at some later date if he is threatened. He has a fear of subjugation or of events whose outcomes he will not be able to control. He develops a pride in his intelligence, in knowing exactly what is going on, where it is happening and who is involved. He develops a pride in predicting what will happen. And in the general atmosphere of pessimism, cynicism, defensive suspicion and paranoia, he becomes a quiet manipulator. He becomes a brooding wolf in sheep's clothing.

Since he is master of all, he is free to *punish* all. He instinctively comes to know the particular vulnerabilities of those around him. He sarcastically points out the foibles of others to them and reminds them of their deprivations, lest they forget their inferior — nay miserable — status in life. He is the past master of embarrassing and humiliating his victims before others by sardonic, often insidious, verbal abuse. How does he rationalize his right to punish those around him? He has developed a self-justified attitude of *intolerance* of others, and once this premise is accepted it covers a multitude of sins. He can always look for the dark side in an individual, and he can always find it. And finding a single stain on another's soul gives him free reign to disparage all of his motives. And if all of that individual's motives are base, he is free to punish him at will.

Since he is master of all, he gives himself the right to *exploit* all others. To the extent that he pictures himself with his back against the wall, facing the rest of the world, he feels justified in taking what should be rightfully his, whether it be from individuals or from institutions. But here he begins to sense an intra-psychic conflict, that something is not quite right. First of all, he finds himself acting furtively, and in his clandestine exploitation of the world around him, he becomes fearful of being discovered, which accentuates his suspicion and his paranoia.

Secondly, in his desire to exploit the world around him, he senses in a deeply repressed but confused manner, that perhaps he is not the master of his environment after all. In fact, he senses, paradoxically, a profound *envy* of those around him. In the same way that he instinctively detects weaknesses in others, so does he see in them all things, large and small, that he himself lacks. And it is in this burning envy of others, which he is loath to admit to himself, that he finds himself from time to time, not the omnipotent master of all, but a lonely outsider in his apparently bustling society of plenty. Thus, we see that, ironically, his sixth sense to exploit others becomes the basis for his furtiveness and for his deepening envy of those around him. And these, in turn, cannot help but lead him to even greater isolation from those around him.

Since he is master of all, he has the inalienable right to *frustrate* others. This is, to be sure, related to his envy of others. He is diffusely stingy and miserly. He begrudges others not only their spectacular successes but also that which is justifiably due them. He is the past master of quietly frustrating others by withholding information that he alone possesses. He is the master of quiet, self-justified sabotage.

Since he is the master of all, he should have the right *to control the emotional atmosphere.* When he intermittently ceases his brooding and comes out of his corner to mingle with the weaklings, then they should — immediately — become cheerful and adulating toward him. And when he is gloomy, he despises even the nuance of any levity and good cheer around him.

Since he is the master of all, the threat of *vindictiveness,* the vindictive rage, or even *violence* is ever present. He is the past master of overreaction and overkill. He will whip a child for stealing an apple from his orchard. If he hears a noise, he will fire his shotgun into the darkness, and whether he hits a burglar, a dog, or his neighbor is almost immaterial to him. He may rationalize that others are masochists, who long for punishment, pain and violence in their longing for attention. He will smile to himself at his maxim, "Beat your wife every night, for if *you* do not know why you are beating her, she will!" Finally, if he finds himself self-righteously devoted to a cause, then violent sabotage or assassination may become the bread and butter of his life. In his mind, his violent acts have taken on the aspect of a just retribution for the events of the past, and in his mind he has pride in the accountability of an elephant that never forgets.

Finally, as a master of all, expressions of sympathy to weaklings are alien to him. He has become emotionally numb, *emotionally dead,* with respect to the plight of others, the plight of society, and the plight of the human race. But a state of abject hopelessness is to be avoided at all costs. It is in the desperation of isolation, of loneliness and of emotional numbness that he attempts to find meaning in life in an "exception that proves the rule." He may be drawn into a morbid dependency; he may attempt to mount to the throne in order to become the power behind it; or finally, he may find meaning in life by devotion to solitude, to poetry, to art, or to periodic demonical thrills of passion.

Q.   And these activities go on without the individual's recognizing the psychodynamics behind his actions!

A.   Alas, it is true. But it is even more true in the spectrum of activities that we call *inverted* sadism. Here, we find subtle, subliminal manifestations of aggressive-sadistic behavior. But it is precisely for their subtlety, for their insidiousness, that they must be exposed for what they are — a not insignificant contribution to the totality of evil that lurks in the human soul.

For example, there may be in the individual a subtle avoidance of physical contact and eye contact with others, as of a master ruling over his subjugated victims. In his relations with others, there may occur a passive obstructionist behavior that may barely be noted. He may "misinterpret" what others say or interpret everything literally. If someone asks him if he has a pencil, he will answer negatively if he has two of them. He will secretly delight in confusion and ambiguity. He will delight in subtly "crying wolf" and then acting as a detached observer. He may find himself wishing for an unfavorable outcome as events unfold around him. He may find himself taking a morbid interest in accidents and natural disasters.

We find that the individual tells "white lies" and breaks promises, which somehow in his mind were not promises at all. There is often a subtle element of stinginess, with his time and with his money. If he finds a wallet, he keeps the money if he is sure that no one will discover his secret. He is selfish. If he finds something that is worth a hundred dollars to someone else but only one dollar to him, he will keep it nonetheless. If given an opening he may be drawn, irresistibly, to exploit it. He may find himself stealing in a kleptomanic manner.

He may persistently and compulsively indulge in raillery as he teases individuals, mocks them, joshes them about their faults, ostensibly in a good-natured, only mildly sarcastic manner. Towards children, he may tickle them, poke them, pinch them — all in fun, of course. He may be callously careless. He will shake out a blanket at the beach and only the devil may care where the sand flies. He will slam the door of an automobile and if someone's fingers are pinched, it is they who should have been more careful. And if a conflict arises, he will delight in self-righteously saying, "Let the chips fall as they may," but if one looks closely at the circumstances, the chips always fall so that they crush someone else's toes.

An individual with inverted sadistic trends would be loath to admit the presence of potential violence in his character structure, but that potential lurks there all the same. It may show itself in his language, as he uses idiomatic expressions such as "penetrate his thick skull," "hit between the eyes," or "picked apart." The latent violence that exists may be seen in the way that he rips open packages, crushes cans, or tears up clothes before throwing them out. He may have a penchant for horror movies, or in a deeply inverted manner he may be horrified at the thought of going to a horror movie.

Latent violence may appear in sexual connotations. It may appear in substitute masturbatory activities such as compulsive vigorous hair combing, or peeling off the label from a bottle at the dinner table. The sexual connotations of latent violence may appear in less disguised forms, such as in the compulsive crumpling up of paper and throwing it into a basket, stuffing paper napkins into a cup, slamming a telephone receiver vigorously into its carriage, or applauding wildly at a basketball game when a powerful dunk shot shatters the glass backboard.

Finally, the individual with inverted sadistic trends is, he too, somewhat emotionally numb. In the same way that he may find it very difficult to give and receive favors, to give and receive gifts, to give and receive thanks, he may find it difficult to reciprocate feelings of tenderness, sympathy and love.

Q.   Are you not exaggerating a bit with these subliminal, so-called inverted sadistic trends? Are they not to be found in normal individuals?

A.   First of all, the expression "normal individual" is not a part of our vocabulary.   All of the character types are prisoners of their unrecognized obsessive-compulsive behavior.   Secondly, the behavior described is nothing in which the human race should take special pride. That the subliminal mannerisms are indeed rooted in sadistic behavior is evident: they are also found in individuals having more manifest sadistic trends, and they are rarely found in N and NP individuals who, of course, do not possess the aggressive trait.

Q.   What is the significance of our recognizing the nature and the extent of sadistic trends in man?

A.   Simply recognizing them for what they are would be a giant step forward. In our recognition of sadistic trends in the human character, we can begin to mobilize our society to render them much less harmful.

Q.   How?

A.   This could easily be the subject of *another* book. Briefly, we believe that as the individual knows his own character better, he is less likely to be frustrated, and less likely, in his frustration, to harm others, and less likely, in his frustration, to harm himself.

With regard to the containment of sadistic trends in society, this is the monumental task that faces mankind, and this task has certainly been recognized in other contexts throughout the ages. As we have seen, aggressive-sadistic trends are not the sole source of evil in the human character. However, the very nature of sadistic trends is such that they are overseen by an ever-present specter of violence, ever ready to shatter the ever so thin veneer of civilization that is the pride of the human spirit. This in itself, we believe, is enough to place sadistic trends in a separate category.

We shall repeat like a refrain: for all of us, each in our own way, there are some things in this life for which we are just not suited. And we must ask ourselves if some of the personages are simply not suited to be world leaders in our age of potential nuclear warfare.

It is now the task of historians to take a fresh look at past history in light of the probable character structures of the individuals who determined its course. We must reexamine the dynamics of the human disasters of the past so that Santayana's admonition to us does not become an epitaph on all of our gravestones.

Finally, if we are to have any hope of containing sadistic trends in our society, then we cannot afford to raise our children, and ourselves, on a pervasive diet of overt to subliminal sadism, from morning to night. Yes, we are referring to American television, where a camouflaged spirit of evil masquerades behind the laugh track — nay claptrap — of what we attempt to convince ourselves is entertainment. If we do not recognize the presence of the sadistic trends in our children's programs, in our news programs and in our Saturday night humor, and if we do not do something about it, then we deserve all that we get, and we will probably get all that we deserve. And more.

Q.   Why are you picking on American television?

A.   It is an easy target!

Q.   I gather that you are absolving the N and NP types from all traces of sadism. Is that so?

A.   Yes and no. If we define sadism in terms of a master-slave relationship based on the behavioral trait of aggression, then this is of course so. However, the N and NP types can, in their self-centeredness and compulsiveness and in a thousand-and-one ways, exhibit behaviorisms that others would brand as frankly diabolical or cruel. Indeed, although this chapter emphasizes the malevolent aspects of the trait of aggression, we should emphasize that the behaviorisms of malignant narcissism, or NPD, can mimic those of sadism and sadomasochism.

In addition, we must mention the unfortunate, poorly adjusted N or NP type who is pushed to the brink of schizophrenia, and adopts a pseudoaggressive antisocial personality. Such an individual can adopt sociopathic behavioral traits that can rival those of the overt sadist. In short, if *any* individual can satisfy the needs of his character structure by overt or covert cruelty toward others, then cruel he shall be. Thus, we absolve the N and NP types of sadism, but not necessarily of selfishness, meanness, cruelty or brutality.

Q.   Could you present a diagram to show in schematic form the occurrence of sadistic trends in the human character structure.

A.   The diagram is shown overleaf in Figure 13.

Q.   Since you are a physician, I don't think that you can close this chapter without saying something about the occurrence of sadistic trends in medicine.

A.   We believe that sadistic trends may come to the fore in any aspect of human relations where one group of individuals establishes a position of dominance over another group. This may occur in a galley slave ship, in a prison, in a family, in a cult, or in the wake of a military victory.

In the field of medicine the dynamics of dominance and submission are unearthed without too much difficulty. Speaking in very general terms, when a nursing student or a medical student enters the scene, he is in a chronic subdued state. He is overwhelmed by the amount of information he must assimilate, overwhelmed by the number of procedures he must learn, and overwhelmed by the hierarchal masters of the nursing and medical staff. He is in a subdued state, thus he is deferential, helpful and sympathetic to all. He is, in particular, loving and compassionate in his relations with patients. His handwriting, at that stage, is the penmanship of a submissive person writing for the benefit of his master. It is neat and legible.

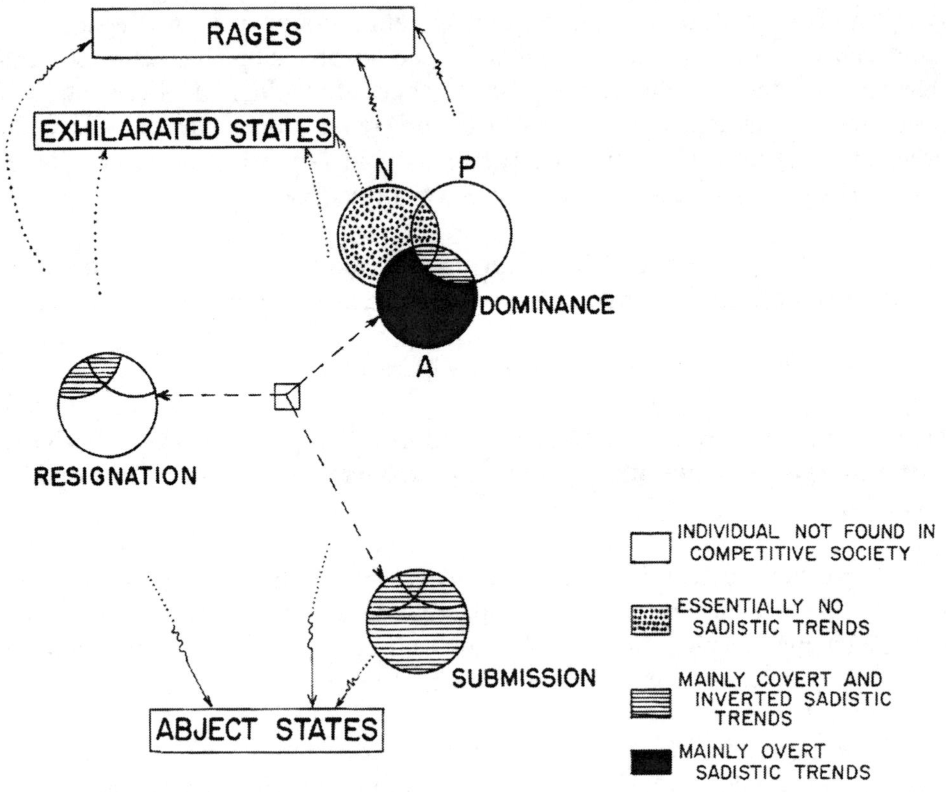

FIGURE 13

Occurrence of sadistic trends in human character structure.

Q.   How about the patient. What is his status?

A.   The hospitalized patient may be completely helpless, as in an operating room or in the intensive care unit. Or he may be overwhelmed by hopelessness, finding himself in an abject state of depression. Hence, he easily adopts the role of a subjugated victim, in whom masochistic trends may come to the fore. He may show limitless sympathy for his masters, the physicians and nurses, and exhibit a limitless tolerance of abuse.

Q.   What happens over time, as a nursing student or medical student advances in his hierarchal structure?

A.   As you may have guessed, as he gradually comes to mingle with the masters, and gradually becomes a master himself, then, as inevitably as the cock crowing twice, sadistic trends come to the fore. His handwriting deteriorates to an abysmal quality. He becomes the unquestioned, omniscient, omnipotent master of all patients.

His self-ordained majesty then may manifest itself in all of the familiar mannerisms of sadistic behavior:

Enslavement — The master is condescending. He may exhibit the characteristic averted eyes of the master presiding over his subjugated victim. Practically anything can be rationalized by his view that the master is, after all, dedicated to helping the patient.

Punishment — The master feels that the patient deserves what he gets. After all, did he not cause his own cancer by his smoking, his gout by his own high living, his diabetes by his obesity, and his heart attack by his own slothfulness?

Exploitation — The master has a right to all of the patient's money. No fee is too high. Sexual subjugation of patient by physician is not unknown, particularly among psychiatrists, who themselves have studied this type of patient-physician relationship.

Frustration — Any demand of the master beyond the call of duty will be automatically and instantly rejected, unless the request is in the context of a special favor to be done.

Emotions — The patient has no right to be happy. Laughter between two patients may be looked upon with disfavor by the staff.   The patient is allowed to be cheerful only with the permission of the masters. At times, in the intensive care unit, the juxtaposition of the jovial staff in an exhilarated state, and the agonizing patients in a deep abject state, takes on a surrealistic quality.

Vindictiveness and violence — "If the patient does not like the way he is being treated, then he can well go elsewhere. And if he bleeds to death on the way, that is his problem!"

Lack of feelings — The weaker the patient, the more inhibited is the master's capacity to show him sympathy. He develops a dullness, a numbness of emotional response toward the patient, a numbness of spirit that masquerades as the proud physician's "professional detachment" from his patient. This explains why patients can be ridiculed *in absentia* during clinical presentations, why staff members can joke to each other while a patient is in pain, and finally why staff members can totally ignore the miserable cancer patient who occupies that last bed at the far end of the ward.

Q.   But are you not exaggerating all of this?

A.   Yes, of course we are… But, then again, no we are not.

CHAPTER 9

INTERACTIONS BETWEEN
CHARACTER TYPES

I do not like thee, Doctor Fell,
The reason why I cannot tell;
But this I know, and know full well,
I do not like thee, Doctor Fell.

Thomas Brown (1680) after
Martial (ca. 100 A.D.)

OUTLINE OF CHAPTER NINE

INTERACTIONS BETWEEN VARIOUS TYPES

Non-love relations

Love relations
    Dominant & submissive
        N types
        NP types
        NA types
        NPA types
        PA types
        A type
        NA– type
        PA– type
        N–P type
    Resigned types

SYMBIOTIC RELATIONSHIPS

The "power behind the throne"
    Non-love relations
    Love relations

The "morbid dependency"
    Non-love relations
    Love relations

THE MONOGAMOUS IDEAL

THE "NEW REALISM"

Q.   What is this chapter about?

A.   Here we would like to consider briefly how the various personages interact with each other. However, note that even if we confined ourselves to only nine different character types taken two at a time (six dominant, one each of compliant and non-compliant submissive types, and a resigned type), we would need to describe forty-five distinctly different combinations of interactions. Rather than attempt such a task, we shall confine our comments to a limited number of generalizations.

Q.   What kind of generalizations can we make?

 A.   The most important point is the following: when two or more personages interact, the overriding principle governing their behavior is their obsessive-compulsive necessities to fulfill the needs of their respective character structures. Everything else is secondary. Human logic may fall by the wayside. The individual's esteemed capacity for human sympathy, compassion and tolerance may become only grotesque lip service. As an automaton, he walks, talks, attacks others and defends himself, but although he may think that he understands what he is doing, he really understands little of why he is speaking or of how he is acting. The truth pales into insignificance as the four horsemen of arrogance, self-righteousness, self-justification and intolerance gallop into view. Group meetings take on a surrealistic quality. Do individuals address the real issues involved? Do they apply themselves to the details of real problems? Yes, sometimes, but only sometimes.

What one observes is the following: in any conflict or problem-oriented meeting involving several individuals, there are actually two separate problems to be resolved.

The first problem is that of the real issue under consideration in all of its complexity. For example: what happened, why it happened, who was at fault, if anyone, and what should be done about it. The second problem is that of resolving the conflicts that the individuals' character structures inevitably generate. These are, in fact, sometimes recognized in somewhat vague terms as "personality conflicts." More often than not, however, they are not recognized as such, but only as other individuals being "unreasonable" or "illogical." What the various parties totally ignore, and indeed are ignorant of, is the fact that they believe they are wrestling with the first problem when in reality they are lost in the second. To a disinterested observer the scene is one of jesters before kings, fools fumbling in the fog, robots repeating by rote and ships passing in the night.

It is all so infuriating. Sometimes one wants to shake these individuals by the shoulders and shout into their ears in order to bring them to their senses. And often, of course, one sees oneself among them.

One recalls Maugham's quotation (Chapter 7) describing in somewhat poetic terms how little individuals understand each other. He has hit the nail on the head! In a single day one can see a dominant NPA type exhorting an NP type to be aggressive, or exhorting a submissive type to speak up at conferences. One can see an aggressive type, blind to his own frailties, berating others as he inflates molehills to the size of mountains. One can see the perfectionist-aggressive type, not able to live up to his own standards of perfection, seething in a stew and then attacking a vulnerable victim in directed vindictiveness. One can see the submissive and resigned types… but we cannot go on.

Ladies and gentlemen: Basta!

Enough!

Q.   Can we make any generalizations or predictions with regard to how the various personages will interact in various relationships?

A.   We certainly can.

In ***non-love relations,*** even just two or three of the character types can interact in exceedingly complicated ways. The reader can easily analyze for himself relationships with which he is familiar and deduce their underlying dynamics with reference to the particular behavioral traits that predominate. Television, too, provides one with a fertile ground for study of interpersonal dynamics. For example, successful comedy teams often comprise a sardonic A or PA type as the dominant individual, together with either a bumbling, unaggressive N type or with a flighty, agitated NA type.

In ***love relations*** we find both stable and unstable relationships, as follows:

- ***Aggressive types.*** Individuals, both dominant and submissive, who have an aggressive component play a "love game" of subjugation. The dynamics of this "game" will be discussed later in this chapter under the heading of "the morbid dependency."

- ***N type.*** An individual of the N type may be profoundly attractive to any of the other personages. However, in his need for continual approbation and reassurance, he may be compulsively promiscuous.

- ***Two N types.*** In a mating of two N types (N × N) we obtain the dazzling duo who alternate between reassuring each other of their glorious qualities and periodically galloping off into the sunset, separately, to lead crusades in search of the fountain of youth or the golden fleece. (As historical examples of matings of this type in Western settings we have Henry of Navarre and Marguerite of Valois, Napoleon III and Eugénie, and George IV and Caroline of Brunswick).

- ***Two NP types.*** Individuals of the NP type often find each other, and the NP × NP mating is often an exceedingly stable one. This is almost inevitable, since the two individuals are duty-bound to themselves and to their partners by their perfectionist requirements. Since these individuals do not "play the game," neither partner is aggressively dominant and the mating has all the appearances of being one between equal partners. And it often is!

- ***Aggressive type & NP type.*** In a relationship between an aggressive type and an NP type (XA × NP), the NP individual may be a defenseless "henpecked husband" or a "courageous, dutiful wife." He may be interminably nagged for not being aggressive enough in the outside world or in the bedroom, and his response to this will be a stubborn impassiveness that is only occasionally punctuated by his narcissistic rages.

There is little doubt that the marriage of Mary Todd to Abraham Lincoln was of the NA × NP kind. In Herndon's biography of Lincoln, we read:

> Mary was quick, gay, and in the social world somewhat brilliant. She loved show and power, and was the most ambitious woman I ever knew. She used to contend when a girl, to her friends in Kentucky, that she was destined to marry a President...

> To marry Lincoln meant not a life of luxury and ease, for Lincoln was not a man to accumulate wealth; but in him she saw position in society, prominence in the world, and the grandest social distinction. By that means her ambition would be satisfied...

Yes, we now understand to what extent Honest Abe was a "henpecked husband":

> Mrs. Lincoln came of the best stock, and was raised like a lady.   Her husband was her opposite, in origin, in education, in breeding, in everything; and it is therefore quite natural that she should complain if he answered the door-bell himself instead of sending the servant to do so... Her frequent outbursts of temper precipitated many an embarrassment from which Lincoln with great difficulty extricated himself.

As in the case of Mary Todd and Abraham Lincoln, when an aggressive NA female chooses a hapless NP type or a submissive type for a mate, these individuals may reveal themselves to be rather unsuited to each other... as in Miss Piggy and Kermit the Frog of television fame.

We note in passing once more the frequent preference of the NA (or N) individual for an older mate: Mary Todd's choice of a man a decade older than she. The dynamics underlying such a preference, based on the narcissistic drive, were discussed in Chapter 6.

- *Two NPA types.* In a mating between two NPA types (NPA × NPA) the sense of duty engendered by the perfectionist behavioral complex will usually make this relationship a stable one. But, oh those NPA rages! There is much volcanic activity in this household, with much apologizing afterward. In addition, since the progeny of an NPA × NPA union are likely to be either NPA or NA types (see Chapter 10), this may be a "loud family" indeed.

- *NPA type & submissive type.* In a mating between an NPA type and a submissive type, the relationship will usually be stable, again due to the NPA individual's sense of duty. If the relationship sours, however, the NPA individual can become a situational sadist, and may attempt to extricate himself.

- *NPA type & aggressive type.* In a mating between an NPA type and another aggressive type, the NPA individual may be the subjugated partner (e.g., NA × NPA). In this case, the NPA individual can be an NPA+ lion at work but an NPA− pussycat at home (if he can control his rages), this being a common example of a "personality split."

- *NA type dominant.* In matings in which the NA partner plays the dominant role, the relationship may be highly volatile, in part due to that individual's hyperactivity and hypersexuality.

- *Two NA types.* In a relationship between two NA individuals, this couple may divide its time between making love and fighting like panthers. The reader can verify for himself that this is often a relationship that does not attain even a semblance of stability!

- *PA type dominant.* The PA individual may interact in several different highly complex ways. If he is confident in his approach to life and feels self-sufficient, then as a dominant partner in marriage his sense of superiority may allow him to dissolve the relationship at his whim. If, on the contrary, he is not so confident in his approach to life and indeed is financially dependent on his partner, then he may attempt to subjugate his partner in the dynamics of "the power behind the throne," which will be discussed later in this chapter.

- *PA type subjugated.* If a PA individual is he himself subjugated in marriage by a stronger aggressive type, for example by an NPA type, then this individual in the PA− state will be schizoid at home, although he may split his personality to that of a dominant PA+ personage in his career setting. In fact, it is possible that the PA individual may sense the danger to his psychic integrity in being subjugated by a stronger aggressive type. He may thus be attracted to an N, NP or submissive type.

- *A type.* In the context of mating, an aggressive A type must at all costs maintain his dominance. Otherwise, he would become an incapacitated, immobile A− personage.

- *Submissive types.* For a submissive type, his ideal partner is either a dutiful NP or NPA type, depending on whether he prefers a quiet or a noisy household. Although this type of marriage is the dream of the submissive type, more often than not he will become involved in one consisting of the morbid dependency. When such a marriage sours, it becomes a gloomy theatrical setting, where the characters act out ceremonial sadomasochistic charades as if predestined to do so.

- *n-C Submissive type.* The non-compliant submissive type can perform as either a dominant or subjugated individual in a love relationship. He may go through life in a single, fairly stable relationship, depending on the exigencies of his partner and on his life situation. Often, however, he is attracted to an NA or PA type, with whom he is hell-bent on alternately soaring to the heights of ecstasy and groveling to the depths of despair.

- *NA- type.* The NA- submissive type, particularly the female, is especially vulnerable to be taken advantage of in love relationships. First of all, she has a strong need to satisfy her narcissistic longings in the context of sexuality, and she may be attracted to a much older male. Secondly, being submissive, she is ever ready to submit in subjugated love to any strong individual. Thus, the vulnerability of the NA- type in a sexual context is determined by two powerful drives acting in synergy: unbridled narcissism and submission.

- *Two Submissive types.* In a mating of two *non-compliant* submissive types, one of the individuals will usually clearly assume the dominant role. Can two *compliant* submissive types live in equilibrium without the dynamics of subjugation intervening? One has the mental image of two confined monkeys clinging to each other, and it is probable that such relationships can exist for long periods of time in a stable state.

- *Two PA types.* Two PA individuals may play a very vigorous "game" of dominance and submission. If a true morbid dependency ensues, then the specter of calculated violence may hover over the relationship.

- *Two paranoid PA types.* Two paranoid PA individuals can exist in a symbiotic state and thus constitute a very stable relationship. We see these individuals, with their backs against the wall, peering suspiciously through their living room curtains at the hostile outside world. If the paranoia assumes psychotic proportions, then this is the state of *folie à deux,* which is well documented in the psychiatric literature.

- *PA- type.* The submissive PA- type in his schizoid loneliness is a timid, often sexually unaggressive individual who spends most of his life teetering on the brink of a schizophrenic break. Indeed, if life circumstances become too constraining, he does go over the brink, with a poor prognosis for full recovery. If the PA- type is subjugated by a stronger individual in the context of mating, his psychic integrity is especially precarious. If his subjugator threatens to leave him, he may make a feeble effort at achieving a submissive triumph, but he may suffer a schizophrenic withdrawal from reality.

- *N-P type.* Similarly, the N-P type is a timid, asthenic, melancholic, often hypochondriacal, sexually unaggressive individual who spends most of his life teetering on the brink of a catatonic state. If his life circumstances do not support him adequately, then he will, all too easily, retire into either the harried nervousness or the immobile mutism of catatonia.

- ***Resigned type.*** Finally, the resigned type tends to avoid marriage like the plague, as if it were the ultimate threat to his independence. If, however, his partner is capable of supporting the individual's detachment, a fairly stable marriage can ensue. Larry, the detached character in Maugham's *The Razor's Edge* is prepared to marry Sophie after she has descended to a state of near-total degradation following the death of her husband. Larry had, of course, consistently avoided any close attachments with all other women, including the predatory NA Isabel.

THE POWER BEHIND THE THRONE

Q.   **What is the basis of the relationship of the "power behind the throne"?**

A.   It is a symbiotic relationship between a perfectionist-aggressive PA type, who plays the role of the *Power* and another individual, who plays the role of the *Throne.*

Q.   **What kind of individual serves as the Throne?**

A.   The individual will often be a trusting, dutiful, ingenuous, perhaps somewhat naive person, who will remain oblivious to the manipulative machinations of the Power. Since the Throne holds the key to success in life, whether it be fortune or fame, he is likely to be one of the dominant personages, or perhaps a *n-C* submissive type.

Q.   **Of the two individuals, who plays the role of dominance in a power-throne relationship?**

A.   This may not be obvious at first, but the Power is actually the dominant individual.

The Power, whatever may be his actual station in life, is deeply insecure. He fears the outside world and is suspicious of it. Indeed he lacks, or senses that he lacks, the natural abilities to become the dominant leader in his competitive society that his character structure demands of him. And in his insecurity he seeks shelter and safety in devoting himself to the Throne. The Power is usually a lonely, somewhat brooding individual, whose few friends fit the "exception that proves the rule" criterion. In fact, the Power's sole bridge to humanity, his sole "exception that proves the rule," may be the Throne.

Somewhat paradoxically, it is in his weakness that the Power mounts to the Throne and takes command in his insidious and barely obtrusive manner. But despite the fact that the Power's security in life is totally dependent on the Throne, it is the Power who is in command. Indeed, in a love relationship of the type PA × NPA the individual assuming the role of the NPA Throne may be subjugated in the dynamics of a morbid dependency.

Q.   How do the Power and Throne perceive the relationship?

A.   We shall consider two situations of a power-throne relationship, namely in a *career setting* and in a *love relationship.* In both situations, but especially in a love relationship, the dynamics underlying the relationship are usually not in the slightest recognized by the individuals involved. In a career setting, perspicacious outsiders often have some grasp of the machinations of the Power behind the scenes, but the kindly king sees only a devoted companion, colleague or mate. In Ibsen's play *A Doll's House,* we see in the following exchange a description of an incipient power-throne relationship:

> NORA. I didn't know that Krog — that this man Krogstad had anything to do with the bank.
>
> RANK. Yes, he's gotten some kind of berth down there. (*To* MRS. LINDE) I don't know if you also have, in your neck of the woods, a type of person who scuttles about breathlessly, sniffing out hints of moral corruption, and then maneuvers his victim into some sort of key position where he can keep an eye on him. It's the healthy these days that are out in the cold.

Q.   Could you summarize the dynamics of a power-throne relationship in a *career setting?*

A.   We see a PA individual, often unmarried, who devotes himself to his boss. The devotion appears to be beyond the call of duty. The Power seems to neglect his personal life in his unceasing attention to the Throne. He does, in fact, seek to make the Throne dependent on him; he seeks to enslave him. He does this in various insidious ways, but the most notable way of establishing dominance is by being the holder of all key information. Thus, whatever be the matter at hand, it is the Power who holds the data, the flow sheets, the minutes of prior meetings, the schedules, and so on. Although others may have access to a portion of the key information, no one else, including the Throne, has access to all of it. In fact, the perfectionist abilities of the Power make him quite adept at juggling times, persons, places and things so that others may not even attempt to challenge him on any point of information. Thus, at the same time that the Power establishes his indispensability, he also maintains his station in life behind a secure defensive perimeter.

The Throne sees only a super-devoted colleague willing to occupy himself with all of those picayune, unpleasant details of running the show. He truly sees the Power as indispensable, so he does not see his manipulative tactics.

Others may notice very little unusual — until they try to make direct contact with the Boss. In that case, they notice that the Power is hovering about, and indeed they may find it impossible to approach the Throne without being confronted by the Power. In fact, the Power lives in secret fear that someone will gain the confidence of the Throne, will bypass the Power, and steal his influence. Any friend of the Throne thereby becomes an enemy of the Power, and the Power will instinctively mobilize all the capabilities available to a perfectionist-aggressive personage to repel the intruder. The attack on the intruder takes on two separate aspects, and is unrelenting until the threat disappears.

First, the Power seizes on some flaw in the intruder, be it in his character, in his philosophy of life or religion, in his personal life, or perhaps on some transgression that he committed in the past. This becomes the basis for unremitting disparagement of the individual, such that the Throne is discouraged from attaining any close personal relationship with the intruder.

Second, when the intruder himself appears on the scene, he may encounter any or all of the techniques of overt or covert aggressive-sadistic activity that we outlined in the previous chapter. Indeed, very little effort may be required on the part of the Power. The intruder may encounter an icy glare from the Power, and simply feel himself unwelcome. If this is not sufficient, the Power will obstruct, embarrass and insidiously humiliate the intruder before others. As we have mentioned, the Power has the uncanny ability to spot vulnerabilities in others. In short, the Power has an almost unlimited range of vindictive tactics, and it is not very long before the intruder senses himself so unwanted that he abandons any hope of close personal contact with the Throne. This type of protective vindictiveness may be very effective, and the Throne may find himself rather isolated. He may recognize this fact but may accept it as a necessary fact of life to make the business at hand run so well.

There is a final important point in the dynamics of the "power behind the throne." The Power, being so dependent on the Throne as his solution to life, must continually prove to himself and to others his indispensability. If the Throne becomes *too* successful, therefore, the Power's influence may begin to fade. Thus, the Power secretly, insidiously, and mostly subconsciously seeks to keep the Throne in a relative state of weakness. If the Throne is in danger of being toppled, the Power will, of course, mobilize all of his energies to defend him. However, a series of major victories on the part of the Throne may constitute grave threat to the Power, as they may lead to events that he cannot control. Hence, the Power may take on the aspect of a brooding wolf in sheep's clothing. He cannot enjoy life because he is anxious and insecure, and his relations with the Throne may not be all that good because of his unconscious competition with him.

The personality of the Power may puzzle others, since at times he may appear as a moody, laconic introvert and at others he may be a good-natured extrovert. His moods may be apparently inappropriate. When the Throne suffers an embarrassment or moderate setback, the Power will offer him commiseration and sympathy, but his mood may be one of *la belle indifférence*. Inwardly, but in a deeply repressed manner the Power's heart is leaping with glee, as he cackles to himself, "See how indispensable I am. See how you need me!"

Similarly, if the Throne achieves great victories, especially if the Power was not directly involved, the Power will offer him his congratulations but may retire from the scene in a brooding schizoid abject state.

Q.   Could you kindly summarize the dynamics of a power-throne relationship in a *love setting?*

A.   In a love setting the dynamics of the relationship are similar to those just described. With an NPA, N, NP or *n-C* submissive individual as the Throne, the marital relationship may be a very stable one indeed.

This couple often appears in a real life situation as an inseparable husband-wife team, in which one of the partners is a PA individual. The key word here is *inseparable*. For example, the couple may comprise a sweet, ingenuous celebrity who is teamed with her stern, overprotective manager-husband. The Power, being profoundly insecure in life, would be utterly lost without his mate. He has little in qualities to embark on a successful career of his own, hence his social and financial aspirations stem from the Throne.

Although the relationship is basically a very stable one — indeed the Throne himself may be deeply devoted to his mate — the Power is in a continual state of anxiety nevertheless. The Power is continually fearful of losing his mate to events that he will not be able to control, hence he will unremittingly attempt to have a complete knowledge of all persons, places, things and events that have even a remote interaction with his mate. He maintains a constant vigilance against any events that may catch him unawares. He screens the friends of the Throne and resents the slightest independence that his mate may show in his relations toward them. Any telephone call is a threat. If the Throne answers the telephone, the Power must know "Who was it?!!" and in the tone of his voice one senses anxiety and desperation. If a friend shows any indication of being able to influence the thinking of the Throne, then this friend is labeled an intruder and will be repelled from the scene by all the overtly and covertly sadistic machinations that the perfectionist-aggressive personality can conjure.

Finally, in a marital situation as well, the Power secretly finds unconscious joy in the frailties and minor failures of his mate. And this cannot help but to accentuate the free-floating anxiety that he intermittently senses but does not succeed in understanding.

Historical examples of the power behind the throne, in the literal sense, are numerous. In the United States, at the presidential level, recent variations on the theme have appeared in the form of wives (Nancy Reagan), secretaries of state (Alexander Haig, Jr.) and vice-presidents (Richard B. Cheney).

THE MORBID DEPENDENCY

Q.   What exactly is a "morbid dependency"?

A.   Individuals who have an aggressive component in their character structure (XA and submissive types) have the innate drive to "play the game" of dominance and submission. When two such individuals enter a long-term relationship, one of them establishes dominance over the other. This is the chronic state of subjugation of which we have spoken, and if both partners accept the status of the relationship, it may appear to be a stable one. However, as we have pointed out previously, the relationship is essentially one of master ruling over slave, and in this context there is always the risk of sadomasochistic elements entering into the relationship. In addition, the master being the omnipotent master that he is, he will feel himself to be well within his rights to terminate the relationship at any time.

The nature of the relationship that ensues in the chronic state of subjugation depends, to a large extent, on the nature and intensity of the sadistic trends in the dominant individual. These were discussed in some detail in the previous chapter, and were seen to lie in the realm of the individual's basic character structure, as well as in other intrinsic and extrinsic forces in life that tend to suppress overt sadistic behavior.

We could describe an infinite number of relationships based on the chronic state of subjugation, each with its particular nuances. On the one hand, we find the most overt sadomasochistic relationship based on outright cruelty and violence. On the other hand, at the other end of the spectrum, we find the self-condemned couple whose life together is a morass of overt and inverted sadomasochistic behavior. This is the silent couple that we see in the restaurant. The master is gloomy, and his eyes are averted from his companion, the slave. The latter is unhappy and looks intermittently to his mate in a deferential manner for recognition. And recognition, when it comes, is only grudgingly pronounced. These two individuals know that something is wrong, but they have only the foggiest notion of what it is.

This subtle state of affairs may involve even the usually dutiful NPA type in either the dominant or submissive role. As a dominant subjugator, the NPA type's sense of duty causes his sadistic trends to be deeply inverted, and if they do break out into overtness, he periodically hates himself for it. In the subjugated role, the NPA type suffers. He suffers because of the dramatic personality split he must assume when in the presence of his master. Furthermore, he is periodically driven to the extremes of frustration, which cannot help but incite him to his NPA super-rages. And in the wake of these rages he senses only a continuation of his confused frustration, and a growing suspicion that just as he is not in control of his emotions, he is equally not in control of his life situation.

A chronic state of subjugation between two individuals can exist, in a seemingly stable relationship, for many years. However, the illusory nature of this stability is revealed time and time again, as we can readily confirm in the real life situations of our friends, of our families and of ourselves. Sometimes the relationship, in all of its apparent placidity, may take on the appearances of a nice, tranquil, uneventful voyage… of the Hindenburg on its last transatlantic crossing.

If the relationship is a fairly stable one, we think that it is appropriate to call it a *dependency of subjugation*. If the relationship is an unstable one, with more pronounced sadomasochistic trends, and the dependent partner is at risk of destroying himself or murdering his mate, then we think that the term *morbid dependency* is appropriate.

The essential dynamics of the morbid dependency have been noted throughout the ages. The American psychologist William James called it "monomania" and presented dramatic examples of in his treatises published at the turn of the 19th century (Appendix B). Hilaire Belloc in his biography of Louis XIV described it in somewhat romantic terms as follows:

> There falls upon some very few human lives an experience
> transcending every other. They that have received it stand
> separate from all their fellows. It has no name. To call it

exalted love or love inspired means nothing. The word "love" is used in every tongue and by all mankind to mean things so different, so varying in degree and quality, that to use it here is meaningless. It has no name.

The thing has no name. For names attach only to things generally known and *this* thing, a revelation, is known to very few and is incommunicable. The only parallel to it is the experience of the mystics, their momentary union with the Divine. This, those who have been so transfigured can never later describe.

But — though what Mary Mancini awoke in him has no name — we can call it a flame of fire. It seized his whole being as from without and from above. It is not of mortality; and in one great English line it has been justly saluted "the ultimate outpost of Eternity." Such a Visitor met Louis in his twentieth year.

Somerset Maugham and Karen Horney were both intrigued by the dynamics of the morbid dependency, in part because of their own life experiences. One can hardly improve on their perspicacious expositions of the subject, and the reader is exhorted, in particular, to read Maugham's *Of Human Bondage* and Horney's *Neurosis and Human Growth*. Horney's descriptions of human psychodynamics are unparalleled, and when removed from their psychoanalytic setting, show remarkable insight.

Maugham, at least in his early years, viewed the morbid dependency as the basis of *all* love relationships. In the words of Philip:

> "I'm afraid that's always the case," he said. "There's always one who loves and one who lets himself be loved."

In our view, the morbid dependency is a fundamental type of behavior limited to personages having an aggressive component in their character structures.

Q.   What are the basic forces involved in the morbid dependency?

A.   It is the role of the master to rule over his slave, using all of the dynamics of sadomasochistic behavior that are available to him. It is the role of the slave to accept all that is meted out to him, so long as the master does not leave him.

As a quick review, the compendium of actions available to the master is:

- to enslave

- to punish

- to frustrate

- to exploit

- to play on emotions

- to use vindictive violence, and

- to show complete lack of feelings toward his companion.

Q.   Are the dynamics of the morbid dependency limited to love relationships?

A.   Definitely not. They may enter wherever there is evidence of dominant-submissive behavior. This may occur in the context of leader and follower, boss and subordinate, priest and penitent, parent and child, physician and patient, or guard and prisoner. It may, of course, appear in a sexual setting, and the overt rituals of sexual sadomasochistic behavior are a well-known fact of life.

Man's relationship with his God may also be interpreted in the light of a sadomasochistic morbid dependency. Is it not true that man calls his God the Lord and Master? Does he not prostrate himself as a slave before his master? Does he not willingly accept punishment for his "sins," even though such punishment may be overtly sadistic, such as the irrevocable condemnation of a human soul to suffering in the fires of Hades for all of eternity? And does he not accept unflinchingly in self-flagellation the violence of natural catastrophes, even going as far as calling natural disasters "acts of God"?

Finally, it should be clear that there is nothing intrinsically sexual about the sadomasochistic relationship of the morbid dependency. If the two partners are of the same sex, homosexuality is not necessarily implied.

It is true, however, that the master may exploit the slave sexually once the morbid dependency is established, and this is without a doubt the basis for the bisexuality of many individuals who adopt the submissive role. In fact, this type of homosexual (as well as heterosexual) exploitation came to the fore in 1978 in the case of the subjugation of hundreds of members of the now infamous Peoples Temple cult by the Reverend Jim Jones. Here we have an example of a morbid dependency, whose basis was certainly initially established in a nonsexual setting, involving a multitude of subjugated individuals who longed only to be cared for. The whole spectrum of overt tendencies of sadistic behavior was present, but of course all of the parties involved were blind to them: the reverend master's need to enslave, to punish, to frustrate, to exploit, to play on emotions, to show complete lack of feelings toward his followers, and in the end to use the ultimate vehicles of sexual exploitation and vindictive violence on them.

Q.   But, in the context of mating, is not the subjugated partner in love with his master?

A.   If we wish to call this relationship "love" we are free to do so. But it has essentially nothing in common with the "love" that two N or NP individuals show for each other. On the contrary, it has many points in common with the "love" that the doomed cultists felt for the good Reverend.

But we make no bones about it; powerful forces are at work here. To the extent that the dependent partner senses the possibility of his being cared for in life, he seeks to be subjugated at all costs. He may resort to seeking subjugation actively in the submissive triumph, described in Chapter 7. This activity is all subjectively interpreted in the context of "falling in love," and once interpreted in these exalted terms, the dependent partner elevates his devotion to a sanctimonious state of selflessness. Once he is smitten as a prisoner of love, it must be "good." And he acquires a feeling that an inevitable destiny awaits him in the consummation of the relationship, be it for the good or for the bad.

Q.   Could you briefly describe how such a relationship unfolds?

A.   We shall describe an extreme, full-blown case in which the two partners embrace upon a path leading to destruction, or at least to a lack of fulfillment. The master might be an NPA, NA, PA, A or n-C type, while the slave might be an NPA, NA, PA, n-C or C type. It goes without saying that neither partner understands the other, neither understands the nature of the relationship, and neither understands himself.

Soon after the initial excitement of a new relationship, in all of its sexual and non-sexual aspects, the master-slave relationship begins to emerge. Even at this early stage, small, barely noticeable horns will begin to grow from the forehead of the master. The slave, though, is hooked: he is devoted to true love until the ends of time.

At a fairly early stage, the master senses his power over the slave. The slave is "in his clutches." He has him "on a string." His partner will "do anything for me." This will be rationalized in the sense that the master enjoys "having things done" for him, or enjoys the constant attention that he receives in a sexual context. But as the initial excitement passes and the subjugation becomes more marked, the inevitable scenario unfolds. The master begins to avert his eyes from the slave, and his initial feelings of attraction turn into gloomy numbness. The master begins to ignore the slave, and the elements of overt, inverted or situational sadism come to the fore. The relationship begins to sour. The illogic of it in its real life setting may be stupefying. The master may consciously or unconsciously, overtly or covertly, begin to search for a new partner. He may make, in similar ways, intimations that the slave simply is not included in his future plans.   In everything that the master says or does, there is a subliminal voice saying, "I am going to leave you… I am going to leave you… I am going to leave you…"

In Kierkegaard's novel *The Seducer* the master's plan is one of conscious, covert sadism. Johannes tells us:

> Another circumstance affecting my *modus operandi* is that I desire nothing which is not, in the strictest sense, freedom's gift. Let crude seducers employ such methods. What, after all, do they achieve? Whoever is unable to captivate a girl to such an extent that she loses sight of everything he does not wish her to see, whoever is unable to pervade a girl's being to such an extent that it is from her that everything emanates as he wishes it, he is and remains a bungler; I do not envy him his enjoyment. Such a person is and remains a bungler, a seducer, something I certainly cannot be called.

> I am an aesthete, an eroticist, who has understood the nature of love and its *pointe,* who believes in love and knows it from the ground up, and the only reservation I make is the private opinion that every love affair lasts half a year at the most and that every relationship is over as soon as the ultimate has been enjoyed.

> All this I know, but I also know that to be loved, to be loved more than anything else in the world, is the highest enjoyment imaginable. To pervade a girl's being is an art, to recede from it is a masterpiece. Still, the latter depends essentially on the former.

Despite all of the illogic in the relationship, even when he sees himself abused, the slave does not give up. On the contrary, he becomes more persistent, more clinging, more groveling, and at the same time more demanding.

Cordelia's reply to Johannes is predictable:

Johannes!

I do not call you mine, I fully realize that you have never been that, and I am severely enough punished for having allowed this thought to delight my soul. And yet I call you mine: my seducer, my deceiver, my enemy, my murderer, the cause of my unhappiness, the tomb of my joy, the abyss of my wretchedness. I call you mine and I call myself yours, and just as these words once flattered your ear which you proudly inclined toward my adoration, so now they shall ring out like a curse upon you, a curse in all eternity. Rejoice not at the thought that it might be my intention to pursue you or to arm myself with a dagger and thus arouse your derision! Flee wherever you will, I am still yours; go to the ends of the earth, I am still yours! Love a hundred others, I am still yours; yes, even in the hour of death I am yours. The very language I use against you must prove to you that I am yours. You have presumed to deceive a human being so that you have become everything to me, so that I would stake all my bliss on becoming your slave.

I am yours, yours, yours, I am your curse.

Your Cordelia

It is at this point that the slave begins to feel himself being torn apart in conflict. On the one hand are his feelings of unlimited devotion to his partner, the desire to be taken in his arms to be merged with him, to disappear into him, so that they may at last be one and the same. On the other hand is the feeling of being abused, of giving everything and receiving nothing. And, in flashes, the subjugated partner sees in himself the first inklings of vindictiveness.

In *Of Human Bondage* Philip is tormented by his love for Mildred:

"Don't be beastly to me, Mildred. I can't bear it."

"You are a funny feller. I can't make you out."

"It's very simple. I'm such a blasted fool as to love you with all my heart and soul, and I know that you don't care twopence for me."

> "If you had been a gentleman I think you'd have come
> next day and begged my pardon."
>
> She had no mercy. He looked at her neck and thought
> how he would like to jab it with the knife he had for his
> muffin. He knew enough anatomy to make pretty
> certain of getting the carotid artery. And at the same
> time he wanted to cover her pale, thin face with
> kisses…

As the relationship deteriorates and the master's intimations of leaving become more apparent, the slave may strike out in an attempt to achieve a full-blown submissive triumph. He implores that he has been so devoted, so faithful, so loving, hence he demands to be loved unconditionally in return. The master loved him before, did he not, hence he must love him forever and ever. And as he clings more and more, and grovels before the master, the latter cannot but shrink away. And as the slave promises in an ever-intensifying manner, warmth, favors, gifts and fidelity, the master cannot but grow colder.

If the attempts at a submissive triumph do not succeed, then the slave may descend into an abject state of suffering, for example in a basement or in a darkened room. The state of suffering is in part a further call to be rescued, by anyone, from being left alone in the cruel world. It is partly vindictive, to shame the master before his children, his friends and his neighbors: "See how rotten I am being treated!" And in fact, the dependent partner breaks out of the abject state periodically in vindictive rages of frightening intensity, frightening to everyone including the individual himself. But in the threats of vindictiveness and in the brief skirmishes of violence, the slave realizes that it is all for naught. For after each spell of vindictiveness he is filled with remorse. And behind the remorse lies the paralyzing fear that his vindictiveness might be the last straw to cause the master to leave.

But for the master to leave? Impossible! Somehow the words have no meaning and do not enter into the reality of the slave. It is just not possible, to the thinking of the slave that this couple who once loved together could ever part.

At this point, the slave may surface from the abject state of suffering, perhaps with the help of a friend, a physician, or a psychiatrist. He will go alone, to some secluded spot, taking a pen and pads of yellow lined paper with him. He will try to figure this out. He is convinced that he will be able to decide in the calm of solitude — either yes or no — whether or not he loves the master. In fact, does he *love* him or does he *hate* him? The question is rarely answered in any definitive terms, but what does emerge on paper is a master plan. The plan is to "turn the tables" on the master, with vindictive elements showing themselves unabashedly. The plan is: "If he does this, then I shall do that." "If he says this, I shall say that." "I shall force *him* to make the moves." "I just don't care about him any more, so if he wants something, *he* will have to ask *me*." And so on.

However, when the time comes to put the plan into effect, all of the careful planning and all of the self-exhortation is for naught! At the first word from the master, all the plans collapse like a house of cards. The slave is shocked to realize that despite all of his planning, all of his logical thinking, the situation is not amenable to logic at all! The master remains the master, and he remains an utterly powerless slave.

In Stendhal's novel *The Red and the Black* the submissive character Julien attempts to "turn the tables" on Mathilde. With the help of a confidant, Korasoff, he is resolved to succeed in his well-rehearsed plan to play "hard to get":

> Surprised, dumbfounded, Julien did not at first comprehend the happiness the scene had in store for him. As he assisted her to a seat, she threw herself with abandon into his arms. At first his joy was boundless, but on second thought he recalled Korasoff.
>
> "I may lose everything now by a single word." His arms became rigid, though indeed the effort was painful.
>
> "I must not permit myself to press this supple and charming figure to my heart; otherwise she would despise me… Julien felt his strength disappear, so frightful was the act of courage which he had outlined for himself. "Those eyes," he said to himself, "will soon express only disdain if I allow myself to be drawn into the allurements of love…"
>
> "My great love and misery, if you do not love me any more," she replied, taking his hands.
>
> The movement removed the shawl for an instant, and Julien saw her charming shoulders, her disordered hair recalling to him a certain pleasing scene; and he was about to yield.
>
> "One imprudent word," he said to himself, "and I begin again that long series of days of despair."

> ...Julien, who was not over-proud over his painful victory, was himself fearing her looks, and had not dined at the house that night. His love and happiness increased rapidly the moment he was removed from the scene of battle. Then again he began to blame himself. "How could I have resisted her? Suppose she does not love me any more? In a moment she can change, and I must admit I treated her shamefully."

But although Julien succeeds, barely, in establishing dominance over Mathilde, indeed forcing her into an abject state of depression, he finds himself in a double bind. If he is submissive to her, he is despised by her; but if he establishes dominance over her, he cannot touch her! In fact, the only way that he would be able to touch her would be to act sadistically toward her. Julien senses that there is no easy solution here, and he feels himself being torn apart.

As a relationship of morbid dependency nears its end, two somewhat opposing forces come increasingly into play:

On the one hand, in the dependent partner, there is an *ever-increasing anxiety* of the ambiguity of the situation, and the situation is, indeed, always ambiguous in the eyes of the slave. For no matter how dismal the outlook is for the success of the relationship, the slave will always have a small hope that, somehow, the master will change his mind, and everything will turn out all right. Or vindictive daydreaming may be prominent: "Just you wait Henry Higgins!" "When you smash yourself to pieces, you'll come to me as a quadriplegic begging me to take care of you for the rest of your life." And, it is true, the master will sense that all of this ambiguity is to his advantage. He has, practically, no need to do anything. Until he decides what he is to do, for the good of all concerned, he can profit from the ambiguity of the situation to play on the emotions of the slave.

On the other hand, there is an *ever-increasing weariness,* a gradually intensifying numbness on the part of the dependent partner. He begins to look forward to the end of the relationship. He thirsts for that intense pleasure, in suffering, of knowing that he is right in being wronged. He will find great satisfaction in being physically maltreated at the end of a relationship. He may have dreams of destroying himself at the doorstep of his offending master in a final act of masochistic martyrdom.

Q.   **What happens after the master leaves?**

A.   He often comes back. In one way or another he returns to the slave.

Q.   He returns??   After all that he did, that good-for-nothing has the nerve to show his face again?

A.   Yes. And the master himself may not really understand why he does so. First of all, he may return in a subdued or even abject state. He may have realized that it is a cold, cruel world out there, and in his inability to cope with it all and in his inability to establish lasting relationships with others, he wonders why he was unable to respond to the tender devotion of the slave. Secondly, he may observe, with burning envy that the slave is doing quite well without him. In fact, he cannot bear to see any independence, resourcefulness or levity in any of his subjugated subjects. In consequence, the master may periodically visit his slaves in order to prove to them, and to himself, that he is the lord and master of all the underlings in his life. Finally, if his sadistic trends are marked, he will derive enjoyment and exhilaration from showing up from time to time in order to play on the emotions of his slaves. The latter, despite all logic, remain helplessly subjugated by the master.

The following example, taken from the advice column* of a daily newspaper, is typical:

> DEAR MEG: I've been in love with a man I'll call George for three years. For various reasons we broke up.
>
> I was pulling myself together and getting ready to start all over again when George called. He didn't want to come back, he just wanted to talk. So we talked. The following week he called again.
>
> Now he calls just often enough to keep my emotions in complete turmoil. This man cheated on me with other women and broke his word when we entered a business deal. Now he's accusing me of giving him the bum's rush.
>
> I admit I still care for him but I know I should put him out of my mind. I can't seem to do it. Would a hypnotist help?
>
> —FRANTIC

---

* Dear Meg, *New York Post,* August 14, 1980.

Q.   **Does a real morbid dependency ever die away completely?**

A.   Philip has one answer for us:

> Philip walked along Parliament Street. It was a fine day, and there was a bright, frosty sun which made the light dance in the street. It was crowded... He crossed Trafalgar Square. Suddenly his heart gave a sort of twist in his body; he saw a woman in front of him who he thought was Mildred. She had the same figure, and she walked with that slight dragging of the feet which was so characteristic of her. Without thinking, but with a beating heart, he hurried till he came alongside, and then, when the woman turned, he saw it was someone unknown to him. It was the face of a much older person, with a lined, yellow skin. He slackened his pace. He was infinitely relieved, but it was not only relief that he felt; it was disappointment too; he was seized with horror of himself. Would he never be free from that passion? At the bottom of his heart, notwithstanding everything, he felt that a strange, desperate thirst for that vile woman would always linger. That love had caused him so much suffering that he knew he would never, never quite be free of it. Only death could finally assuage his desire.

A somewhat different answer is quoted by William James (1908) in this account:

> For two years of this time I went through a very bad experience, which almost drove me mad. I had fallen violently in love with a girl who, young as she was, had a spirit of coquetry like a cat. As I look back on her now, I hate her, and wonder how I could ever have fallen so low as to be worked upon to such an extent by her attractions.
>
> Nevertheless, I fell into a regular fever, could think of nothing else; whenever I was alone, I pictured her attractions, and spent most of the time when I should have been working, in recalling our previous interviews, and imagining future conversations. She was very pretty, good humored, and jolly to the last degree, and intensely pleased with my admiration. Would give me no decided answer yes or no, and the queer thing about it was that whilst

pursuing her for her hand, I secretly knew all along that she was unfit to be a wife for me, and that she never would say yes... My own conscience despising me for my uncontrollable weakness, made me so nervous and sleepless that I really thought I should become insane...

The queer thing was the sudden and unexpected way in which it all stopped. I was going to my work after breakfast one morning, thinking as usual of her and of my misery, when, just as if some outside power laid hold of me, I found myself turning round and almost running to my room, where I immediately got out all the relics of her which I possessed, including some hair, all her notes and letters, and ambrotypes on glass. The former I made a fire of, the latter I actually crushed beneath my heel, in a sort of fierce joy of revenge and punishment. I now loathed and despised her altogether, and as for myself I felt as if a load of disease had suddenly been removed from me.

That was the end. I never spoke to her or wrote to her again in all the subsequent years, and I have never had a single moment of loving thought towards one who for so many months entirely filled my heart... From that happy morning onward I regained possession of my own proper soul, and have never since fallen into any similar trap.

We note, however, that the narrator above had the good fortune of being able to separate himself physically from his subjugator. Maugham, in fact, recommends this highly. In *The Razor's Edge* he states:

"You know, when one's in love," I said, "and things go all wrong, one's terribly unhappy and one thinks one won't ever get over it. But you'll be astounded to learn what the sea will do..."

"Do you speak from experience?"

"From the experience of a stormy past. When I suffered from the pangs of unrequited love I immediately got on an ocean liner."

Q.   This is all rather depressing. "Playing the game" of subjugation seems rarely to lead to relationships of real stability. How can this be? Why does it not work?

A.   There is no easy answer to this question, but we must face the fact that, at least until the present time, it has very often not worked. The enormously high divorce rates in many Western countries attest to this state of affairs, not to mention the innumerable morbid dependencies that occur without attaining the status of wedlock, or end in murder or suicide. How indeed are we, who have a measure of the trait of aggression, to respond to Philip's conclusion that "there's always one who loves and one who lets himself be loved"?

Part of the answer may be found in three observations:

First, one is reminded of G.B. Shaw's comment regarding Christianity, to the effect that the only trouble with it is that it has never been tried! Similarly, we can say that the only trouble with human relations based on a mutual under-standing of the needs of the involved partners is that it has never been tried!

Second, we must re-examine the question of whether we are all suitable for marriage and for raising a family. We have, most of us, been instilled with a concept of an inflexible *monogamous ideal.* It is a concept that few of us have dared to question. Hilaire Belloc describes it as follows:

> Man is monogamous: even men as individuals are normally monogamous. It is of their nature, especially in and after maturity. The life of all the world, its social terms and institutions prove the thing. This does not mean that men are necessarily rigid in morals or unexceptional in their commerce with one mate, but it does mean that the nature and general course of life lead men to this condition of monogamy and maintain them in it.
>
> A man's mate is his wife: the man, his wife and the child are the cell and unit of human affairs. The departure from such a norm may be rare or frequent, but norm it remains. Custom at the least, and at the most some obscure deep-seated instinct creates that social thing, monogamy, and so roots and nourishes a bond that all permanent rupture of it is disastrous. All but a very few, very imperfect men, take monogamy for granted even when they least profess it, even when they least know it.

The question remains whether we should without reservation inflict this concept on our children.

Finally, we must think the unthinkable and ask ourselves the unmentionable: is the dependency of subjugation based on aggression a vestigial remnant of our animal past? For, as we shall see in the next chapter, many societies — in particular some Oriental ones — are comprised mainly of NP individuals, where the morbid dependency based on aggression is virtually nonexistent.

Q.   Let's say that this book reaches me while I am in the throes of a morbid dependency, as the dependent partner. What should I do?

A.   It would be presumptuous of us to tell you what to do. However, we can make the following observations:

First, your recognizing the dynamics of a morbid dependency and intellectualizing about it will not make it disappear. The forces operating are powerful ones indeed!

Second, in recognizing the true nature of the relationship — and in some cases the monstrosity of it — you may trade the high anxiety of ambiguity for a low abject state of hopelessness. But we predict that the state of hopelessness will not last long. There is, in fact, a whole world out there! And as most evidence points to the conclusion that we are on this earth for just a short journey, it is all to our advantage to make the best of it.

Finally, out of all this should emerge a spirit of healthy resignation toward life. We do not, of course, mean resignation in the sense of detachment from society. Rather we mean a healthy realization that we all cannot do everything in life, that some things are just beyond our comprehension and capabilities, and that sometimes our goals and expectations of ourselves in love relations are frankly outlandish.

And as part of this "new realism" we recognize the plight of the master. Is he evil, with all of his brooding sadistic machinations? If we say "yes" and declare him "guilty," we must nevertheless recognize the extenuating circumstances that have plagued his life. For in doing what he has done, he has been carrying out a ritual that stems from the deepest instincts of his human soul. We forgive him for he knows not what he does.

PLATE 23. Psychiatrist Karen Horney: a highly-motivated NPA– type who succumbed to cancer. [*K. Horney Institute*]

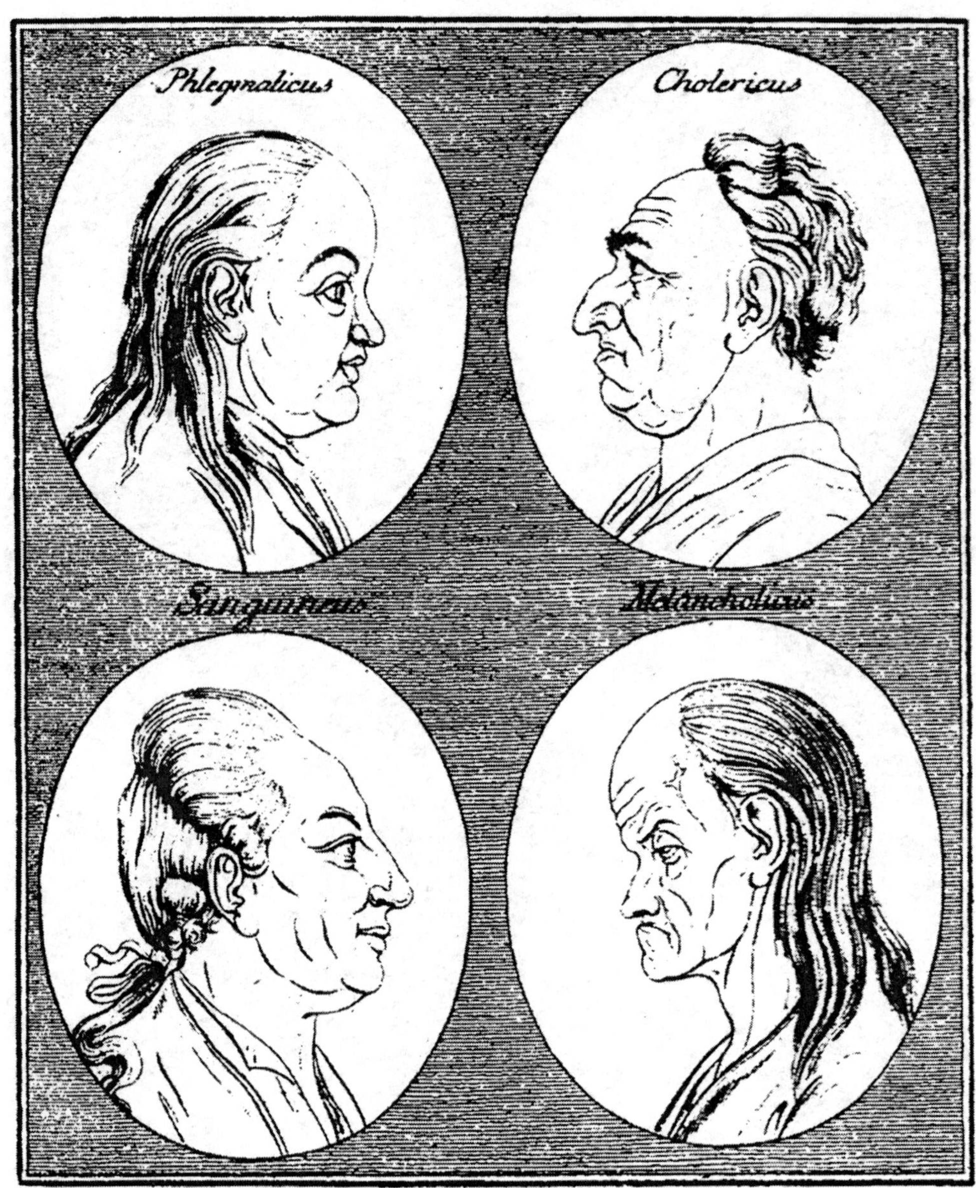

PLATE 24. The four character types according to the ancient theory of "humors": *Phlegmaticus, Cholericus, Sanguineous* and *Melancholicus.*
 [*J.K. Lavater, ca. 1775*]

Annabel Lee.
By Edgar A. Poe.

It was many and many a year ago,
    In a kingdom by the sea,
That a maiden there lived whom you may know
    By the name of Annabel Lee; —
And this maiden she lived with no other thought
    Than to love and be loved by me.

She was a child and I was a child,
    In this kingdom by the sea,
But we loved with a love that was more than love —
    I and my Annabel Lee —
With a love that the wingèd seraphs of Heaven
    Coveted her and me.

And this was the reason that, long ago
    In this kingdom by the sea,
A wind blew out of a cloud by night
    Chilling my Annabel Lee;
So that her high-born kinsmen came
    And bore her away from me,
To shut her up, in a sepulchre
    In this kingdom by the sea.

PLATE 25. The behavioral complex of perfectionism at work in a poet, demonstrating a powerful need for neatness, order, symmetry and repetition. [*Courtesy of Houghton Mifflin Co.*]

PLATE 26. Character types having a measure of the trait of aggression often have "spirited eyes." Here, Franz Ferdinand of the Habsburgs, an NPA type known for his irascibility and his prominent eyes.
[*Heeresgeschichtliches Museum*]

PLATE 27. The eye contact of dominance and submission: Peter the Great and his doomed son, Alexis. [*Nikolai Gyé*]

PLATE 28. A dominant type, especially an A or PA, may pompously look away from the camera. Here, Allied leaders of the World War II conflict pose for a group picture. [*Imperial War Museum*]

PLATE 29. A camera-shy submissive type may use a prop to avoid eye contact with the camera. Here, Somerset Maugham, an NPA– type.
[*The Rank Organization*]

PLATE 30. Character traits can often be read from an individual's countenance. Here, the physiognomies of two introverts: Charles I (an NPA– type) and the Cardinal de Richelieu (an NP type).

[*By courtesy of H.M. the Queen & the National Gallery, London*]

PLATE 31. American President Andrew Jackson in his later years: a rambunctious narcissistic N type. [*Mathew Brady*]

PLATE 32. Marie-Antoinette: an ingenuous, somewhat scatterbrained N type. [*Kunsthistorisches Museum*]

PLATE 33. Prince Regent, the future George IV: a pleasure-loving N type. [*National Portrait Gallery, London*]

PLATE 34. Catherine of Aragon: a dutiful, obdurate NP type. [*Kunsthistorisches Museum*]

PLATE 35. Abraham Lincoln in 1860: a tall, slender, melancholic NP type.
[*Chicago Historical Society*]

PLATE 36. Mary Todd Lincoln at age 28: an ambitious, spirited NA type. [*Library of Congress*]

PLATE 37. Kaiser Wilhelm II: a vainglorious N type.
[*Bettmann Archive*]

PLATE 38. Friedrich Wilhelm I: a sanguine NPA autocrat known for his irrational rages. [*Charlottenburg Castle Authority*]

PLATE 39. Henry VIII: a sanguine autocrat of the N type.
[*Thyssen-Bornemisza Collection*]

PLATE 40. Napoleon I: a pseudoaggressive N type, easily mistaken for an A type. He had the capacity for a smile "that would light up his entire countenance." [*National Gallery of Art, Washington*]

PLATE 41. Catherine de Médicis: a matriarchal sanguine N type. She was an asset to her weak and probably psychotic son, Charles IX, but not the strong, scheming "power behind the throne" of popular lore. [*Louvre*]

PLATE 42. The Archduchess Sophie of Bavaria: an authoritarian NP type, easily mistaken for a PA "power behind the throne." [*Austrian National Library*]

PLATE 43. Count Kaunitz of Austria: an eccentric NP type who served as foreign minister to Empress Maria Theresa. [*Archiv fur Kunst und Geschichte*]

PLATE 44. Henry III of Valois: a sexually ambivalent narcissistic exhibitionist (N type), known for his contingent of "minions."
[*N.Y. Public Library Picture Archive*]

PLATE 45. Caligula of Rome: a narcissistic extrovert (N type) who developed a psychosis and showed pseudoaggressive qualities. [*Ny Carlsberg Glyptotek*]

PLATE 46. King Ludwig II of Bavaria: a schizoid, camera-shy PA– type who became frankly psychotic. [*Culver Pictures*]

PLATE 47. Vincent van Gogh: an N–P type who developed a psychosis with pseudoaggressive characteristics. [*Musée Orsay*]

PLATE 48. Empress Maria Theresa: was she an NP type… or a dutiful, resolute NPA– type afflicted with rheumatoid arthritis? [*Akademie der bildenden Kunste*]

CHAPTER 10

GENETICS

Heterogeneous personality has been explained as the result of inheritance — the traits of character of incompatible and antagonistic ancestors are supposed to be preserved alongside of each other. This explanation may pass for what it is worth — it certainly needs corroboration.

William James (1908)

OUTLINE OF CHAPTER TEN

TERMINOLOGY OF GENETICS
    Dominance & recessiveness
    Pleiotropic genes

BEHAVIORAL COMPLEXES
    Narcissism: recessive gene n
    Perfectionism: dominant gene P
    Aggression: recessive gene a

SYMPATHETIC-PARASYMPATHETIC NERVOUS SYSTEM
    A rage
    N rage
    Combined rages

PREDICTIVE ASPECTS OF GENETIC MODEL
    Genotypes corresponding to phenotypes
    Transmission of traits N, P, A
    Infertility: P and null progeny
    Phenotypes of children according to phenotypes of parents

POPULATION GENETICS
    Hardy-Weinberg approach
    Genotype and phenotype frequencies
    Polymorphism of character types
    Theoretical habitancies
        U.S.A. (Polymorphic)
        Punctilious
        Sublime
        Demonstrative
        Authoritarian
        Militant

CHAPTER SUMMARY

Q.   What is this chapter about?

A.   Our model proposes that the human character structure hinges primarily on the existence of three behavioral complexes, namely those of *narcissism, perfectionism,* and *aggression.* We propose that each of these behavioral complexes is controlled by its respective genetic locus, and each is normally expressed with *complete penetrance* in dominant individuals. That is, if the genes are present in individuals in accord with the minimal requirements for their expression (heterozygosity for a dominant gene and homozygosity for a recessive one), then all such individuals will exhibit the corresponding *behavioral complex* or *trait.*

In addition, we propose that if more than one behavioral complex is expressed, the resulting phenotype will be characterized by a resultant *behavioral syndrome* that is manifested by a recognizable interaction of the component behavioral complexes. The behavioral complexes may be thought to interact in either a *synergistic* or *antagonistic* manner. Individuals exhibiting either a single behavioral complex, or a behavioral syndrome based on more than one behavioral complex, will be recognized as *discrete character types.*

For example, the N and A behavioral complexes act synergistically to produce the unbridled ambition of an NA individual. The P and A traits act somewhat antagonistically to produce the passive aggressiveness of the PA individual.

Q.   Before you go on, would you kindly review some of the basic principles of human genetics to which you will be referring.

A.   We summarize below some of the concepts that we will be using. If you are interested in this chapter, we would suggest a reading of one of the many available books on the fascinating history of genetics.

Chromosomes are the cell constituents that carry the hereditary material, *DNA.* Somatic (body) cells contain the *diploid number* of chromosomes, while gametes (germ cells) contain one-half the diploid number, called the *haploid number.* The diploid number of chromosomes in man is 46, consisting of 22 pairs of *autosomal chromosomes* and two *sex chromosomes* (XX in the female, XY in the male). In the process of *meiosis,* the female produces *gametes* called *ova* in which the haploid number of chromosomes (22 plus X) is randomly assorted from the 23 pairs. Similarly, in meiosis the male produces *spermatozoa* in which the haploid number of chromosomes is randomly assorted (22 plus X, or 22 plus Y).

Genes are sections of DNA in the chromosomes. They are encoded with information to produce molecular chains of amino acids called *polypeptides* (an intermediate step involving RNA is involved). Many complex polypeptide molecules act as *enzymes* in the various biochemical reactions on which life processes depend.

Alleles are alternative forms of a gene at a given locus on a chromosome. A *genotype* is *homozygous* with respect to a locus if the alleles are identical on both *homologous chromosomes.* It is *heterozygous* if the alleles are different at the locus. If a trait (behavioral complex) is expressed when the responsible gene is present in the heterozygous state, then it is said to be *dominant.* If the trait is expressed only when the responsible gene is present in the homozygous state, then it is said to be *recessive.*

We shall assume *Mendelian genetics.* Mendel's first law is that of *segregation of alleles.* That is, only one of each pair of homologous chromosomes is present in any gamete. Mendel's second law is that of *independent assortment* of genes existing on separate chromosomes. That is, there will be a random distribution of non-homologous chromosomes to the daughter cells (gametes) during the process of meiosis.

If two genes are close together on the same chromosome, they are said to be *genetically linked.* Two characteristics having a genetic basis may thus exhibit *association* in a family if their corresponding genes are linked. The association may be loose, however, for if two genes are relatively far apart on the same chromosome, *recombination* (crossing over) may occur between homologous chromosomes during the process of meiosis. (However, the association of two phenotypic characteristics in a population does not necessarily imply genetic linkage as the basis of the association). In a separate sense, the term *X-linked* simply refers to a trait determined by a gene present on the X chromosome.

Q.   How can a behavioral complex depend on the expression of just a single gene?

A.   Just how this occurs is not known, but other examples in human genetics show that, indeed, a single major gene can control a whole syndrome of various phenotypic characteristics. The action of such a gene, having multiple effects, is known as *pleiotropism.*

In addition, there is little doubt that different genes for the expression of several different modes of behavior can act as complementary genes and lead to intermediate forms of behavior. For example, Dilger succeeded in hybridizing two different species of the parrot genus *Agapornis.* The two species had very different, stylized modes of nest-building behavior. The hybrid had a resultant syndrome of nest-building behavior that incorporated aspects of each of the parental types. In the resultant mode of nest-building behavior, some of the bird's mannerisms were mutually antagonistic, and in fact the adult birds came to suppress some aspects of the resultant behavioral syndrome through the process of learning.

Even more to the point, investigators working in the field of aggression in animals have shown that different strains of rats exhibit widely variable tendencies for muricidal (mouse-killing) behavior. Furthermore, by inbreeding a colony of rats the frequency of muricidal individuals could be dramatically increased.

In short, there is to our mind little doubt that pleiotropic genes can control rather complicated behavioral complexes. The behavioral complexes, like any phenotypic expression having a genetic basis, are subject to being mediated by other "modifier" genes and by environmental influences. In addition, we believe that at the level of a population of individuals, the behavioral complexes are not immune to the processes of artificial and natural selection.

Q.   How can the principles of genetics be applied to populations?

A.   We shall utilize the principles of population genetics later in this chapter.

Briefly, if one assumes numerical values for the frequencies of the genes under consideration and assumes random mating of individuals, all of whom give rise to viable progeny, then it is easy to calculate the frequencies of genotypes and phenotypes in an isolated population. This type of analysis, known as the *Hardy-Weinberg approach,* reduces the problem of genetic variability in a population to probabilistic terms.

It will be seen that we have good reason to believe that the above underlying assumptions of the Hardy-Weinberg approach are not rigorously valid for the character traits N, P and A. Nevertheless, we believe that the approach will provide us with a good first order approximation to the occurrence of the various character types in widely different populations.

Q.   You mentioned that the gene for the behavioral complex of perfectionism is expressed as a dominant, while those for narcissism and aggression are recessive. How did you arrive at this?

A.   Consider the following example that we have observed:

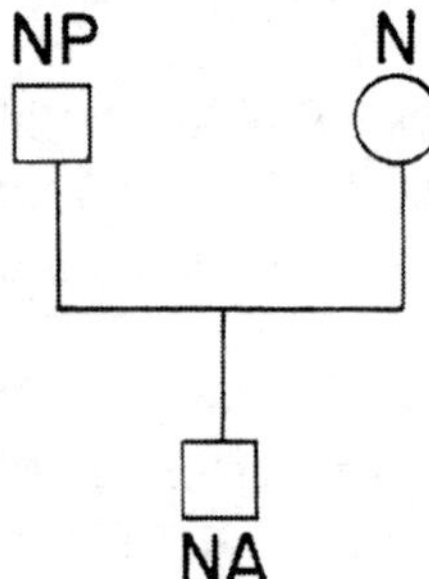

This depicts the mating of NP and N types, producing a child, a son, having an NA phenotype. (A square denotes a male, while a circle denotes a female). This example suggests that the phenotypic trait A is expressed as a recessive (gene **a**), since the trait A is not present in either parent.

We consider a second example that we have observed:

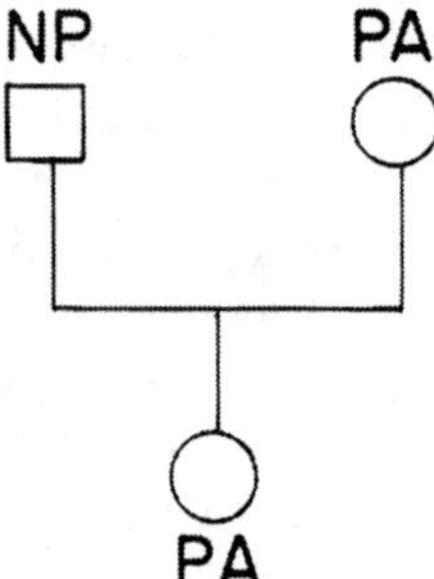

This indicates that the gene for trait A is not an X-linked recessive, since otherwise the father would exhibit the trait.

As a third example, we consider:

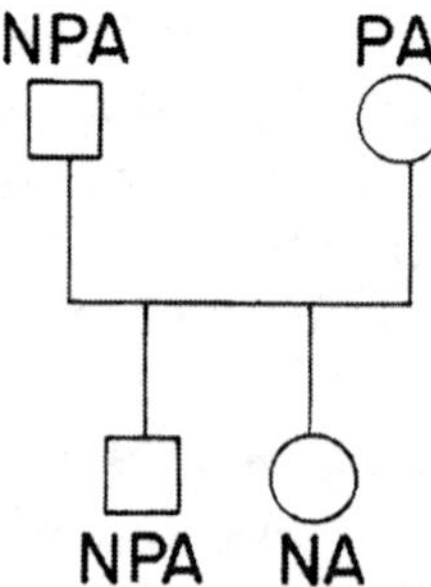

The preceding example shows that the trait P is expressed as a dominant, since the trait is lacking in the NA daughter. It also shows that the gene **P** is not X-linked, since otherwise the NA daughter would obligatorily have received the father's dominant gene **P**. The above example shows, in addition, that the gene for trait N is not an X-linked dominant, since the NPA son receives his only X chromosome from his mother.

Consider a fourth example that we have observed:

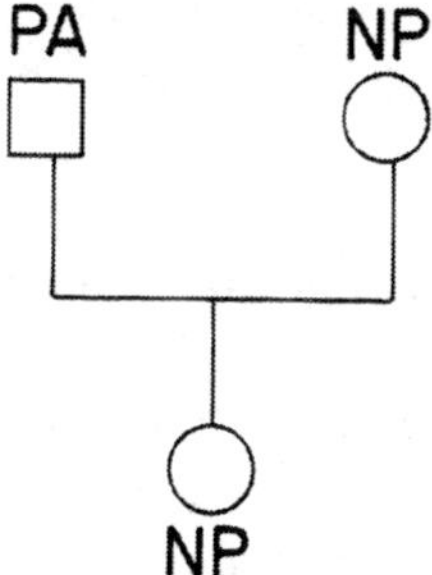

This shows that the gene for trait N is not expressed as an X-linked recessive either, since otherwise the father would exhibit trait N. Therefore, we infer that the gene for trait N is autosomal.

It still remains for us to determine if the gene for N is a dominant or high-frequency recessive one. In fact, when the frequency of a gene is high in a population, it is difficult to differentiate between dominant and recessive modes of transmission without accurate family pedigrees. If the gene for N were expressed as an autosomal dominant, then we would expect to find some families of the types:

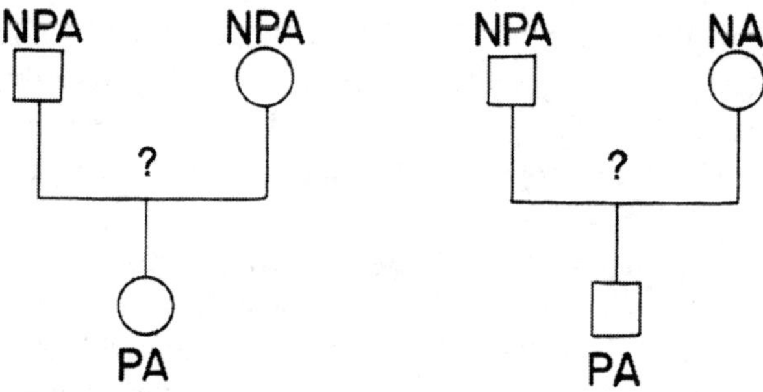

However, in our experience we have not been able to unequivocally verify such examples. (It is difficult to do so without asking embarrassing questions!) It is our impression that PA individuals ordinarily have at least one parent who is also a PA type (or possibly an A type), or in other words, a non-sanguine individual must have at least one parent who is also non-sanguine. This point will be discussed in the next chapter in connection with schizophrenia. We therefore infer that the gene for trait N is transmitted as an autosomal recessive.

In summary, we have circumstantial evidence that the behavioral complexes are controlled by autosomal genes. The genes for the traits of narcissism and aggression appear to be recessive, while that for perfectionism is dominant.

Q.   What are the implications of this result?

A.   First of all, the result would have a pleasing symmetry to it. That is, the two components of ambition, i.e., narcissism and aggression, are both transmitted by recessive genes, while their mediator, perfectionism, is transmitted in the dominant mode. From the biochemical viewpoint this could mean that the alleles corresponding to the absence of **n** and **a** produce *enzymatic inhibitors* of the N and A traits. If this is true, then mutations in these alleles could lead to the spontaneous appearance of a variety of different **n** and **a** genes. These mutations would have different DNA nucleotide sequences, and might thus lead to a variety of inactive or partially active inhibitory enzymes. Hence, we might be alert to the possibility of the spontaneous appearance of N and A phenotypes based on mutations.

Since the gene for the P trait is expressed as a dominant, we might suspect that it codes for an *activating enzyme*. Hence, mutations in **P** might be expected to lead to gametes lacking the ability to confer the P trait to the individual's progeny. On the other hand, if an individual lacked the P trait because of two different defects in the **P** genes, at different locations on the two homologous chromosomes, then it is possible that *intragenic crossing over* during the process of meiosis would lead to an intact **P** gene. Thus, the dominant P trait could appear spontaneously in the progeny of this individual, as if by a mutation.

We recognize that in genetics things always turn out to be much more complicated than they first appear. We have glossed over the possibility that the phenotypic expression of the trait P may be somewhat different depending on whether the gene P is present in the homozygous or heterozygous state. Also, in assuming that the behavioral complexes are expressed "with complete pene-trance," we are undoubtedly making a simplification. That is, we are glossing over the whole area of *expressivity,* or the degree to which a trait will be present in an individual having the character vector of dominance. However, we are doing so for a good reason: we have few objective data regarding possible biological variability in the traits of narcissism, perfectionism and aggression. Here, we cannot afford to forget Lord Kelvin's cautionary admonition:

> "When you cannot measure it, when you cannot express
> it in numbers, you have scarcely, in your thoughts,
> advanced to the stage of Science, whatever the matter
> may be."

Obviously, it would be of great interest to be able to identify and to characterize the various personages in quantitative biophysical and biochemical terms. With regard to the behavioral complex of *aggression,* much work has already been done in elucidating its basis in the sympathetic nervous system. The "flight or fight" response to stress and the aggressive rage have been extensively studied, but little is known of the physiologic bases of the behavioral responses of *narcissism* and *perfectionism.*   But we have little doubt that these bases will be found once they are looked for.

As an example of the quantitative characterization of a character type, we might look to the NP individual. We noted earlier that of all the character types, the N and NP individuals are the most conducive to blushing. This may imply an enhanced susceptibility to peripheral vasodilation in response to a wide variety of vasoactive agents in these individuals. For example, Wolff, in a study of the effect of alcohol ingestion in various groups of individuals, found that Japanese, Taiwanese and Korean subjects exhibited a markedly higher incidence of facial flushing than Caucasian subjects. Since in our estimation some (but not all) of the Oriental groups were in the majority NP individuals, we believe that Wolff's results were in part due to an increased frequency of N and NP types in some of the Oriental groups.

If it is shown that N and NP individuals tend to exhibit an enhanced response to vasoactive agents, then this would have immediate application in medicine. For example, some N and NP individuals may be especially susceptible to vascular collapse and the state of shock that accompanies an anaphylactic reaction (e.g., the allergic reaction following a bee sting). Similarly, N and NP individuals may be especially sensitive to the therapeutic vasoactive agents that are administered intravenously to patients in the intensive care unit.

Finally, there are at least three other general areas in which it would be of great interest to be able to characterize quantitatively the behavioral components of the various character types:

- First, it would be of immense value to be able to identify by a biochemical test the character type (N, P, and A traits) of the *newborn*. This would, for example, allow the pediatrician to identify those infants at higher risk for the development of infantile autism and childhood schizophrenia. (We would imagine that the astute pediatrician will be able to determine the N, P and A traits in the infant by the age of a few months from the infant's behavioral characteristics, for example from his narcissistic smile, his perfectionistic head-banging, and his aggressive sucking).

- Second, although we acknowledge the possibility that some *submissive* and *resigned types* may have their bases in environmental conditioning, we must nevertheless look for biophysical and biochemical accompaniments to these behavioral states. In addition, we cannot rule out the possibility that there exists a well-defined genetic predisposition to an individual's assuming a submissive or resigned state (see Chapter 11). This information could be obtained from comparative studies of identical and fraternal twins who were separated in early infancy.

- Finally, in being able to identify the character types of individuals, psychiatrists should be able to study on a more firm ground the biophysical and biochemical bases of *behavioral disorders.*

Q.   Do we have any idea how the genes n, P, and a could control human behavior?

A.   We do not know, but it is easy to hypothesize that these genes act by controlling biochemical neurotransmitters at the level of the central nervous system. It has long been known that emotions and "personality" are determined to a large extent by neural activity in the hypothalamic-limbic system in the deeper recesses of the brain, with the existence of large "association areas" in the frontal lobes of the neocortex. Indeed, by suitable experimental stimulation of areas of the brain in the region of the hypothalamus, at least two different types of rages may be induced in animals. One is called the "sham rage" of aggression, while another is termed the "septal rage" of hyperactivity and withdrawal.

It is, thus, tempting to hypothesize that the **a** gene controls the release of inhibition of one portion of the autonomic nervous system. That is, an individual having the A trait is able to effect a mass discharge of the sympathetic nervous system, hence to be incited into an *aggressive rage.* As is well known, the prodromal physiologic response of the aggressive rage includes pupillary dilatation, blanching of the skin, piloerection, diaphoresis and so on. An individual lacking the A trait would have this aspect of behavior genetically suppressed. He would simply not be able to effect this mass discharge of the sympathetic nervous system.

Similarly, we hypothesize that the **n** gene controls the release of inhibition of a different portion of the autonomic nervous system. That is, an individual having the N trait would be able to effect a mass discharge of the parasympathetic nervous system, hence to be incited into a *narcissistic rage.* The prodromal response to this, we surmise, might be a complex autonomic response involving in part blushing, flushing and lacrimation. Again, an individual lacking the N trait would not be able to effect a mass discharge of the narcissistic behavioral complex, this being genetically suppressed.

Of course, individuals having both the N and A traits would be able to effect mass discharges of both behavioral complexes, either selectively or in synergy, the latter comprising the previously described NA and NPA rages.

Q.   You have established, somewhat tentatively, that the n, P and a genes are autosomal ones.   Do they assort independently?   That is, are they genetically linked or are they on separate chromosomes?

A.   This could be determined from careful pedigree studies. For example, it would be of interest to study matings of the kind NP × NP that have NA types among the progeny. If the genes **n**, **P** and **a** assorted independently, then the character types of the progeny would be in the ratio NA:N:NPA:NP = 1:3:3:9. On the other hand, if the three genes were closely linked, the character types of the progeny would be in the ratio NA:N:NPA:NP = 4:0:0:12, that is, essentially devoid of N and NPA types.

On the basis of a limited number of family pedigrees, we have concluded that the genes **n** and **a** assort independently. Hence, these genes are probably on separate chromosomes.

Q.   Do we have any idea on which chromosomes the genes for the traits N, P and A are located?

A.   Not with any certainty. But in recent years a compendium of "genetic markers" has been gathered. These are phenotypic traits that have been deduced to correspond to definite locations in the human complement of chromosomes. For example, certain genes determining tissue type (histocompatibility) are known to be located on the short arm of the 6th chromosome. By studying in families the association of the traits N, P and A with known genetic markers, we should be able to deduce the location of the genes **n**, **P** and **a** in the chromosomal complement.

It is possible that the condition known as *Down's syndrome* ("mongolism") may provide us with some clues to the manner in which the behavioral complexes are expressed phenotypically. It is now well known that Down's syndrome arises as a result of a condition called "trisomy-21," that is, the presence in the child's somatic cells of a third, or extra, autosomal chromosome. The extra chromosome is, in fact, of the number 21 kind, and in most cases arises due to the non-disjunction of the homologous pair of number 21 chromosomes at the time of formation of the gametes (oöcytes) during the process of meiosis in the female. As we know, the occurrence of Down's syndrome is strongly related to maternal age.

Now, the personality type of the Down's syndrome child is eerily reminiscent of a child of the NP type. These children are often described as *imitative, good-tempered, amiable, responsible, well-mannered, unexcitable, playful, unaggressive, relaxed,* and *scrupulous.* Sometimes the phrase "stereotyped personality" is used. Now, if the Down's syndrome child is indeed often an individual of the NP character type, then we might expect that racial and ethnic groups having high frequencies of NP types would have high incidences of individuals with Down's syndrome, and vice versa. However, the limited data available on the epidemiology of the occurrence of Down's syndrome do not allow us to come to this conclusion.

What is the Down's syndrome child trying to tell us? Are N or NP mothers more susceptible to having a Down's syndrome child? Does the extra "dosage" of genes located on the number 21 chromosome cause suppression, by a process of epistasis, of the expression of gene **a** located on some other chromosome? It is apparent that the origin of the gentle personality of the Down's syndrome child remains a puzzle to us.

Q.   On the assumption of autosomal genes n and a being recessive and P being dominant, what are the possible genotypes corresponding to the various phenotypes?

A.   They are shown in Table 1:

TABLE 1

Genotypes Corresponding To Phenotypes

Phenotype	*Genotype*
N	**(nn) (nna)**
P	**(P)* (PP)* (nP)* (nPP)* (Pa)* (PPa)* (nPa) (nPPa)**
A	**(aa) (naa)**
NP	**(nnP) (nnPP) (nnPa) (nnPPa)**
NA	**(nnaa)**
PA	**(Paa) (PPaa) (nPaa) (nPPaa)**
NPA	**(nnPaa) (nnPPaa)**
0 (null)	**(o)* (n)* (a)* (na)**

Note the genotypes in Table 1 above that are marked with an asterisk (*). Such genotypes could not appear in viable zygotes, since one or both parents would have to be of either the null or P phenotype. Since null and P phenotypes are not found in competitive society, we assume that they are not capable of reproduction, being either non-viable or severely stunted.

Note that only one possible genotype exists for the NA individual, namely the doubly homozygous state (**nnaa**).

Q.   With the same assumptions, what are the possible phenotypes of the progeny in various matings?

A.   Table 2 overleaf shows the possible progeny of all possible combinations of matings, according to the phenotypes of the parents. Note that this does not mean that a *given* set of parents, with given genotypes, would always necessarily yield *all* of the possibilities.

In Table 2 and in much of the discussion of this and the following chapter, we have implicitly lumped *submissive, resigned* and *non-aggressive withdrawn types* with dominant types. We have done so in order to keep our development relatively simple, not because we believe that these types are always genetically identical to *dominant types*.

According to our genetic model, we note the following:

TABLE 2

Possible Phenotypes of Children According to Phenotypes of Parents

MOTHER \ FATHER	N	A	NP	NA	PA	NPA
N	N — — NA — — — —					
A	N — — NA — — 0 A —	— — — NA — — — A				
NP	N NP — NA NPA — — —	N NP P NA NPA PA 0 A —	N NP — NA NPA — — —			
NA	N — — NA — — — —	— — — NA — — — A	N NP — NA NPA — — —	— — — NA — — — —		
PA	N NP P NA NPA PA 0 A —	— — — NA NPA PA — A —	N NP P NA NPA PA 0 A —	— — — NA NPA PA — A —	— — — NA NPA PA — A —	
NPA	N NP — NA NPA — — —	— — — NA NPA PA — A —	N NP — NA NPA — — —	— — — NA NPA — — —	— — — NA NPA PA — A —	— — — NA NPA — — —

- N and A individuals need not have N or A parents. Such individuals can arise *de novo* so long as at least one of the parents is an NP and PA individual, respectively.

- PA individuals must have at least one parent who is of either the PA or A type.

- NP individuals must have at least one parent who is of either the NP or N type.

- NA individuals can arise *de novo* from any combination of phenotypes.

- The mating of two NA types can yield progeny of only NA types.

- The mating of an NPA type with an NA type can yield progeny of only NPA or NA types.

- Certain combinations of parental genotypes may lead to zygotes that have only the P trait. The possible phenotypes of the parents are: N × PA, A × NP and NP × PA. Thus, the model predicts that some of these matings will fall into the category of *relative infertility,* on our assumption that most P individuals are not viable. For example, one such parental combination is a mating of the genotypes **(nP/nP) × (nPa/Pa),** which would correspond to a mating of phenotypes NP × PA. Roughly fifty percent of the zygotes would be of genotype **(nP/Pa),** hence non-viable.

- Certain combinations of parental genotypes may lead to zygotes that are lacking in all three behavioral traits. These zygotes will be called *null progeny,* denoted by the symbol "0" in Table 2, and are also assumed to be non-viable. For example, one such combination is a mating **(n/nP) × (a/Pa),** which would correspond again to mating of phenotypes NP × PA. Such a union falls into the category of *complete infertility,* since the progeny can be only of genotype **(na),** i.e., null zygotes, or **(nPa)** and **(nPPa),** i.e., P zygotes.

    In summary, an examination of Table 2 shows that matings leading to *relative infertility* may be of the combinations N × A, N × PA, NP × A and NP × PA. In addition, some of these combinations may be characterized by *complete infertility.*

Q.   **Where do we go from here?**

A.   We will now see what insight we can obtain from the area of *population genetics,* which has advanced rapidly in the past fifty years. We shall commence with the classic *Hardy-Weinberg approach.* We assume random mating of individuals, with independent assortment of the relevant genes, so that the occurrence of a particular allele in an individual is reduced strictly to probabilistic terms.

We denote by the symbols *n, p* and *a* the gene frequencies for the traits N, P and A respectively. By a *gene frequency* we mean the probability that a gene locus will be occupied by the appropriate allele. On the assumption of traits N and A being expressed as recessive and the trait P as dominant, we present in Table 3 below expressions for the distribution of phenotypes in the progeny of the population.

TABLE 3

Relative Incidences of Phenotypes on Basis of Gene Frequencies

Phenotype	*Relative Incidence*
N	$n^2 (1-p)^2 (1-a^2)$
P	$2n(1-n)\ p(2-p)\ 2a(1-a)$
A	$(1-n^2)\ (1-p)^2\ a^2$
NP	$n^2 p(2-p)\ (1-a^2)$
NA	$n^2 (1-p)^2\ a^2$
PA	$(1-n^2)\ p(2-p)\ a^2$
NPA	$n^2 p(2-p)\ a^2$
Null	$2n(1-n)\ (1-p)^2\ 2a(1-a)$

Q.   **Where did these expressions come from?**

A.   We assume that the probability of a genotype's occurring is equal to the product of the probabilities of the individual alleles assorting in the required manner. For example, considering gene **P**, we obtain expressions for the following probabilities:

$$\textbf{P} \text{ on both chromosomes} = p^2$$
$$\textbf{P} \text{ on neither chromosome} = (1-p)^2$$
$$\textbf{P} \text{ on one or neither} = 1-p^2$$
$$\textbf{P} \text{ on one or both} = 1 - (1-p)^2 = p(2-p)$$
$$\textbf{P} \text{ on one only} = (1-p^2) - (1-p)^2 = 2p(1-p)$$

Similar expressions are obtained for genes **n** and **a**. Note that in the expressions for the incidences of P and null progeny (Table 3) we do not allow for genotypes that were marked with an asterisk in Table 1, since such genotypes could not arise from the mating of viable individuals. This explains why the sum of all of the incidences in Table 3 is algebraically not quite unity.

Q.   How can we estimate the gene frequencies *n*, *p* and *a*?

A.   We choose a group of individuals and characterize them by their behavioral phenotypes.

As an example, our group is a sample of 103 individuals, aged 17 +/– one year, who attended secondary school together in the environs of New York in the year 1956. The author was acquainted with all of the individuals for at least one year's duration. The sample consists of 52 males and 51 females. Studio portraits taken by the same professional photographer were available for 102 of the individuals, and handwriting samples were available for 100 of the individuals.

We found that from memory we could, with some degree of certainty, characterize the phenotypes of 75 individuals of the 103 total, yielding 42 males and 33 females. The distribution of these phenotypes is given in Table 4 under the heading "observed." We shall refer to this population as "the U.S.A. habitancy."

TABLE 4

Phenotypes of U.S.A. (or "Polymorphic") Habitancy

Frequency of Phenotype

Phenotype	*Observed*	*Theoretical*	*Theoretical (viable)*	*Theoretical (viable, %)*
N	4	5	6	8
P	--	3	--	--
A	2	3	3	4
NP	20	14	15	20
NA	15	12	12	16
PA	12	7	7	10
NPA	22	30	32	42
Null	--	1	--	--
Sum:	75	75	75	100

Summing the individuals having the traits N, P and A, we obtain respectively 61, 54 and 51. We calculate the gene frequencies from the equations:

$$n^2 = 61/75$$
$$p(2-p) = 54/75$$
$$a^2 = 51/75$$

giving

$$n = 0.90$$
$$p = 0.47$$
$$a = 0.83$$

Substituting these values into the expressions for the phenotype frequencies of the progeny (Table 3) and normalizing to a total of 75 individuals, we obtain the values that are given in Table 4 under the heading "theoretical."

We can make the following observations:

1. The gene frequencies for **n** and **a** of this sample are quite high, on the order of 0.9 and 0.8 respectively.

2. The theoretical distribution correctly predicts that the most common individuals should be NPA and NP types, followed by NA types and PA types. It also correctly predicts that N and A types should be fairly uncommon.

3. If we conduct the Chi-square statistical test, to test for a significant difference between the observed distribution of individuals and the theoretical distribution of viable individuals in the progeny (Table 4), we find that we cannot reject the hypothesis that the theoretical and observed distributions of phenotypes are identical.

4. The theoretical calculation predicts that about 5 percent of zygotes should be of the P and null types. Is this reasonable? It is, indeed, not unreasonable if we assume that P and null individuals are not viable. McKusick (1969) indicates that at least 15 percent of all zygotes are naturally aborted, while another 3 percent lead to stillborn infants.

5. Finally, we note that the predicted distribution of character phenotypes is valid only as a first approximation because the assumptions of the Hardy-Weinberg approach are not strictly valid. In fact, the appearance of non-viable P and null types in the progeny means that in order to maintain constant gene frequencies from generation to generation, one would require either that non-

random (assortative) mating occurred, or that certain combinations of phenotypes were conferred reproductive advantages. However, since the number of non-viable types is small in relation to the total, we believe that the preceding calculation of phenotype frequencies of viable individuals is nevertheless valid as a good approximation for the first generation of progeny.

Q.   Why should the traits of aggression and narcissism persist in the human species?

A.   They are the primary traits that determine an individual's ambition, or motivation. Consequently, they act synergistically with other basic instincts for self-preservation and would tend to be powerful driving forces in an individual's struggle for survival.

The *unbridled* traits of aggression or narcissism (i.e., unaccompanied by the trait P) would confer a direct reproductive advantage to an individual:

- The trait of *aggression* would allow the individual to dominate a potential sexual partner, either subtly or not so subtly.

- The trait of *narcissism*, flashing its alluring smile and flaunting its motto: "I am great, beautiful, charming... irresistible" would be a major factor in the attraction of potential sexual partners and would also confer to the individual a decided reproductive advantage.

Q.   Why would a population be comprised of different character types? If natural selection were operative, why would not just a single character type have emerged?

A.   This is a question that the population geneticists have considered in other contexts over the years and leads to the concept of *balanced polymorphism.* Most geneticists quote C.E. Ford's definition:

> "Polymorphism may be defined as the occurrence together in the same habitat of two or more discontinuous forms of a species in such proportions that the rarest cannot be maintained merely by recurrent mutation."

An example frequently cited is that of the ABO system of antigens related to human erythrocytes (red blood cells). In that case, the balanced polymorphism has its basis in multiple alleles existing at a single gene locus. In our model of discrete character types, the polymorphism is based on the combined action of three separate genes.

Q.   What are the population dynamics that would lead to a balanced polymorphism of character types?

A.   This is an extremely complex problem that we are not going to be able to solve here. Much data will be required before we will be able to make even the most tentative of conclusions. Nevertheless, we offer the following observations:

1. The polymorphism of character types in various areas of the world may not be in a state of balanced equilibrium at all. Indeed, during the past few hundred years very significant migrations of races, nationalities and religious groups have occurred.

2. In the face of natural selection against certain character types, polymorphism could be maintained by non-random, or assortative, mating. We have noted earlier that a character type may be intensely sexually attracted to one personage yet be rather indifferent to another.

3. Although a population may appear to be a "melting pot" of individuals in a given habitat, there may exist macrocosmic or microcosmic isolates within the population. These isolates could have their bases in race, religion, even somatotype, and so on. In fact, if we believe the old adage that "birds of feather flock together," then we see that the character type of the individual may itself be a driving force for the formation of an isolate within a population.

4. Some human societies appear to require a diversity of character types. Consider some of the harrowing alternatives. If all the individuals of a community were N types, there would be much posturing and much rushing to the limelight, but very little real work would get done. If all were A types, the individuals would spend all of their time fighting, and the hospitals would be filled. If all were NPA types, the individuals would spend all day talking and shouting at each other, while vying for leadership, and again, very little real work would be done. And so on.

    Conversely, we could imagine (somewhat tongue-in-cheek) a balanced society consisting of a polymorphic distribution of character types. The NPA's would be the group leaders, N's the creative visionaries, NP's the quiet workers, NA's the hunters, A's the warriors and defenders, PA's the intimidating enforcers, submissives the homemakers, and resigned types the artists.

5. There is evidence that a relatively stable human society of a *monomorphic* character type is possible. In fact, we propose that approximations to such a society already exist in some parts of the world. We are referring, in particular, to societies having very high frequencies of either N or NP individuals.

Q.   The conclusion that stable monomorphic societies of NP individuals seem to have arisen appears to be striking one. Do not the implications of this conclusion merit serious attention?

A.   Yes, indeed!

Q.   Can we infer anything else from the principles of population genetics?

A.   In order to see what further insight we can gain from the principles of population genetics, we present in Table 5 (overleaf) theoretical frequencies of phenotypes in six different hypothetical habitancies. These distributions represent the prevalences of behavioral phenotypes in the progeny of the populations on the basis of assumed gene frequencies. The computations were done on the basis of the expressions presented in Table 3. For each habitancy the theoretical frequency of character phenotypes is given per 100 viable individuals.

Q.   I note that you have attached labels to the various habitancies, namely *U.S.A.*, *Punctilious*, *Sublime*, *Demonstrative*, *Authoritarian*, and *Militant*. Where do these labels come from?

A.   The *U.S.A.* (or *"Polymorphic"*) *habitancy* is the same as that presented earlier in Table 4. The gene frequencies for this habitancy were found to be $n = 0.9$, $p = 0.5$ and $a = 0.8$, approximately.

Q.   How about the *Punctilious habitancy*?

A.   We note that the gene frequencies of **n** and **a** in the U.S.A. habitancy were quite high, being 0.9 and 0.8 respectively. We inquire into the nature of the population that would ensue if the aggressive trait were less common, and the perfectionist trait more common. With the use of the values $n = 0.90$, $p = 0.80$ and $a = 0.30$, we obtain the distribution of phenotypes given in Table 5.

We note the following:

1. The incidence of NP individuals is very high, being over ten times greater than that of any other viable type.

TABLE 5

**Theoretical Frequencies of Phenotypes in Six Hypothetical Habitancies
(per 100 viable individuals)**

HABITANCY

Phenotype	U.S.A.	Punctilious	Sublime	Demonstrative	Authoritarian	Militant
N	8	4	78	2	0.7	1.3
A	4	0.08	0.07	2	18	36
NP	20	86	18	7	2	1.3
NA	16	0.4	3	20	6	12
PA	10	2	0.02	7	55	37
NPA	42	8	0.8	62	18	12
Gene	$n = 0.9$	0.90	0.99	0.95	0.50	0.50
frequencies	$p = 0.5$	0.80	0.10	0.50	0.50	0.30
	$a = 0.8$	0.30	0.20	0.95	0.95	0.95

2. The frequency of N types is even lower than that of the U.S.A. habitancy.

3. The frequency of A types is extremely low, on the order of 8 per 10,000 individuals.

4. The frequency of NA types is also very low, on the order of 4 per 1000.

5. The frequency of PA types is about 2 percent. This is five times less frequent than that of the U.S.A. habitancy, but not entirely negligible.

6. The frequency of NPA types is about 8 percent, which is more than five times lower than that of the U.S.A. habitancy.

Thus, the model permits the existence of this society of predominantly NP individuals, hence a society of quiet perfectionist achievers. They would be known for their unaggressiveness, their affable gingival smiles of recognition and their narcissistic bows.

Q.   What are some of the underlying forces that would maintain a high frequency of NP types in the Punctilious habitancy?

A.   Such forces would stem, on the one hand from behavioral characteristics generated by the character types themselves, and on the other hand from specific cultural mores of the society.

We might conjecture as follows:

First of all, NP types would need to have a natural attraction for each other. This appears to be the case, as has often been observed in our own society.

Secondly, in such a society of non-aggressive individuals we might expect that aggressive tendencies in children would be severely suppressed by parents. Hence, some NPA children, as well as the rare NA and A individuals, might appear as submissive types on the basis of environmental conditioning.

Thirdly, we would expect that there would be no particularly strong sexual attraction between NP types and either A or NA types, which from our observations appears to be the case. If there did exist such a strong attraction, the appearance of an aggressive type in the society would raise sexual havoc, and his promiscuity would spread the **a** gene at the expense of the **P** gene. In fact, one can see that the "positive feedback" of this situation would cause the **a** gene to spread rampantly through the population. Thus, maintenance of the Punctilious habitancy requires that the NP individual be sexually relatively unresponsive to predatory aggressive types, or that promiscuity be strongly repressed by cultural mores.

Finally, we make the observation that the Punctilious habitancy tends to be a stable one because of two other socio-cultural factors. One factor is the very low rate of crime based on aggressive acts, which can hardly be otherwise due to the low frequency of the **a** gene. The other factor is the low rate of overt marital discord and divorce. We have pointed out earlier the inherent stability often observed in the NP × NP marriage, based on a mutual sense of duty.

Q.   **How about the *Sublime habitancy?***

A.   Referring to the Punctilious habitancy, we noted that this unaggressive society revealed itself to be highly narcissistic and perfectionistic. We ask ourselves what the attributes of the society would be if it were less perfectionistic. Assuming gene frequencies of $n = 0.99$ (highly narcissistic), $a = 0.20$ (unaggressive) and $p = 0.10$ (non-perfectionistic), we obtain what we have labeled the *Sublime habitancy.*

We note the following:

1. The incidence of N individuals is very high, corresponding to over three-fourths of all individuals.

2. The next most common personage is the NP type, who is still fairly common (about 18 percent).

3. The NA type is uncommon, but not rare (about 3 percent).

4. The NPA type is rare (about 8 per thousand individuals).

5. The A and PA types are exceedingly rare (less than one individual per thousand).

Therefore, our model permits the existence of this society of predominantly N types, or a society of friendly, beatific, perennially helpful individuals. Isolated from the harried world of aggressive types, they might live on their own "special island." They would not be especially known for their perfectionistic work activities, but in fact, enough NP individuals exist to satisfy these requirements of their society. They are known better for their relaxed manner at the ocean shore, their immodesty, their colorful dress, and their ethereal ceremonial rites. When stressed, they scold, quarrel and have tantrums, but the incidence of violent crime is low indeed in this peaceful society where "live and let live" is the law of the land.

Q.   **May we proceed to the *Demonstrative habitancy?***

A.   Intuitively, our version of the stereotyped *Demonstrative habitancy* consists of at least 80 percent of the individuals who are loud, hyperactive and colorful. These must be mainly NPA, NA and A types. But despite its vigor and its vivacity, this is not a particularly aggressive nation, and it would only infrequently think of marching beyond its borders. Thus, we might expect that the frequencies of A and PA types would not be especially high. As everyone knows, the demonstrative swashbucklers of this habitancy are constantly shaking hands with each other, are great lovers, and have the ability to dance indefatigably until the wee hours of the morning.

In order to arrive at a distribution of phenotypes having a greater number of NPA and NA types than the U.S.A. habitancy, the gene frequencies of **n** and **a** must be higher than those of the latter. Since these gene frequencies in the U.S.A. habitancy are already quite high ($n = 0.9$, $a = 0.8$) we are forced to assume very high values of n and a for the Demonstrative habitancy. The distribution of phenotypes in Table 5 assumes values $n = 0.95$, $a = 0.95$, with p maintained at 0.5, i.e., the same value as for the U.S.A. habitancy.

We make the following observations:

1. The frequency of NPA individuals (62 percent) and NA individuals (20 percent) is high.

2. Contrary to what one might expect, this noisy society has few A types (about 2 percent).

3. The incidence of N types is low (about 2 percent).

4. The types NP and PA, although less frequent than in the U.S.A. habitancy, are not rare (each about 7 percent).

Q.   **How did the *Authoritarian habitancy* present itself?**

A.   Referring to the Demonstrative habitancy, we noted that this was a highly narcissistic-aggressive, moderately perfectionistic society. We ask ourselves what the attributes of the society would be if it were less narcissistic. Assuming gene frequencies of $a = 0.95$ and $p = 0.50$ (i.e., identical to those of the Demonstrative habitancy) and $n = 0.50$ (significantly lower than that of any other habitancy), we obtain what we have labeled the *Authoritarian habitancy.*

We note the following:

1. Compared to the other habitancies, there is a markedly higher incidence of PA and A individuals (55 and 18 percent respectively).

2. The types NPA and NA are not uncommon (18 and 6 percent, respectively). They are, however, much less frequent than in the Demonstrative or U.S.A. habitancies.

3. The types NP and N are very uncommon (about 2 and 1 percent, respectively), being significantly less common than in the Demonstrative habitancy.

Thus, the model now reveals the possibility of a society in which over 70 percent of the individuals are non-sanguine PA and A types. This is evidently a strict, no-nonsense, somewhat paranoid society. Behind its quietly aggressive policies lies the will of perfectionist domination. It is known for its marching militia and not unknown for its iron fist.

Q.   Finally, how did you arrive at the *Militant habitancy*?

A.   Referring to the Authoritarian habitancy, with its relatively high degree of perfectionist aggressiveness, we ask ourselves what the attributes of the society would be if it were less perfectionistic. Assuming gene frequencies of $n = 0.50$ and $a = 0.95$ (i.e., identical to those of the Authoritarian habitancy), and $p = 0.30$ (significantly lower than that of the Authoritarian habitancy), we obtain what we have labeled the *Militant habitancy.*

We note the following:

1. Compared to the Authoritarian habitancy, the sum of the incidences of PA and A types is the same (73 percent), but now these two types are present in about equal numbers. Therefore, of the non-narcissistic aggressive types there is a marked shift from perfectionistic aggression to unbridled aggression.

2. Compared to the Authoritarian habitancy, the sum of the incidences of NPA and NA types is about the same (just under 25 percent). However, these two types are now present in about equal numbers. Therefore, of the types having both the traits N and A, there is a marked shift from bridled to unbridled ambition.

3. The types N and NP are relatively uncommon (about 1 percent each).

Therefore, the model presents us with a non-sanguine, aggressive society having a high degree of unbridled ambition. If the aggressive A types take command, then any semblance of perfectionist restraint evaporates into the winds of naked aggression, barbarism, overreaction, overkill and the self-justified preemptive strike.

Q.   Can we obtain any other points of insight from comparisons of the various hypothetical habitancies?

A.   Yes, and these will be brought out in the next chapter when we consider disorders of human behavior.

Q.   Have you not been neglecting submissive and resigned types in this chapter?

A.   For simplicity we have confined our quantitative estimations to dominant types. However, we can safely predict that some societies will prove to have much higher frequencies of submissive and resigned types than others. In the following chapter we shall provide evidence that the character vector of submission, for example, may be determined by clearly delineated genetic factors endogenous to the individual.

Q.   What if it turns out that the trait N is transmitted according to dominant rather than recessive inheritance?

A.   If trait N reveals itself to be dominant, then certain quantitative aspects of our model would be somewhat different (See Appendix C). However, the differences would be rather minor. In particular, the frequency distributions of character phenotypes for the hypothetical habitancies presented in Table 5 would be unchanged.

Q.   Could you kindly summarize the main points of this chapter?

A.   Here is a summary of the main points of our genetic model:

- On the basis of the identification of discrete character types in dominant individuals, we propose the existence of three behavioral complexes, namely those of narcissism, perfectionism and aggression.

- The behavioral complexes are determined by separate major pleiotropic genes that follow the laws of Mendelian genetics.

- The genes determining man's ambition, namely the traits of narcissism and aggression, are expressed in a recessive manner. The gene determining perfectionism is expressed in a dominant manner.

- In a given individual having a character vector of dominance, the traits of narcissism and aggression are either present or they are not. The traits allow the expression of narcissistic and aggressive rages, respectively.

- If the phenotypes of parents are known, then the possible phenotypes of the progeny can be deduced (Table 2). If a family pedigree is available, then the genotypes of the individuals can often be deduced from knowledge of the phenotypes.

- Certain combinations of parental genotypes may lead to zygotes that are lacking in both the N and A traits. These parental combinations will be prone to miscarriages and stillbirths. Some combinations will not be able to produce any viable progeny and would fall into the category of complete infertility.

- With the use of a simple probabilistic model (similar to the classic Hardy-Weinberg approach) we could predict the distributions of character phenotypes that would occur for selected combinations of gene frequencies. Calculations were given for six different "societies" termed the *U.S.A.* (Polymorphic), *Punctilious, Sublime, Demonstrative, Authoritarian*, and *Militant* habitancies.

- The model proposes, not only that the character structure of any given individual is largely genetically determined, but also that the distribution of character types in a society is determined by mechanisms of Mendelian population genetics. We therefore propose that differences in personality between various peoples of the world, formerly ascribed to cultural or environmental factors, can now be analyzed on the basis of a model of quantitative genetics.

CHAPTER 11

DISORDERS OF HUMAN BEHAVIOR

I have always wondered whether one should suspect, behind these cases, heredity.

Sigmund Freud (1896)

We are standing at the threshold of an era in which the entire proud edifice of medicine — including psychiatry as a whole as well as psychoanalysis — will rest on genetics.

Sandor Rado (1961)

OUTLINE OF CHAPTER ELEVEN

PERSONALITY THEORIES

GENETICS & ENVIRONMENT

PERSONALITY DISORDERS

SCHIZOPHRENIA
> Classes & types (either N or A trait)
>> Class I, NP type: infantile autism
>> Classes II & III, NP type
>> Class I, PA type
>> Classes II & III, PA type
>> N type
>> A type
>
> Quantitative genetics of schizophrenia
> Folie à deux
> Schizophrenia in NA, NPA types (both N & A traits)
> Conclusion

PERIODIC PSYCHOSES
> Manic-depressive illness
> Other periodic psychoses

ENDOGENOUS DEPRESSION & EUPHORIA

EXOGENOUS DEPRESSION & EUPHORIA

ALCOHOLISM

STATES OF DEPRESSED AMBITION

HYPERACTIVE CHILDREN

SPEECH DISORDERS

ENDOGENOUS & EXOGENOUS ETIOLOGY

PSYCHOSOMATIC DISEASE
> Genetic basis for Submission & Pseudonarcissism
> The Introspective Habitancy

CHARACTER TRAITS IN OTHER PRIMATES

EUGENICS

Q.   What is your approach going to be in this chapter?

A.   We will try to be brief. We will attempt to enumerate the various "disorders" of human behavior that have heretofore been described in psychiatric terms, and show how they fit our model. We acknowledge that a genetic basis for the various disorders of human behavior has become increasingly recognized over the past twenty years.

Q.   You mentioned earlier that Freud's tripartite theory of id, ego and superego was of little use to us. Are any of the other theories of human personality that have been advanced in the past helpful?

A.   We note that *Henry Murray* in his "theory of needs" utilizes some of the motivational drives that we ascribe to genetic and environmental factors in our model. Murray arrived at a tentative list of twenty "needs" and among them we find the three basic character vectors ("Dominance," "Abasement" and "Autonomy") as well as the three behavioral complexes of dominance ("Achievement," "Order" and "Aggression"). But Murray's hierarchy of needs includes fourteen other motivational factors as well.

Karen Horney in her theory of neurosis also arrived at the three basic character vectors and the three behavioral complexes of dominance. Her theory, though, was based almost entirely on an individual's "neurotic" behavior, whose essential roots lay in his faulty conditioning as a child during his critical developmental period. However, Horney's break with Freud was a clear and courageous one and took on all of the aspects of a Protestant Revolution. One has the impression that if Horney had lived another ten years, she would have arrived at essentially the same model that we present here.

Q.   It appears that the polymorphism of discrete human character types passed unnoticed simply because it was not looked for. Is this correct?

A.   We think that this is essentially correct. We, as individuals, had all been so proud of ourselves, so proud of our "uniqueness," that the possibility of our personalities having been struck from only a few molds just did not occur to us.

But the clues have been with us since the time of Hippocrates (ca. 400 BCE), whose theory of "humors" led to a description of character types that became known as *phlegmaticus, cholericus, sanguineous* and *melancholicus.* Since those days of yore, workers who have attempted to quantify the human personality have diverged in two opposite directions. On the one hand, they have optimistically developed complicated empirical "character profiles" based on innumerable questions concerning everything from alpha to omega (such as the Minnesota Multiphasic Personality Inventory). On the other hand, they have abandoned all hope and attempted to distill the human personality to only two types (the "Type A" and "Type B" of the medical literature).

Q.   Some years ago W. H. Sheldon postulated that there exists a correlation between body habitus (somatotype) and personality. Did you find any such correlation?

A.   Sheldon's point of departure was the occurrence of three basic somatotypes, namely the *endomorph* (obese), the *ectomorph* (lean) and the *mesomorph* (muscular). It is true that certain types do appear frequently, for example the diminutive A type, the portly NPA type, and the ectomorphic NP type — sometimes of Marfanoid proportions. However, exceptions abound, and at best there probably exist loose statistical correlations between character type and somatotype.

Correlations between character type and somatotype could stem from two major sources. First, the genes **n**, **P** and **a** could themselves directly influence morphologic attributes on which somatotype depends. Secondly, in a given population the genes **n**, **P** and **a** could be in linkage disequilibrium with the genes determining somatotype.

There is little doubt, however, that the so-called "jovial endomorph" is usually an N or NA type. We have noted earlier the propensity of the NA type to binges of overeating, alternating with periods of abstinence. This appears to be related to the individual's cyclical hypomanic-depressive behavior, and we find that some of these individuals gravitate to a state of obesity.

Q.   Are some of the character types more intelligent than others?

A.   In developing our model we attempted to divorce human intelligence from the human character structure. We had assumed that intelligence was a continuous function and was a foundation on which the character structure was overlaid.

But, what is intelligence? If we define intelligence by the I.Q. scores derived from standard tests, then we believe that some character types will tend to score better than others. To be specific, we think that the methodical, perfectionistic NP and PA types may often prove to be the most "intelligent," as measured by some of the tests of cognitive ability that are routinely in use today.

We might reexamine the question of intelligence with reference to the behavioral components of the human character structure. We might ask ourselves once more: what exactly are we trying to measure in an intelligence test? One can, without too much difficulty, imagine separate tests that would favor, in turn, individuals with pronounced narcissistic, perfectionistic, aggressive, submissive, or detached trends.

In addition, we have over the years assumed that intelligence is a continuous function. Perhaps, here too, we have been in error. We might consider the alternative: that there exists a discrete polymorphism for intelligence and that this is partially rooted in the three genetically determined behavioral complexes.

Q.   What exactly constitutes a "normal" individual?

A.   This apparently simple question opens a Pandora's box, and we cannot hope to arrive at definitive answers here. As we have mentioned, none of the characters of our model (and none in real life!) remotely resembles the mentally healthy, well-balanced individual that, in our own minds, we have tried to emulate during our lives. All of the characters exhibit profound obsessive-compulsive behavior, usually without the slightest understanding of their motivations.

We have seen that all of the characters must fulfill a pride structure, so that they may justify their own existence on the basis of feelings of self-esteem. In so doing they attempt, sometimes desperately, to find life situations that are compatible with their needs. As they go through life, in particular if they are in a competitive society, they assume exhilarated states, abject states, and incite themselves into rages. If they "play the game," they frequently undergo personality splits. But who is to say that all of this psychic activity is not normal?

But, you will say, if an individual's psychic state deteriorates to a point at which he can no longer function in his society, then he must be labeled as "abnormal." For example, if an individual retires from active life to an autistic-like or schizoid state of abject depression, should we not call him "abnormal"? Yes, but is the *individual* abnormal or is his *life situation* abnormal? If an elephant, at long last, runs amok at a circus, is it the elephant who is acting abnormally? Or is his behavior perhaps a result of his being forced into a life situation that is totally inappropriate for him?

Q.   Your model incorporates aspects of genetics on the one hand, and environment on the other. Could you clarify the roles of these two very different determining factors?

A.   These two factors, *heredity* and *environment,* or "nature and nurture," have been widely recognized in the past, in a qualitative way, as determinants of the human character structure. The present model, we believe, is a step toward the quantification of the two factors.

The primary *genetic bases* of the human character structure lie in the genes coding for the behavioral complexes of narcissism, perfectionism and aggression. Of course, many other "modifier genes" are also involved. The *environmental bases,* we acknowledge, are enormously complex. Great gaps lie in our understanding of the precise genetic and environmental bases of the character vectors of submission and resignation.

What indeed are the exact factors that are responsible for an individual's adopting a submissive, resigned, or schizophrenic state? Is some "failure" in nurturing involved on the part of the parents? Does an infant have to be imprinted with certain behavioral patterns at several critical periods in his life? Are there definite alleles that predispose an individual to adopt a submissive, resigned, or schizophrenic state? What are the precise roles of learning and conditioning in the determination of the human character structure at maturity? All of these questions, and many more, await answers.

Q.   Are you implying that parents are at fault for raising submissive or detached children?

A.   In no way do we imply culpability on the part of the parents. The precise factors, both genetic and environmental, that lead to submissiveness or resignation are simply not known at the present time. Parents should certainly not flagellate themselves for imagined failures.

We envision a future, perhaps a not too distant one, when the genetic counselor will have many of the above answers. And we speculate that he will have to be not only a *genetic* counselor in the pure sense of the word, but an *environmental* counselor as well.

Q.   I gather that the term "neurotic" completely loses its meaning in the framework of the present model. Is this correct?

A.   Yes. We suggest that we should strike this word from our vocabulary. If anything, it is not we who are neurotic, but rather certain aspects of the competitive society that we have forged for ourselves.

Q.   Are you implying that present day psychiatry is useless?

A.   No.

We do suggest, however, that classic psychiatry has gone astray in over-emphasizing the importance of environmental factors (especially the influence of childhood experiences in a psychosexual context) in relation to genetic factors. Similarly, behaviorists have gone astray in overemphasizing environmental factors, in the context of learning and conditioning, as eventual determinants of the human character structure at maturity.

Many individuals have real emotional problems, and many need direction and guidance. But the starting point for both psychiatrist and patient must be a thorough understanding of their own genetically determined character structures. Indeed, if this is not understood, then endless ruminations with respect to psychosexual factors cannot possibly yield insight into the patient's current problems.

The psychiatrist of the future will be above all a genetic counselor. His task will be to assist the patient in matching his real life situation to the requirements of his basic character structure. And in so doing the psychiatrist will assist the patient to narrow the gap between his outlandish expectations in life and his more modest capabilities. And in establishing this equilibrium it is just possible that the individual will achieve at least a modicum of "happiness" in his life, that elusive intangible state of mind that for so long has been a mystery to so many of us.

Q.   Could you briefly summarize how the present model interprets the various classic states of abnormal human behavior that have heretofore been catalogued in a rather random fashion?

A.   We present below a brief summary of some of these states, as outlined in the beginning of this chapter. We shall emphasize the genetic bases of the disorders.

PERSONALITY DISORDERS

The so-called "personality disorders" of the psychiatric literature have been, by the psychiatrists' own admission, a poorly defined group of diagnoses. Many authors have commented on their impressions that personality traits are determined in part by heredity. Our model permits us to approach in the precise terms of genetics the modes of inheritance of these so-called "personality disorders." We summarize below our interpretation of the diagnostic categories of the psychiatric literature (see *Chapter References*).

- *Narcissistic personality disorder (NPD)* — This is usually a poorly-adjusted narcissistic (N or NA) character type who comes to the attention of the mental health professional.

- *Arrogant-vindictive personality* — This is simply our aggressive (A) character type.

- *Obsessive-compulsive personality* — All of the character types are in reality "obsessive" and "compulsive." Most of the descriptions in this category correspond to individuals having the P trait. These are most commonly our NP and PA character types. However, NPA–, NPA= and even NPA types can exhibit marked obsessive-compulsive trends (obsessive thinking, compulsive cleanliness, etc.). It is quite humorous that the NP personage is often referred to as an "anal sadistic" type in the older psychiatric literature.

- *Perfectionist personality* — This is most often our NP character type (see *obsessive-compulsive personality*, above).

- *Cyclothymic* or *hypomanic (hypomanic-depressive) personality disorder* — This is usually an NA character type. However, a harried N or NP type can sometimes mimic a hypomanic NA type.

- *Hysterical* or *histrionic personality* — This is often a poorly adjusted N, NA or NA– character type.

- *Explosive personality disorder* — This is one of our NPA types during the throes of his NPA super-rage.

- *Anorexia nervosa personality* — The young people afflicted with anorexia nervosa (a pathologic aversion to eating) seem to be drawn from the ranks of several character types. Major subgroups appear to consist of perfectionistic individuals, as well as cyclothymic young females of the N and NA types.

- *Multiple personality disorder* — This characterizes a labile individual who readily undergoes dramatic changes to distinctly different compartmentalized personality states. Elements of hysteria and amnesia are sometimes prominent. Subgroups of this disorder appear to include cyclothymic individuals of the NA type as well as borderline schizophrenics.

- *Schizoid personality type* — This term is used very loosely in the psychiatric literature to denote a withdrawn, aloof individual. Thus, this category could include NP, PA as well as submissive, resigned and non-aggressive withdrawn types.

- *Paranoid personality* — This is usually a PA type. However, sometimes a vigilant aggressive (XA) or resigned type, a negativistic N or NP type, or a sullen *n-C* submissive type may show intense paranoid traits.

- *Passive-aggressive personality* — This is usually a PA type. However, other character types having negativistic qualities may also act in a passive-aggressive manner.

- *Antisocial* or *sociopathic personalities* — These are probably drawn from the ranks of all of the character types and deserve further study.

- *Passive-dependent personality* — This is our compliant submissive type.

- *Self-effacing personality* — This is also our compliant submissive type.

- *Inverted sadistic personality* — This corresponds most closely to some of our non-compliant submissive types.

- *Detached neurotic personality* — This corresponds to our basic character vector of resignation.

- *Asthenic* or *neurasthenic personality* — This corresponds to our N– or N–P type. However, other submissive and withdrawn types may also be characterized by weakness, timidity and a low level of energy.

SCHIZOPHRENIA

The present model attempts to explain the genetic and environmental bases of schizophrenia in definitive terms. If the model is verified, this information will be of practical use to the genetic counselor.

The schizophrenic syndromes are devastating mental disorders involving major disturbances in thought, mood (affect) and behavior. They are not at all rare, with a worldwide prevalence on the order of tens of millions of cases. We shall, in this section, refer to the various clinical syndromes of schizophrenic withdrawal, but we shall not describe them in detail.

We have noted earlier that even so-called "normal" individuals lead a somewhat precarious existence as they blunder through life, in part because of the lack of insight into the limitations of their character structures. We have noted that their lives are ruled by states of anxiety, confusion, exhilaration, rage and depression. As an individual begins to slip into a state of schizophrenia or affect disorder, his anxiety, confusion and depression may become accentuated. He may desperately try to understand what is happening to him as he descends into a twilight zone where the rules of logic hold no longer. As he withdraws farther and farther away from close relationships with others, he may begin to wonder aloud if he is insane and if he is the victim of a hereditary taint, as in this recent letter* to a newspaper columnist:

> DEAR ANN LANDERS:
>
> I am a college student, only 20 years old, but I feel as if I've been 100 years in the grave. Before I got in this shape I was as popular and happy as any of my friends. Now I am miserable.
>
> These last two years I have been unable to communicate with anyone. Lately, I've been spending all my time with books. I go from being severely depressed to feeling as if I am on top of the world... I am aware that there are some loons in my ancestry... I can't figure out why I am like this. Can you help me — or am I already crazy?
>
> — Monroe, La.

Turning to our model, we advance the hypothesis that chronic schizophrenia may be viewed as an altered behavioral state that ensues when an individual's drives of ambition are suppressed:

- The *genetic basis* of an individual's susceptibility to schizophrenia is assumed to lie in the precise allelic structure of the genes **n, P** and **a**, as well as in other "modifier genes" and in genes that confer specific susceptibilities.

- The *environmental constraints* that may eventually lead to a schizophrenic syndrome are, no doubt, quite varied. They probably range from the most overt stresses of outright physical

---

* Ann Landers, Gannett Newspapers, *The Daily Item,* June 18, 1981.

trauma to the most subliminal, insidious psychological stresses imposed on an individual as he plays the "mating game," or as he attempts to succeed in his competitive society. We shall not attempt to list the types of environmental constraints that an individual may feel imposed on him. Rather, we shall confine ourselves to the general statement that, whether consciously or unconsciously, the individual senses that he is unable to cope with the outside world. It becomes more and more difficult for him to conduct meaningful interpersonal relations with those around him, especially in a competitive society. He feels the walls of life closing in around him, and his only path of escape is the one leading into himself.

In the context of our model, if an individual's narcissistic and/or aggressive behavioral complexes are not permitted to flourish, he may be left in the confused state of pure perfectionism with no drives of ambition to perfect, or lacking the behavioral complex of perfectionism, in a state of aimlessness with no functioning behavioral complex at all.

If an individual's character structure is on the verge of disintegration, or is undergoing acute changes, we might expect to see grotesquely distorted states of narcissistic and/or aggressive rage in the appropriate individuals.

Since predisposition to schizophrenia may be considered in terms of suppression of the drives of ambition, we surmise that the most vulnerable individuals would be:

- Those having only *one component of ambition,* namely the withdrawn types N–, A–, N–P and PA– and the dominant types N, A, NP and PA.

- Next in line in vulnerability would be *submissive* types (NA=, NPA=, NA– and NPA–) and resigned types (N–A and NP–A). These latter individuals might be somewhat vulnerable to schizophrenic withdrawal at maturity, since they already carry a muted behavioral complex of aggression.

- Finally, we speculate that the types least likely to succumb to schizophrenia would be the *dominant* NPA and NA types.

Since A types are relatively uncommon in most societies (Chapter 10), we suggest that the bulk of schizophrenic cases worldwide will be found in the N, NP and PA categories.

Confining ourselves for the moment to individuals having only *one component of ambition* in their character structures, we have outlined below our model's prediction of states of chronic schizophrenia. Note that the model predicts three major classes of schizophrenia:

Class I: Juvenile onset. This kind of schizophrenia pattern may be described by the following notation:

$$N \quad \rightarrow \quad N= \quad \text{[N schizophrenia]}$$

$$NP \quad \rightarrow \quad N=P \quad \text{[NP schizophrenia,}$$
$$\text{"infantile autism"]}$$

$$A \quad \rightarrow \quad A= \quad \text{[A schizophrenia]}$$

$$PA \quad \rightarrow \quad PA= \quad \text{[PA schizophrenia,}$$
$$\text{"childhood schizophrenia"]}$$

The Class I schizophrenic is severely limited in the area of interpersonal relations in childhood, and he may find himself institutionalized at an early age. His limitation may, in fact, have been present from birth. Hence, the premorbid state of dominance, as denoted above, is a hypothetical one for infants who are congenitally afflicted. In this simplified categorization, Class I schizophrenics include children with both schizoid and dominant premorbid personalities.

Class II: Maturity onset with a schizoid premorbid personality

$$N- \quad \rightarrow \quad N= \quad \text{[N schizophrenia]}$$

$$N-P \quad \rightarrow \quad N=P \quad \text{[NP schizophrenia]}$$

$$A- \quad \rightarrow \quad A= \quad \text{[A schizophrenia]}$$

$$PA- \quad \rightarrow \quad PA= \quad \text{[PA schizophrenia]}$$

The Class II schizophrenic is an individual whose basic character vector at maturity is submission (A-, PA- types) or withdrawal (N-, N-P types). These individuals are schizoid but not psychotic at maturity. With time, however, these individuals decompensate as they yield to stresses with which they cannot cope, becoming frankly schizophrenic with either a gradual or acute onset of psychotic behavior. For the A and PA types, this process corresponds to submissive individuals who are non-compliant at maturity but who become compliant. Thus, the Class II schizophrenic may often be described as an individual who undergoes a "schizophrenic withdrawal having a smoldering onset."

Class III: Maturity onset with a dominant premorbid personality

$$N \quad \rightarrow \quad -N \quad [\text{N schizophrenia}]$$

$$NP \quad \rightarrow \quad -NP \quad [\text{NP schizophrenia}]$$

$$A \quad \rightarrow \quad -A \quad [\text{A schizophrenia}]$$

$$PA \quad \rightarrow \quad P-A \quad [\text{PA schizophrenia}]$$

In this class the individual has a basic character vector of dominance at maturity and functions more or less adequately in interpersonal relations. He may be somewhat schizoid (aloof), or he may not. With environmental stress, however, this individual undergoes a dramatic "schizophrenic break" from the dominant personality type to an incapacitated psychotic one.

The relations between the three classes of schizophrenic patterns may be seen schematically in Figures 14 and 15 (overleaf), where we consider separately schizophrenias due to suppression of the A and N traits respectively.

Classes and Types of Schizophrenias

In the following sections we will consider our interpretation of the various schizophrenias. According to our nomenclature, *class* delineates the premorbid state of the individual (dominant or withdrawn, juvenile or maturity onset), while *type* refers to the genetically determined character type of the individual.

We believe that many of the schizophrenic states have been, in various contexts, extensively described in the psychiatric literature. We shall, in particular, refer to the studies of Karl Leonhard of Berlin, who has painstakingly classified, on the basis of clinical and behavioristic criteria, the various psychoses that he observed. In particular, Leonhard accepted, in part, Emil Kraepelin's classical subdivisions of schizophrenias into *catatonic, hebephrenic* and *paranoid* forms, and subdivided them further. He proposed that the various forms have distinct genetic bases. Leonhard's work in the form of a book entitled "The Classification of the Endogenous Psychoses" has recently become available in an English translation.

Class I **NP schizophrenia (*infantile autism*).** In the NPA model the autistic child ("Kanner's syndrome") is an NP individual whose behavioral complex of narcissism has not been fully expressed during infancy. We describe this state by the notation N=P, and we realize that we have here an individual who is deprived

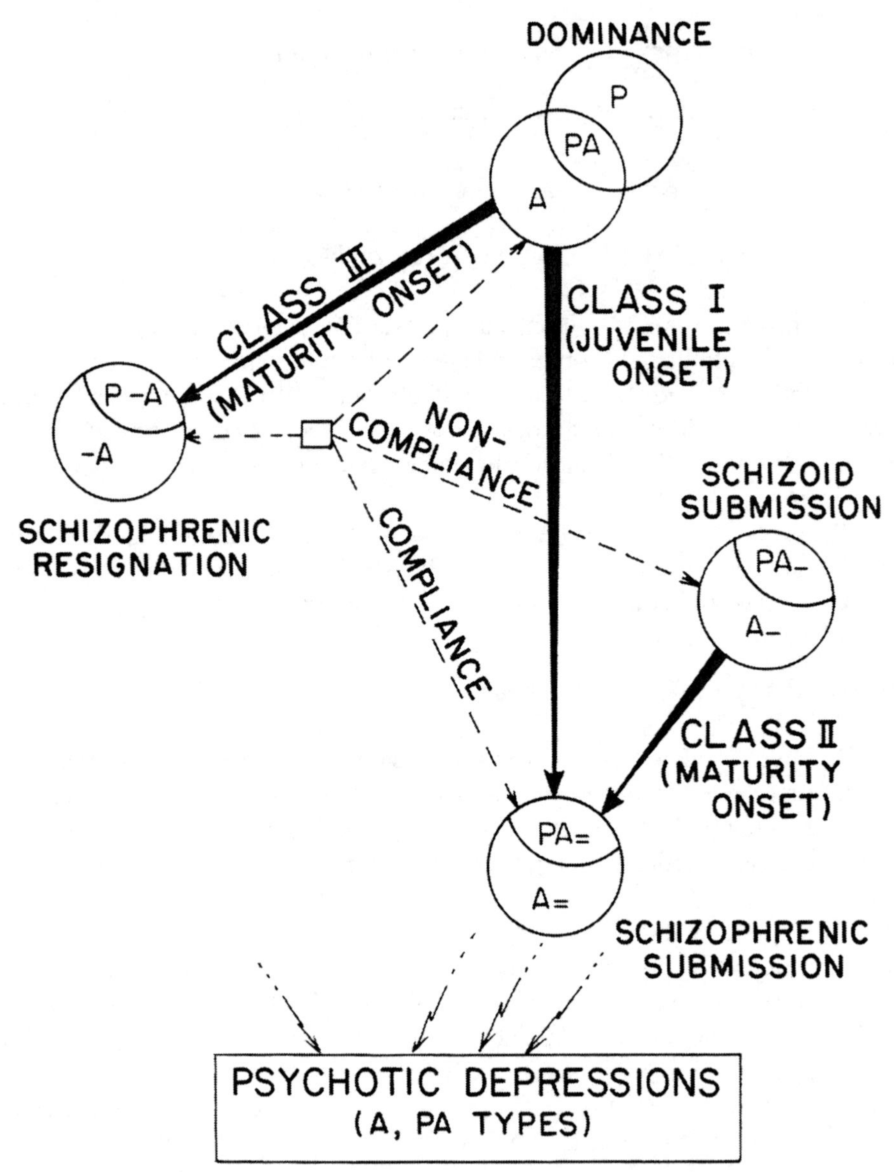

FIGURE 14

Schematic representation of *schizoid* and *schizophrenic states* for character types
having the trait A. The psychotic depressions correspond to situational abject
states (reactive depressions).

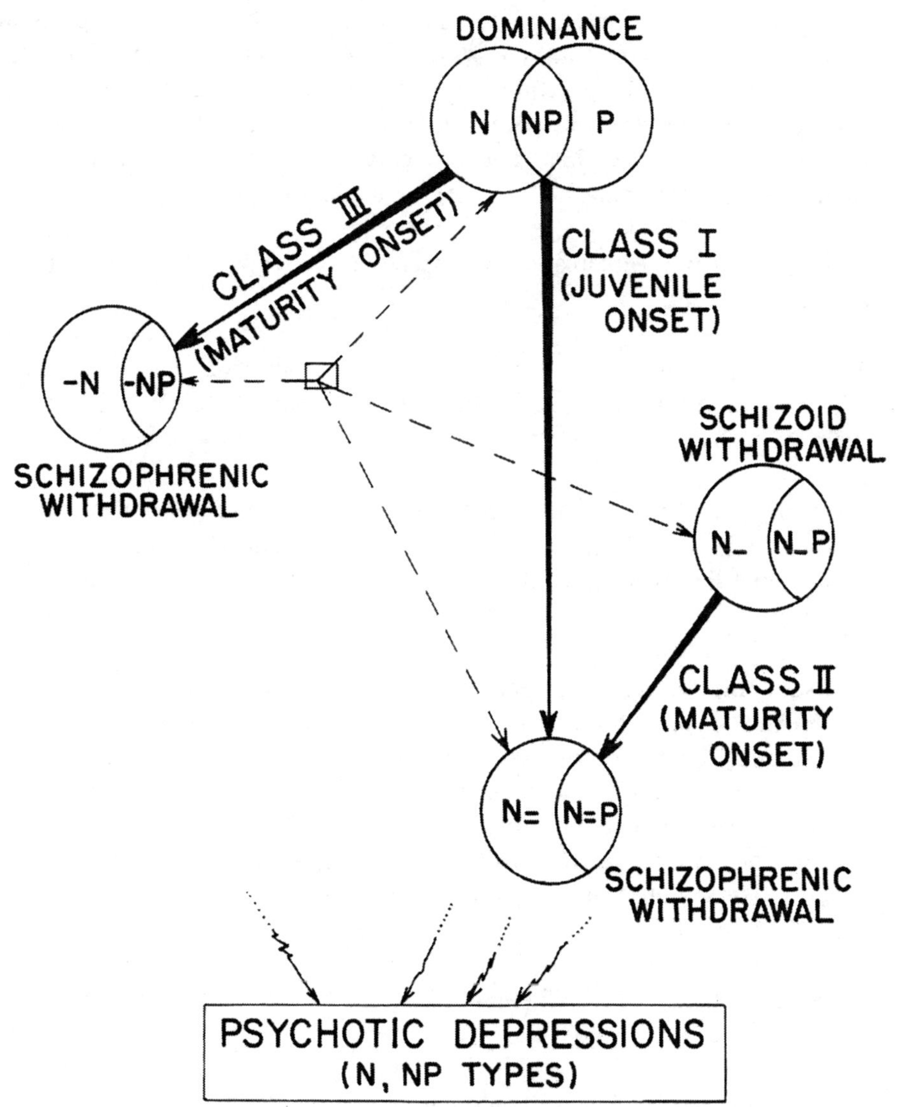

FIGURE 15

Schematic representation of *schizoid* and *schizophrenic states* for character types having the trait N. The psychotic depressions correspond to situational abject states (reactive depressions).

of his only possible source of ambition, narcissism, and who is left in a state of pure perfectionism with nothing tangible to perfect.

Another way of interpreting the association of NP character type with infantile autism is that NP children with a wide variety of developmental disorders tend to be placed by practitioners into clinical categories of "autism." In this interpretation NP children are not at risk for developing autism. Rather, NP children with existing or incipient developmental disorders are at high risk of being placed into the heterogeneous category of autism by practitioners.

In support of our conclusions we offer the following information culled from the medical literature:

- The behavior patterns of the autistic child describe well the ritualism and desire to preserve order that we would expect from the pure perfectionist.

- In much of the older medical literature autistic children have been confused with other child schizophrenics, and for good reason: we believe that the majority of the latter are pure perfectionists of the PA= type.

- Autistic children have no strong family history of psychosis, almost as if they appeared in a random fashion among NP individuals. There appears to be no link between infantile autism and the more common schizophrenias of the PA type. On the contrary, many parents of autistic children are described as being *narcissistic, intelligent, obsessive, perfectionistic,* and *humorless.* Since an NP infant must have at least one parent who is an NP type (or N type, see Chapter 10), we see that the personality characteristics of parents in infantile autism are easily explained. A character sketch of an NP parent of an autistic child was given in Chapter 6.

- The extreme rarity of infantile autism (an incidence of about 4 per 100,000), as well as its very low recurrence rate in siblings (about 2 percent), may be explained by the supposition that the genetic and environmental factors that lead to infantile autism are far more critical than the ones that lead to other forms of childhood schizophrenia.

But why is the sex ratio of autistic children predominantly male (about 2 to 4 boys for every girl)? The answer to this question is not clear. Perhaps hormonal factors play a strong role in the etiology of infantile autism.

- Lacking the ability to express the narcissistic behavioral complex, autistic infants do not exhibit the characteristic smile of recognition that normally appears by the age of about two months.

- Autistic children have no natural instinct for aggression. If they do develop "aggressive" tendencies these are of a mechanistic, retaliatory nature, and are a learned behavioral response. (The highly stylized, ritualistic mannerisms of the arts of self-defense of certain Oriental societies come to mind). With the occurrence of poorly directed hyperactivity, the so-called "aggressive" tendencies of autistic children may be interpreted in the context of the narcissistic rage, or tantrum. That is, the child is not so much being "aggressive" as he is reacting in a diffuse fashion in a desperate attempt to compensate for the repression of his narcissistic ambition.

Richard, a less severely impaired autistic German adolescent describes in his own poetic words his tendency to unaggressiveness in this account from Bosch's book *Juvenile Autism:*

> "Resteten" was the name I gave to my dream world, a world full of harmony and peace in which nothing evil happened, a world rotating round some distant sun far out in the universe. When I was a small boy, whenever life here on earth seemed difficult and incomprehensible, I liked to retire there to its majestic mountain scenery through which the fast-running Olympia River flowed. This happened very often.

> For from my earliest childhood on I was different from the others of my age, and for children this is of course sufficient reason to mock, punch, and torture. I was unable to defend myself, because a deep-rooted feeling prevented me from raising my hand to hurt another. As a result mixing with schoolmates, which is for most people an enjoyable part of going to school, was torture for me.

Nevertheless, on the very first day I found a friend who was always ready to "roll his sleeves up" for me. His name, Inno, is more distinctive than he himself ever was; he is a practical, down-to-earth boy who (unfortunately for me) went back recently to live in America where he was born and where he fits in well.

The "Idiot Savant." These individuals can perform phenomenally complicated numerical computations mentally, often without being able to explain how they arrive at their answers. Various authors have related the so-called "idiot savant" to the autistic child, hence we are drawn to the conclusion that this mental wizard is also of an NP character type.

Class II and III* NP *Schizophrenias. From an examination of the psychiatric literature, including case studies and descriptions of premorbid personalities, we come to the conclusion that the *mature* NP individual is vulnerable to several different subtypes of schizophrenic decompensation. We believe that the subtype is determined by the premorbid NP personality subtype mainly through inheritance. This is in accord with the views of Leonhard.

A clue to the nature of the Class II and III schizophrenic decompensation at maturity in the NP type may be found in the case history of Deiter, an autistic child presented in Bosch's monograph. (A sketch of Deiter's father was given in Chapter 6). When Deiter was about 14 years old...

... the boy complained more and more of various physical ailments: in his intestines, in his genitals, and in 1962 he said he thought he was being poisoned; consequently his parents, who believed his story, took him away from the new home in which, in the meantime, he had been placed. At home with his parents, however, he soon became so aggressive and destructive that they immediately had to put him once more in the psychiatric clinic of the local university...

When he was admitted he reported hallucinations. At home he had heard the voices of his ex-schoolmates who insulted him and called "filthy things" after him; according to him they spoke to him through the air. He also reported that the voices had continually called to him telling him he was being poisoned, that there was poison in his coffee. He therefore spat everything out when he was at home. He complained about pain in his genitals and said, "My 'Peter' is killing me. I wish I was a girl." During a later stay in the psychiatric clinic of Marburg University it was also observed that he suffered from vivid acoustic hallucinations. Most of the progress in making contact that he had shown during his 2 years in the children's home had been lost again. More and more often he was aggressive and destructive and obviously much preoccupied with his bodily sensations, particularly those centered on his genitals and anus...

The case of Deiter indicates that at maturity the normally quiet NP individual can undergo an acute schizophrenic-like decompensation in which the characteristics of hyperactivity, hypochondria, diffuse "aggression," acute paranoia, and auditory hallucinations are prominent. Such a picture is also present in some of the adult schizophrenic patients studied by M. Bleuler in Switzerland. If we speculate that Bleuler's patients came from a population somewhat resembling the Punctilious habitancy presented in the previous chapter, then we may surmise that the majority of Bleuler's Class II and III chronic patients were schizophrenics of the NP type.

If we read between the lines in Leonhard's text, we come to the conclusion that his "systematic schizophrenias" correspond mainly to individuals of the NP type. Leonhard further subdivided the classic schizophrenic subtypes of Kraepelin (catatonic, hebephrenic, paraphrenic) into a total of 16 categories, and not surprisingly it is possible to correlate some of Leonhard's various subtypes with subtypes of the NP character structure that we have observed in our own society. Thus, NP individuals with tics and epileptoid mannerisms would decompensate into the category of "parakinetic catatonia," those with speech hesitation into "sluggish speech catatonia," those with pronounced negativistic or autistic traits into "negativistic catatonia" and "autistic hebephrenia," those with elfish traits into "affected catatonia" and into the hebephrenias, and so on.

Many of Leonhard's patients fell into the categories of "systematic paraphrenias," which are *paranoid types.* This is very important since this indicates that, like Deiter above, many NP schizophrenics are characterized as paranoid types. This is an unexpected finding, since the premorbid personality of most NP and N–P types is not at all paranoid. It explains why the schizophrenias from two very different personality types (NP and PA) have been lumped together throughout the years. It seems that the agitated paranoid state of the NP schizophrenic is more closely related to the narcissistic rage, in which the individual diffusely lashes out at people surrounding him, seeing them only as shadowy beings intent on frustrating him from attaining a sense of order in his life or from reaching an indistinct limelight flickering in the distance.

Are the various subcategories of Leonhard's systematic catatonias, hebephrenias and paraphrenias genetically determined? We believe that the answer is affirmative in the following sense. Since we did not find another major gene determining an obvious fourth behavioral complex, we surmise that Leonhard's systematic schizophrenias may correspond to NP subtypes that are determined by allelic differences at the **n**, **P** and **a** loci, and by "modifier genes." Since there are probably many of these, we tentatively suggest that the subdivisions of his "systematic schizophrenias" are somewhat arbitrary, and from a genetic point of view, somewhat overlapping.

Class I* PA *schizophrenia.   Most juvenile-onset psychotic children do not resemble the clinical syndrome of the NP autistic child. In addition, many of these children have a family history of psychosis, which is usually absent in infantile autism.

Some investigators, such as Rutter, have pointed out that the onset of schizophrenia occurs in a bimodal pattern. The first peak of incidence occurs during the ages of 0 to 6 years, and we speculate that it is comprised mainly of schizophrenic children of the juvenile onset PA= type (the N=P autistic child being relatively rare). The second broader peak, as reported in the literature, begins at about 7 to 8 years of age. This, no doubt, begins to include *maturity onset* (Class II and III) schizophrenics of the PA types, but the various reports in the literature appear to describe admixtures of schizophrenics of various character types. Thus, it is difficult to glean quantitative data from the literature relevant to only the PA schizophrenic.

***Class II and III* PA *schizophrenias*.** What is the evidence that many individuals who become schizophrenic at maturity are individuals of the PA type? Let us offer the following conjectures:

- Several investigators, in fact dating back to Kraepelin, have over the years distinguished a separate group of *paranoid schizo-phrenias* having a smoldering onset and a prolonged course. This is the *paraphrenia systematica* group of Kraepelin, and is termed "affect-laden paraphrenia" by Leonhard. This group was thought by Leonhard to be genetically distinct from the "systematic (paranoid) paraphrenias," which, as we have surmised, correspond mainly to categories of NP individuals. In fact, Leonhard considered that "affect-laden paraphrenia" was related to the *chronic paranoid condition,* which is in concordance with a common premorbid character trait of the PA individual.

- Some individuals of the PA type, even those functioning as leaders in competitive society, may exhibit markedly schizoid tendencies. Hence, it is not difficult to extrapolate the behavioral states of dominant PA individuals, both schizoid and non-schizoid, to the more severely withdrawn states of Class II and III schizophrenia.

- Individuals in the presumed categories of PA schizophrenias apparently are not known for their ability to display a warm smile of recognition. This is in contrast to some schizophrenics of the N and NP types, in particular the hebephrenics, who often break out into silly, sheepish or self-conscious smiles.

- Some paranoid schizophrenics are described as having colorless, pale or sallow countenances, in contrast to others, presumably mainly NP types, who are described as having sanguine or flushed complexions.

- This paranoid schizophrenic, presumably of the PA type, may be violently aggressive.

Narcissistic* N *schizophrenias.   The N character type is not uncommon in Western societies, hence it should be possible to identify individuals in this category without great difficulty.

We speculate that this schizophrenic might be a withdrawn individual who would periodically come out of his shell in an aura of grandiose megalomania. We believe that some of these N= and –N types may have been included in Leonhard's category of "expansive paraphrenias" (a possible historical example being the mother of Charlie Chaplin). These individuals often greet the visitor with a deep narcissistic bow and sweeping hand, dress up in outlandish costumes, speak in loud unctuous voices, sing to the heavens, are apparently unaggressive, and claim intense personal relationships with the Creator.

Aggressive* A *schizophrenias.   Given that individuals of the A type are uncommon in most societies, these would be infrequently seen disorders. We speculate that this would include paranoid, withdrawn A= and –A types who would periodically become violently aggressive individuals, or loud, intrusive individuals in the institutional setting.

Quantitative Genetics of Schizophrenia

If our model is essentially correct, then it should be able to explain the occurrence of schizophrenia not only in qualitative terms, but also in quantitative ones. In fact, the precise genetic basis of schizophrenia has puzzled geneticists and psychiatrists for decades. It is of note that several workers had concluded that some sort of dominant mode of inheritance with incomplete penetrance and variable expressivity was involved.

For example, Heston came to the conclusion that a dominant gene was involved, with nearly complete penetrance but with an expressivity spectrum that varied from a schizoid personality to outright schizophrenia. This is not far from our model's description of the NP and PA schizophrenias. For example, the PA phenotype is transmitted in a near-dominant manner (a PA individual must have a parent who is either a PA or A type), with the expressivity of the PA personality type on the scale of schizoidia to schizophrenia being determined by the exact structure of the **n**, **P** and **a** alleles, by modifier genes, by specific susceptibility genes, and by environmental factors.

In addition, Karlsson, on the basis of a meticulous study of an Icelandic pedigree, came to the conclusion that at least two genes were involved. He surmised that one was dominant and that a second was probably dominant also. Again, this is consistent with the present model, which predicts that susceptibility to schizophrenia can be determined by two separate genes (see later in this chapter: *Psychosomatic Disease*). For example, in the PA– phenotype at risk, lack of the trait N ("non-narcissism") would be transmitted in the dominant mode, since the complementary allele of a recessive gene codes for a dominant trait. In addition, lack of full expression of the trait A on the basis of an allele **a-** would also be transmitted in the dominant mode in a habitancy having a high prevalence of the trait A.

Next, let us approach the question of the *incidence of schizophrenia in the relatives of the chronic schizophrenic*. There exists an enormous body of data in the literature, but it is very difficult to interpret since most workers appear to have combined in their studies not only N, NP and PA types, but also patients with periodic psychoses having distinctly different genetic etiologies.

If we assume that most of Leonhard's "systematic schizophrenias" correspond to chronic schizophrenias of the NP type, then his statistics for this category indicate that the risk of psychosis in parents and siblings of an affected individual is on the order of 1–2 percent. Interestingly enough, this is about the same incidence as that of infantile autism in siblings of an autistic child.

For Leonhard's group of "affect-laden paraphrenia," which we assume to include PA types, the risk of psychosis in parents was in the range of 0-2 percent, and in siblings it was 1-9 percent. Although the numbers of patients involved were relatively small, Leonhard's data indicate that relatives of a PA chronic schizophrenic may be at a somewhat higher risk of psychotic illness in comparison to those of an NP schizophrenic. The higher prevalence of schizophrenia in siblings as compared to parents, if shown to be statistically significant, may be explained by the supposition that schizophrenics themselves are less likely to marry and have children than other more healthy schizo-phrenogenic PA types.

In studies of identical twins with schizophrenia, neither is the concordance 100 percent (it is more like 50 to 80 percent), nor is the clinical course in concordant pairs always similar. This is evidence for environmental factors playing a role in the chronic schizophrenias.

Finally, let us consider the occurrence of *schizophrenia in children whose parents are both schizophrenic.* Most studies report a frequency of schizophrenia of about 30 to 60 percent in these children. Since both parents are schizophrenic, most of the matings will be of types (1) NP × NP, (2) PA × PA, and (3) NP × PA in a locale where the P trait is prevalent. Matings of the first and second categories would be likely to yield at least 75 percent NP and PA types respectively, while those of the third category would yield at least 66 2/3 percent both NP and PA types (at least 25 percent of the zygotes would be non-viable). Thus, taken together, the progeny of all such unions might yield about 20 to 30 percent who are primarily NA and NPA types. This group of individuals would have a low risk of succumbing to schizophrenia, although we speculate that it might have a relatively high prevalence of the submissive character vector. The remaining 70 to 80 percent would be NP and PA types, and if we accept for the entire group the frequency of schizophrenia of 30 to 60 percent quoted above, we conclude that the risk of schizophrenia in these NP and PA children is well over 50 percent. This is about 10 times the risk of schizophrenia in siblings of a schizophrenic, and probably 50 to 100 times the risk of chronic schizophrenia in the general population. This is, once again, not inconsistent with the supposition of a contributory role of environmental factors in the genesis of the chronic schizophrenias. It is, in fact, difficult to imagine a more unfavorable environ-mental setting for a growing infant than one emanating from a situation in which both parents are schizophrenic.

We realize that the above are back-of-the-envelope calculations at best. They would have to be verified on homogeneous groups of patients once the precise modes of genetic transmission of the traits N, P and A, and of the character vectors of submission and resignation, are elucidated by careful pedigree studies.

We note that, according to the model, it may be not at all unusual to encounter schizophrenics of more than one character type in the progeny of a given family. For example, in a mating of the NP × PA kind, this couple besides being vulnerable to miscarriages and stillbirths (see Chapter 10) may produce progeny of both the NP and PA character types. Thus, it would certainly be possible for schizophrenias of both the NP and PA types to appear in the offspring.

Folie à Deux

This so-called "double insanity" or "psychosis by association" is, we propose, the symbiotic relationship between two poorly adjusted PA individuals. To this conclusion we offer the following evidence:

- A genetic basis is plausible from the fact that the two individuals are often members of the same family.

- There is often a family history of schizophrenia.

- The relationship is usually described in terms of a dominant and a submissive individual, sometimes in the context of sado-masochism."

- The disorder is characterized by paranoid delusions.

- One of the individuals is usually described by the term "schizoid," "reclusive," or "suspicious."

In the case of the two individuals being mates, if both are indeed PA types, then according to the discussion of the previous section the probability of their children also being PA types is high, on the order of 75 percent or greater. Since these parents may not be well suited for the nurturing of children, we might expect a high risk for schizophrenia in the progeny of such couple because of unfavorable environmental factors.

Schizophrenia in NA and NPA types

Since NA and NPA dominant types have two components of ambition, we speculate that schizophrenia in these types would be rare. But it is probable that such cases (e.g. –N–A and –NP–A types) will eventually be identified. Similarly we believe that cases of schizophrenias will be identified in individuals having the N trait with premorbid character vectors of submission and resignation.

Conclusion

Chronic schizophrenia shows itself to be a devastating disorder that prevents the individual from functioning in a competitive society. It has appeared throughout history as a mental disorder that has all too often been dismissed under the vague term of "madness." The list of famous individuals from all walks of life who have succumbed to schizophrenia is a long one, and we shall resist the temptation to compile such a list. We do suggest, however, that analysis of the family pedigrees of these individuals will, without too much effort, reveal the basis for the *damnosa hereditas* that has plagued families at all levels of society since time immemorial.

Finally, we note that our model of chronic schizophrenia is not based on unusual "genetic defects" or "inborn errors of metabolism." Rather, it is based on environmental stresses that are imposed on vulnerable individuals possessing very frequently occurring genes. This is in basic agreement with the views of Manfred Bleuler, who recently stated:

> The most likely explanation for the fact that the frequency of the incidence of schizophrenia remains the same in spite of reduced fertility among schizophrenics, lies in the assumption that the hereditary predisposition for schizophrenia does not consist of individually mutating, pathogenic genes, but rather in a disharmony of the congenital tendencies for personal development.

> According to my concept, the schizophrenic and his family may well rid themselves of the tormenting idea that they are carriers of an evil heritage, or that they are, as they have been accused of being, "eugenically undesirable individuals." It is very likely that their hereditary factors are not at all pathological; they may simply possess normal traits in unfavorable combination.

> Primarily, the concept of the nature of schizophrenics, as it is depicted here, should serve to rekindle anew the flames of enthusiasm for their treatment. It shows us that we are acting correctly when we continually expand with increasing care and indefatigable persistence those treatment methods which our experience shows to be most effective, and which is the treatment that reintegrates the patient into an active community, which sometimes shocks

him to awaken his dormant faculties, and which at other times tranquilizes him in his agitation and helps him to reestablish his inner consciousness.

Later in this chapter we will present a more precise hypothesis. Namely, we shall propose that in some family pedigrees the predisposition to chronic schizophrenia may lie in particular alleles at the **n** and **a** genetic loci acting in concert with other "high susceptibility" genes.

PERIODIC PSYCHOSES

The literature contains many accounts of periodic or remitting forms of schizophrenic-like illness, most of which seemingly have a much better prognosis than the chronic schizophrenias that we have described in the previous pages. These remitting forms have led to much confusion, since they have often been lumped with the chronic schizophrenias.

Manic-Depressive Illness

There exists an enormous literature on the subject of manic-depressive disease, which we have not attempted to peruse in depth. The literature is somewhat confusing, but we shall try to present the main thrust of what is known about the condition.

1. The occurrence of chronic schizophrenia and manic depressive disease appears to be mutually exclusive in a given individual or in identical twins.

2. According to Leonhard, the illness occurs in two types, namely the unipolar form (depression only) and the bipolar form (both manic and depressive episodes).

3. In the families of bipolar individuals only about one-half to one-third of all affected relatives will show the bipolar form of the disease, while the remainder will exhibit depressive but not manic symptoms.

4. In families of unipolar individuals there is usually no strong history of bipolar disease. (Unipolar disease is much more prevalent than bipolar disease).

5. The studies of Winokur and colleagues (1969) with bipolar probands showed: a) a female to male ratio of about 2:1 in affected family members, b) apparently no transmission of the disease from father to son, and most important, c) the existence of families in which the disease was transmitted in association with color blindness. Thus, these studies provided evidence for a dominant X-linked gene as a determinant of the disease in families of bipolar individuals.

6. Bipolar individuals are reported often to have a baseline personality of the cyclothymic hypomanic-depressive type.

From this information, we can distill the following hypotheses:

- The bipolar individual may be an NA type who also has the X-linked gene for manic-depressive disease. Since the female has two X chromosomes, while the male has only one, the probability of an affected female in a population will be roughly twice that of an affected male, in agreement with the observed female to male ratio. Since most of the first-degree relatives of the bipolar individual will not be NA types (in a habitancy with a fairly high frequency of gene **P**) this could explain why the majority of affected relatives will not exhibit manic symptoms. In addition, as Winokur and colleagues point out, the X-linked determinant for the disease may be expressed with less than complete penetrance.

  Therefore, we propose that at least one form of the bipolar form of the disease is expressed when the X-linked gene occurs in conjunction with the NA phenotype, the latter being autosomally determined. In the same family, periodic forms of the disease without symptoms of mania would be expressed when the X-linked gene occurred in conjunction with phenotypes other than the NA character type.

- Some reports in the psychiatric literature indicate that bipolar manic-depressive disease may be transmitted from father to son. Assuming that the diagnoses were correct, this would imply a genetic mechanism of inheritance other than X-linkage. Hence, we cannot rule out the possibility that bipolar disease in the NA type is a genetically heterogeneous group of diseases. Alternatively, we note that some investigators describe the premorbid personality type in bipolar manic-depressive disease as being "melancholic."

Therefore, we can hypothesize that the periodic agitated states of NP schizophrenic individuals (sometimes described as "catatonic excitement") have in the past been classified in the category of manic-depressive illness. Similarly, the chronic schizophrenic of the PA type can also periodically display agitated states of megalomania that might have been categorized under the rubric of "manic-depressive disease." In fact, various authors have pointed out that perhaps one out of every ten individuals initially misdiagnosed as manic-depressive eventually degenerate into a state of chronic schizophrenia.

Other Periodic Psychoses

If we surmise that a classic picture of bipolar manic-depressive illness is found in the NA character type, we might expect to find distinctly different clinical syndromes of depressive illness in affected relatives having the various other character types. Thus, as the X-linked gene assorted itself with different character types as it propagated through a family, a wide variety of periodic psychotic illnesses would be observed, with a preponderance in females.

We consider first the X-linked gene in association with the NP type. Carrying over what we learned from the highly variable clinical syndromes that occurred in the chronic NP schizophrenias, we might expect the X-linked forms to be characterized by *periodic* catatonic, hebephrenic and paraphrenic (paranoid) states. Leonhard, in fact, describes remitting illnesses of this type under the groupings of "confusion and motility cycloid psychoses," "cataphasia," and "periodic catatonia." Could these be the X-linked variants of manic-depressive disease in NP individuals?

Considering next the X-linked gene in PA individuals, we would expect to find remitting psychotic states resembling the paranoid PA schizophrenias. Could this correspond to some of the cases included in the "schizo-affective psychoses" of the psychiatric literature? Mendlewicz, in fact, indicates that there is some evidence for X-linkage in the "schizo-affective psychoses."

Some evidence for the supposition that the "schizo-affective psychoses" of the psychiatric literature are not a genetically homogeneous entity may be gleaned from studies that have considered patterns of aggression in psychotic patients. In one group there were patients with *internally-directed aggression* and a second group with *externally-directed aggression*. Those in the first group were found in several studies (reviewed by Beck) to have a better prognosis for remission. For example, according to Beck, Phillips and Ziegler postulated that...

> ... the person who assumes a "turning against the self" role has incorporated the values of society and, consequently, experiences guilt when he does not successfully meet these values.

This is consistent with our notion of the dutiful NP individual. (In addition, we may interpret internally-directed and externally-directed modes of aggression in the context of narcissistic and aggressive rages, respectively, in NP and PA individuals). Thus, we are led to the supposition that in remitting psychoses the NP individual may have a better prognosis for remission than a PA individual.

Finally, in family members of individuals with X-linked manic-depressive disease we would expect to find remitting psychoses also in the N, A and NPA character types. We suspect that descriptions of such psychoses may be scattered about in the psychiatric literature.

ENDOGENOUS DEPRESSION

Leonhard considered that *unipolar depression* or "pure melancholy" was an illness genetically distinct from bipolar manic-depressive disease. In addition, the older psychiatric literature describes "involutional melancholia" as a psychotic reaction occurring in the middle life period, usually being accompanied by depression. One characteristic pre-psychotic personality is sometimes described as *obsessive-compulsive, meticulous, worrisome, thrifty, intolerant, stubborn, scrupulously honest,* and *restricted in interests.* We postulate that this category of endogenous depression is rooted in the NP character type. The mechanism of genetic transmission is unclear, but from the current literature does not appear to be X-linked. Later in this chapter we shall raise the possibility that some subgroups of endogenous depression are rooted in allelic polymorphism at the loci of genes **n** and **a**.

Other involutional syndromes have been described in the context of paranoid states, perhaps rooted in the PA character type. Leonhard lists five categories of endogenous depression. Some of them apparently describe individuals of the NP type, but the genetic bases of these categories are not clear.

ENDOGENOUS EUPHORIA

Leonhard describes five categories of psychotic "pure euphorias" and a category of "pure mania." The category of *pure mania* probably refers to NA types (or harried NP types?), but it is unclear whether this is a genetically distinct group of individuals or whether these correspond to individuals with X-linked manic-depressive disease who have not yet had a depressive episode.

With regard to the *pure euphorias,* most of the patients appear to have been NP types. However, the unctuous, ecstatic, altruistic megalomaniacs in the category of "enthusiastic euphoria" may have included N types. Leonhard describes the familial occurrence of "enthusiastic euphoria" in a father and son, which is, of course, not inconsistent with our genetic model if both of these individuals were N types.

EXOGENOUS DEPRESSION AND EUPHORIA

According to our model, all of the personages can assume *situational exhilarated* or *abject states,* depending on how well they have adjusted to their life situations. In individuals having only one functioning component of ambition, we suppose that a situational abject state (or "reactive depression") could, if severe, drive an individual into a profound stuporous or harried state of psychosis (Figures 14 and 15). Similarly, in individuals with unbridled ambition, lacking the behavioral complex of perfectionism (N, A and NA types), we suppose that a situational exhilarated state might degenerate into a psychotic euphoric state of hyperactivity, exaltation, hysteria, mania, or megalomania (see Figure 16).

Therefore, we suggest that it would be of special importance to differentiate between situational exhilarated and abject states (psychotic reactive euphorias and depressions) on the one hand, and endogenous psychiatric illnesses having separate and distinct genetic components on the other hand.

ALCOHOLISM

From a cursory reading of the medical literature it is apparent that there is no single character type corresponding to the alcoholic individual. In fact, the literature describes alcoholic personality types in the categories of schizophrenia, manic-depressive disease, obsessive-compulsive personality, passive-dependent personality, and so on.

Insight into the genetic basis of alcoholism may be found in the studies of Cruz-Coke (1971), who reported that at least one form of alcoholism was associated with genetically determined color blindness in Western *populations* (not just in closely related individuals). Other investigators have pointed out that the prevalence of alcoholism is much greater in males than in females. In fact, the prevalence in males in some populations is very closely equal to the square-root of the prevalence in females, corresponding to male-to-female ratios of 10-30 to 1. These data provide a hypothesis that alcoholism is determined in part by an X-linked recessive gene. It is apparent that such a heritability would cut across the various divisions of character types of our model, hence it is consistent with the heterogeneity of observed alcoholic personality types.

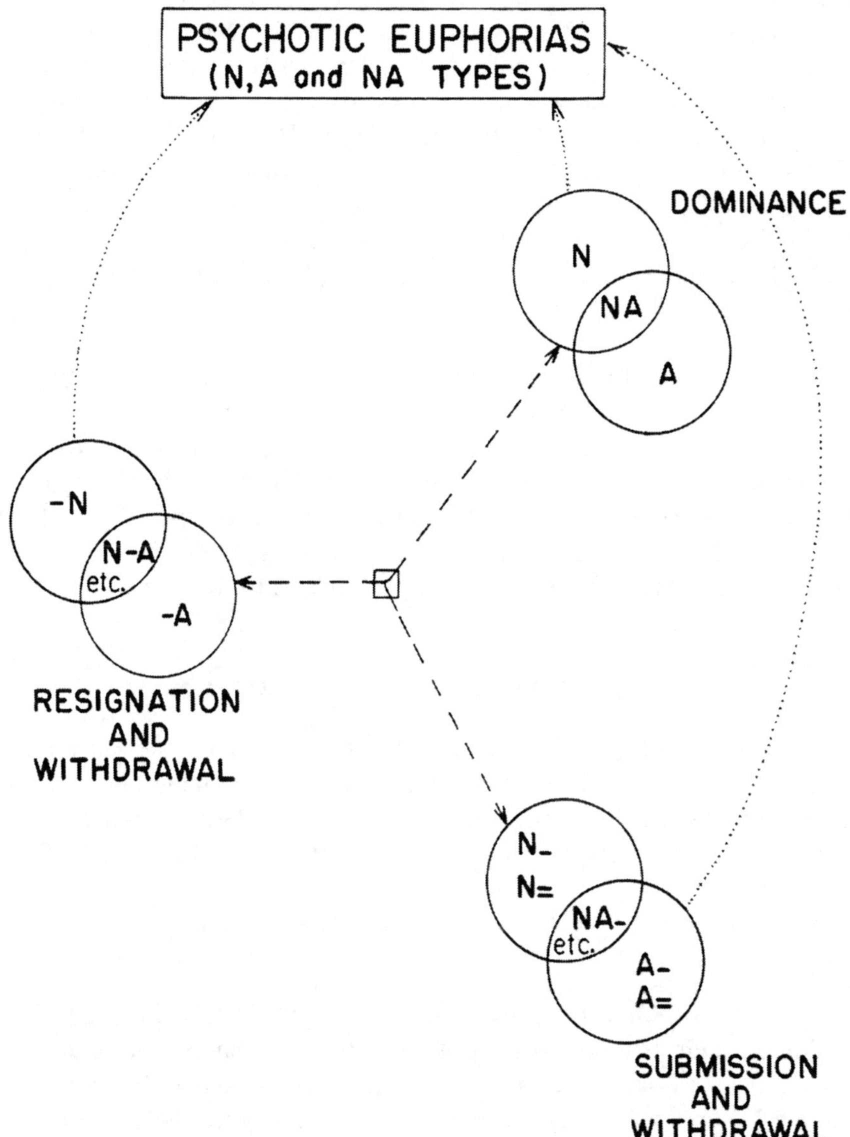

FIGURE 16

Schematic representation of exogenous psychotic euphorias for character types having unbridled ambition (trait N, A or both, but no P trait). The euphorias correspond to the situational exhilarated states.

A particularly intriguing question is raised by Cruz-Coke's tentative finding. Why, indeed, should a gene for alcoholism be linked to the genes for color blindness (on the X-chromosome)? Is this gene being now expressed in modern society in a manner in which it was never "intended"? It is interesting that it has been reported that mothers of alcoholics and color-blind individuals exhibit higher fertility rates, these individuals possibly being asymptomatic heterozygous carriers of the gene for alcoholism. How the character type of the individual would interact in this confusing state of affairs is not clear. Cruz-Coke and co-workers have hypothesized that the genes for color blindness and alcoholism have contributed in some unknown way to human survival, hence that color blindness and alcoholism may be related phenotypic traits belonging to a common system of genetic polymorphism.

Other workers, for example Winokur and colleagues, have concluded that all forms of alcoholism are not transmitted by X-linkage. In fact, they found a high prevalence of familial alcoholism in a subgroup of patients with unipolar depressive disease, who we postulate were mainly NP types. In addition, another indication of the occurrence of alcoholism in some subgroups of the NP type is found in the studies of M. Bleuler. He found a high prevalence of familial alcoholism in his studies of Swiss schizophrenics, presumably mainly of the NP type.

STATES OF DEPRESSED AMBITION

Referring to our model, we have identified various psychic states for which we could interpret the individual's behavior in terms of an inhibition in his ability to express ambition, i.e., the behavioral complexes of narcissism and aggression (exhilarated states were characterized by an accentuated expression of ambition).

We summarize below our interpretation of the various states of depressed ambition:

Acute personality splits. Individuals who "play the game," having the aggressive component of ambition, can acutely change their behavioral state from a dominant one to a subdued one, or from a submissive one to an energetic one. We previously used the notation:

$$XA+ \;\rightarrow\; \leftarrow\; XA-$$

Chronic state of subjugation. Individuals who "play the game" are subject to being subjugated by other stronger individuals. For a dominant individual (XA) who assumes the behavioral characteristics of subjugation, we would again use the notation:

$$XA \;\rightarrow\; \leftarrow\; XA-$$

Basic character vector of submission. We used the notation XA– and XA=, including individuals who are functioning members of society as well as individuals who would be classified as mentally ill. The character vector of submission merges with schizophrenia, in particular for individuals of the PA– and A– types.

Basic character vector of resignation. Somewhat simplistically we did not subgroup detached individuals according to their degree of detachment, and we used the notation X–A. Again, the character vector of resignation merges with schizophrenia for the PA and A types.

N *and* **NP** *withdrawn types.* Examples of these include a withdrawn narcissistic type (N–), narcissistic schizophrenics (N= and –N), a borderline autistic individual (N–P), and NP type autistic children and schizophrenics (N=P and –NP).

Other withdrawn types. Examples of these would include former submissive types who become resigned or who lose their narcissistic self-esteem after maturity, as well as other reclusive types (N=A=, etc.).

Trends of pseudoaggression and pseudonarcissism. Our interpretation of pronounced trends of pseudoaggression and pseudonarcissism, in individuals lacking the A and N traits respectively, was that these trends appeared as compensatory mechanisms of defense against the suppression of ambition, hence against schizophrenia. If the compensatory mechanism is effective, then the individual evolves a pattern of motivation that preserves his function in his society. If the compensatory mechanism is only marginally effective, then, according to the model, the individual barely functions in his competitive society, spending most of his time in a schizoid-autistic twilight zone of submission or withdrawal.

Abject states. In individuals having the behavioral complexes of N or A, we may think of the abject states of depression in terms of states of depressed ambition. In an individual whose baseline state is a submissive or withdrawn one, a depression into a situational abject state might scarcely be noticed by the individual's acquaintances. It would, of course, be intensely perceived by the individual himself. Conversely, in a vigorous individual whose baseline state is an extroverted dominant one the appearance of an abject state would be a dramatic event and would be readily noticed by all.

Senility. If we avow that schizophrenia, or *dementia praecox* as it was called by Kraepelin, may be thought of as a suppression of ambition, then we may hypothesize that some of the personality types having two fully functioning components of ambition (NA and NPA types) may be somewhat less vulnerable to some of the syndromes of the dementia of old age.

Other states. Ambition can also be suppressed to various degrees in:

- *anaclitic depression* of the first year of infancy (prolonged separation of the infant from the mother),

- *concentration camp syndrome,* and

- *states of acute dementia and delirium* (e.g., drug psychoses, postpartum dementia, as well as the delirium of the agonal state).

Again we postulate that individuals having only one functioning component of ambition would be at higher risk of succumbing to these syndromes.

Perfectionism. Finally, we ask ourselves: if the components of ambition (N and A) can be suppressed by adverse environmental factors, why not also P? Are there any disorders in which the behavioral complex of perfectionism is selectively suppressed? We do not know, but our impression is that the behavioral complex of perfectionism is from a behavioral point of view a more primitive one, hence one that is not easily suppressed by environmental constraints.

HYPERACTIVE CHILDREN

We believe that hyperactive children will prove to be a heterogeneous group of various personality types. Children of the N, A, NA and NA– types (unbridled narcissism and aggression) may be especially prone to hyperactivity and attention deficit disorders (ADD, ADHD). Non-compliant submissive NPA-types may also contribute to this loosely defined behavioral category.

SPEECH DISORDERS

We hypothesize that the child in whom the *onset of speech is delayed* is often an N or NP individual. Some of these individuals might be regarded as "successful" or "borderline" autistic children.

The speech disorders known as *stammering* or *stuttering* appear often to be rooted in an incapacity to express fully the trait of aggression. We believe that most stutterers will turn out to be NP, PA and submissive individuals. Some famous stutterers of the past include Charles I of England, his son James II, Louis XIII, George VI, Claudius of Rome, and Somerset Maugham.

ENDOGENOUS AND EXOGENOUS DISORDERS

Q.   What point do you wish to make here?

A.   We would like to clarify a bit out notion of the bases underlying behavioral anomalies, and in particular the notions of *endogenous* and *exogenous* disorders.

According to our concept, there may be no clear dividing line between one condition that we would call "normal" and another that we would tend to call "abnormal." Often we call a condition "abnormal" simply because it occurs infrequently in a given population. But whether a condition occurs frequently or infrequently, and whether we decide to call it "normal," "a disorder," or a "disease," our concept of the condition rests on the premise that there exist metabolic and neurologic mechanisms that can explain the condition on the basis of non-mystical principles of biophysics and biochemistry.

Q.   To what kind of conditions are you alluding?

A.   We are referring to just about any condition that would be labeled a "deviation from the norm" in a given society. This would include not only the various systemic diseases, such as diabetes mellitus and rheumatoid arthritis, but also all of the behavioral disorders to which we have referred in this chapter. Concerning disorders of behavior, whether it be anorexia nervosa or schizophrenia, it no longer suffices for us to accept, for these conditions, explanations based solely on psychodynamic theory. Rather we must clarify both the endogenous and exogenous bases for these conditions, and the relations between them.

Q.   What exactly do you mean by *endogenous* and *exogenous?*

A.   By an *endogenous* disorder we mean one that is engendered by a specific genetic constitution. For example, a clearly delineated endogenous disorder may be caused by a mutation at a specific genetic locus. This can induce a specific "inborn error of metabolism" and can lead to a familial disease that may be recognized easily, especially if it is expressed with complete penetrance. Another example of an endogenous disorder is Down's syndrome, since virtually all individuals having an extra number 21 chromosome will exhibit the syndrome of mongolism, irrespective of environmental factors.

At the other end of the spectrum, we refer to an *exogenous* disorder as one in which the exact nature of the individual's genetic constitution is of negligible importance in comparison with environmental factors predisposing to the condition. For example, if nearly all individuals of a population are susceptible to a particular infection during an epidemic, then we would certainly analyze the epidemic from the point of view of the clearly identified extrinsic cause. We would not immediately think to analyze susceptibility to the infection from the point of view of inheritance.

However, it appears that for many diseased states it is not appropriate to describe the multifactorial etiology of the condition simply by the use of the pithy terms endogenous and exogenous, as if they were mutually exclusive. In fact, in our previous discussions, when we hypothesized that there may exist a *genetic predisposition* to an individual's adopting the character vector of submission or resignation, we implicitly acknowledged the possibility of both endogenous and exogenous factors acting together. Similarly, in our discussion of *incomplete penetrance* and *variable expressivity* of an inherited trait, we acknowledged that exogenous factors may be the deciding ones in the etiology of the condition.

Recent advances in medicine are beginning to show that the above discussion is more than just an academic one. For example, so-called *endogenous diseases* such as juvenile onset diabetes mellitus may in fact require an individual's exposure to an infectious agent in order for the clinical syndrome to ensue. Conversely, so-called *exogenous diseases* such as rheumatic fever, tuberculosis or leprosy may require certain genetic susceptibilities in given individuals in order for the underlying infection to assume clinical proportions.

Q.   How does the above discussion apply to behavioral disorders?

A.   In the development of our model we originally assumed that the character vectors of submission and resignation, as well as the states of non-aggressive withdrawal, chronic schizophrenia, pseudoaggression and pseudonarcissism, were determined by environmental (exogenous) stresses on "susceptible" individuals. Here, we would like to examine the possibility that these states may, in fact, be predisposed by clearly identifiable (endogenous) genetic mechanisms.

Q.   What is the evidence that these states may be predisposed by endogenous genetic mechanisms?

A.   We had assumed that these states can be brought about by environmental factors. This conclusion followed from our consideration of the results of studies in various areas of animal behavior and medicine. For example, studies with non-human primates indicate that isolation during infancy predisposes to irreversible states resembling submission in the adult animals. Similarly, prolonged separation of a human infant from the mothering individual during the first year of life may induce in the infant a condition of withdrawal known as anaclitic depression. Finally, in the area of schizophrenia, the occurrence of infantile autism has been correlated with perinatal trauma, while in studies of identical twins one does not find complete concordance for the occurrence of either infantile autism or the other schizophrenias. These lines of evidence support the importance of *environmental factors* in the genesis of these conditions.

Nevertheless, there exists some evidence for the occurrence of *definite genetic mechanisms* determining these states in certain families. For example, in studies of identical twins there appears to be a paucity of twins, even those raised apart, in which one individual is a dominant type and the other is a submissive type. With regard to schizophrenia, an endogenous mechanism is plausible from recent information showing, in certain patients, anomalies in brain morphology revealed with the use of X-ray scanning techniques (computerized axial tomography). Finally, in certain family pedigrees, including those of royal lineage, the character vector of submission can be traced from generation to generation, as if it were transmitted by a dominant gene.

Q.   So you are implying that certain behavioral states, such as the character vector of submission, may have primarily *either* endogenous or exogenous roots?

A.   Yes. And if this is indeed the case, then we would expect to find subtle phenotypic differences in individuals, depending on the exact etiologic mechanisms underlying the condition. Thus, subtypes of the character vector of submission could have their bases in either endogenous or exogenous factors (i.e., either in genes or environment). But whatever the underlying causes, we believe that behind the observed behavioral phenotypes lie biophysical and biochemical mechanisms that await our discovery.

Q.   What would be the philosophical implications of the conclusion that these states, for example the character vectors of submission and resignation, may be endogenously determined?

A.   This would imply that our free will, as individuals, may be an even more elusive entity than we had originally supposed. It implies further that if we as parents had, over the years, thought that we could raise our children — by "good upbringing" — to fulfill our notions of some preconceived ideals of ambition, motivation and decorum, then this has, once again, been little more than the most fantastic of delusions.

Q.   With regard to these states to which you have been referring (submission, resignation, non-aggressive withdrawal, schizophrenia, pseudoaggression and pseudonarcissism), how, specifically, might they be genetically determined?

A.   We could propose a variety of complex genetic mechanisms. We shall consider below the simplest hypothesis, namely that these states can be engendered by mutations at the genetic loci **n** or **a**. This hypothesis follows from our consideration of *psychosomatic disease,* which we shall consider in the next section.

PSYCHOSOMATIC DISEASE

Q.   Why should character type have anything at all to do with disease?

A.   If we believe that the behavioral complexes N, P, and A are controlled by major genes acting at the level of the central nervous system, then we are drawn to the supposition that individuals of the various character types will have vulnerabilities to particular diseases. This follows, in part, from what we know about mutations at other genetic loci. Namely, a mutation at a genetic locus will, in general, engender an endogenous condition in the individual that will be interpreted by the physician as a "deviation from the norm." The great body of such conditions, most of which occur very infrequently, is what constitutes the thick catalogue of the inherited diseases known to modern medicine.

Q.   What are psychosomatic diseases?

A.   Over the years, psychiatrists and other physicians have noted that certain somatic disorders seem to be related to personality type or to patterns of emotional factors. These disorders include asthma, eczema, psoriasis, rosacea, dermatitis herpetiformis, rheumatoid arthritis, lupus, Graves' disease, ulcerative colitis, peptic ulcer, essential hypertension, glaucoma and many others.

For many years the various factions of internists and psychiatrists have been engaged in a running controversy regarding the primacy of "the chicken or the egg" in the etiology of the psychosomatic diseases. On the one hand, one group of investigators has maintained that it is the psychic or emotional factors that cause the somatic disease, with the disease sometimes being regarded as a defense mechanism against psychic disintegration. On the other hand, another group has clung to the belief that the emotional and personality factors are simply the behavioral reactions of an individual afflicted with a chronic disease. As things stand in the present day of modern medicine, the psychic elements of the psychosomatic diseases are regarded by the physician with a wink and a chuckle, and at most the physician will grudgingly admit that, yes, perhaps in some diseases the symptoms and signs may be induced or exacerbated by the patient's personality or emotional state.

Q.   What does your model predict with regard to the origins of the psychosomatic diseases?

A.   We focus first on the trait of aggression. Lack of the A trait in an individual implies an incapacity to effect a mass discharge of the sympathetic nervous system. We might expect, therefore, that certain individuals lacking full expression of the A trait (e.g., N, NP, submissive and resigned types) may be especially prone to disorders against which the sympathetic nervous system normally acts as a defense. Such disorders might include asthma, allergic reactions and anaphylactic shock.

Q.   Is there a more precise way by which your model can predict the origin of psychosomatic diseases?

A.   Yes. We postulate that mutations at the **n**, **P** and **a** loci can be the link between a certain character subtype and a systemic disease causing somatic symptoms. Thus, at the same time that we are proposing that mutations at the **n**, **P** and **a** loci can cause disease, we propose that allelic polymorphism at these loci can be the origin of character subtypes to which we have referred in this and preceding chapters.

Q.   What would we expect to find in the case of a mutation at one of the loci n, P or a?

A.   Focusing first on the gene **a**, we would expect to find the following:

1. We would find a familial disease, which might be transmitted according to either dominant or recessive inheritance.

2. If the mutation occurs at a critical point in the gene, it could lead to inhibition (or accentuation) of the ability to express the trait A. If

we assume that the allele for non-aggression ordinarily complementary to **a** (to be called **a₀**) codes for a fully active repressor enzyme, then we might expect that a mutation in **a₀** (to be called **a-**) would be expressed as dominant with respect to **a** but recessive with respect to **a₀**. If the mutation leads to a partial inhibition of the ability to express the trait A, then the phenotypic expression of the allele **a-** would be A–, or non-compliant submission.

3. As the mutant allele **a-** assorted itself with other alleles at the **n, P** and **a** loci, a set of discrete character subtypes could be generated. A systemic disease engendered by the allele **a-** could thus be associated with a set of discrete character subtypes.

4. If a systemic disease is engendered by the complementary action of the gene **a** and one of the other character genes (say **n**), then the disease would be associated with only a few relevant character types. It would be said to be transmitted according to "polygenic inheritance."

5. If a disease is thus identified with a particular character type, it would be associated with mental disorders to which that character type is predisposed. Conversely, mental disorders to which the character type is not predisposed would not be found in association with the systemic disease.

6. The systemic disease engendered by gene **a-** might be one of a spectrum of similar diseases, with overlap existing between some of the signs and symptoms of each. According to the model, this spectrum of diseases could have its roots in at least two mechanisms. First, allellic polymorphism at the **a** locus could give rise to similar diseases having partial overlap in their clinical presentations. Second, a disease could have altered forms as a mutant gene **a-** acted in complementarity with other genes and with environmental factors.

7. An easily identified gene **H** closely linked to gene **a** could serve as genetic marker for the entire spectrum of diseases.

8. The various alleles of gene **H** could serve as a general genetic marker for gene **a**, hence for the different character types and subtypes, as well as for character subtypes prone to various categories of mental disorders.

Q.   Are there any psychosomatic diseases that seem to fit the picture that you have outlined?

A.   Yes. As an illustrative example we shall choose one, namely *rheumatoid arthritis (RA)*. We note the following:

1. *RA* is a familial disease which has been thought by some to be transmitted according to polygenic inheritance.

2. The personality type in *RA* has been consistently described in terms of an inhibition of the ability to express aggression. These individuals are often described as *dependent, self-sacrificing, masochistic, forgiving, righteous, conforming, self-conscious, shy,* and *inhibited but active*. This most closely resembles a subtype of our non-compliant submissive character type. Historical examples of this subtype include Pierre-Auguste Renoir, the American President James Madison... and perhaps even the Egyptian Pharaoh Amenhotep II.

3. Individuals having *RA* appear to be well motivated, apparently having full expression of the trait N. This might imply that *RA* is limited to the character types NPA– and NA–. In fact, individuals with *RA* of the presumed NPA– subtype tend to have sanguine complexions and, when not intimidated, have the sparkling, spirited eyes* of their NPA cousins.

4. If *RA* is in fact limited to individuals of the NPA– and NA– types, then the occurrence of schizophrenia in *RA* should be a very rare event. This, in fact, appears to be the case. It has long been known among psychiatrists that the occurrence of arthritis in schizophrenia is very unusual.

5. *RA* is one of a spectrum of "autoimmune" or "connective tissue" diseases showing overlap in many of the signs and symptoms. These diseases include lupus, systemic sclerosis, polymyositis, periarteritis nodosa, and others.

6. The occurrence of *RA* and the spectrum of diseases with which it has been allied on clinical grounds has been shown to be linked to a common genetic marker, namely the HLA histocompatibility loci on the short arm of chromosome 6.

7. Subgroups of patients with schizophrenia and depressive disease have also revealed linkage of the disease to the HLA loci.

---

* Sometimes the eyes appear proptotic, suggesting a relation to Graves' disease.

Q.   So what conclusions do you draw from the example of rheumatoid arthritis?

A.   We conclude that there is indeed some evidence that:

- Some systemic diseases may be endogenous to mutations at the **n**, **P** and **a** loci.

- In some of the so-called psychosomatic diseases a mutation at the **n, P** or **a** loci may be responsible for both the somatic disease and the characteristic personality or emotional picture.

- The character vectors of submission, resignation and non-aggressive withdrawal, states of pseudoaggression and pseudo-narcissism, as well as certain categories of schizophrenia and endogenous depression, may be genetically predisposed on the basis of allelic differences at the **n** or **a** loci.

- Allelic differences at the **n**, **P** or **a** loci may be the origin of specific character subtypes (e.g., NPA–).

- The **a** locus, or a gene interacting with the **a** locus by a process of epistasis, may be located on the short arm of chromosome 6 in the vicinity of the HLA histocompatibility complex of genes.

Q.   What are the medical implications of your discussion of psycho-somatic disease?

A.   We speculate that some of the various diseases whose metabolic bases are not well understood might possibly be related to the genes **n, P** and **a.** This would include:

- *psychosomatic disorders, allergic disorders, anomalies of the connective tissue and of development* (e.g., the "collagen vascular diseases," congenital heart diseases, the arthritides, sarcoidosis, Marfan's syndrome)

- *disorders of the endocrine and immune system* (e.g., Addison's disease, Graves' disease, subacute bacterial endocarditis, acquired immune deficiency syndrome)

- *disorders associated with flushing* (e.g., hives, rosacea, urticaria, lupus, dermatomyositis, rheumatic heart disease, gout, maturity-onset diabetes mellitus)

- *dermatologic disorders* (e.g., psoriasis, atopic dermatitis, dermatitis herpetiformis)

- *infectious disorders* (e.g., acute rheumatic fever, tuberculosis, leprosy, mononucleosis)

- *neoplastic and associated disorders* (e.g., cancer, Hodgkin's disease, acute leukemia, myasthenia gravis)

- *disorders conducive to episodes of acute psychosis* (e.g., lupus, ICU psychosis in post-rheumatic heart disease)

- *disorders related to "personality type"* (e.g., coronary artery disease, urticaria, gout, migraine headache, carcinoma of breast and lung)

- *disorders linked to the HLA and immune response genes* (e.g., juvenile-onset diabetes mellitus, hemochromatosis, rheumatic fever, celiac disease, pernicious anemia).

Thus, it would be of particular interest to sort out how the genes coding for the above disorders relate to those determining the traits N, P and A. It is of interest that predisposition to some of the diseases linked to the HLA genes (e.g., Hodgkin's disease, multiple sclerosis, juvenile-onset diabetes mellitus) is thought to be associated with exposure to an infectious agent. Therefore, we should not rule out the possibility that predisposition to some of the mental disorders, and in particular to schizophrenia, may in some cases be linked to an infectious agent acting in endogenously determined vulnerable phenotypes.

Q.   If non-compliant submission is primarily endogenously determined in a given population, can its incidence be quantified?

A.   Yes. We assume three possible alleles at the **a** locus, namely $\mathbf{a_0}$, **a-** and **a**. The alleles $\mathbf{a_0}$ and **a** are those for non-aggression and aggression, respectively; they correspond to the two-allele model that we used previously. The allele **a-** is the mutant allele. Thus, if we denote by a_o, a- and a the gene frequencies for the three alleles, we obtain the following:

Phenotype	*Genotype*	*Phenotype frequency*
Aggression (A)	**(aa)**	a^2
Non-compliance (A–)	**(aa-)**, **(a-a-)**	$a\text{-}^2 + 2aa\text{-}$
Non-aggression	$\mathbf{(a_0a)}$, $\mathbf{(a_0a\text{-})}$, $\mathbf{(a_0a_0)}$	$1 - (a + a\text{-})^2$

Thus, we may use the Hardy-Weinberg approach as outlined in the preceding chapter. As a basis, we begin with the U.S.A. habitancy, for which the reader will recall the gene frequencies were $n = 0.90$, $p = 0.47$ and $a = 0.82$. To obtain our modified habitancy, which we call the *Introspective habitancy*, we maintain the same values of n and p as above, but we replace a with two separate alleles having the gene frequencies $a- = 0.25$ and $a = 0.57$. Thus, the sum of the gene frequencies *(a + a-)* is 0.82, the same as that for the U.S.A. habitancy. The calculated prevalences of the expected phenotypes for the U.S.A. and Introspective habitancies are presented in Table 6.

We note the following:

- The allele **a-** generates four additional phenotypes for the Introspective habitancy, namely the non-compliant types A–, NA–, PA– and NPA–.

- The sum of the phenotype frequencies XA + XA– for the Introspective habitancy is the same as that for the U.S.A habitancy.

- The Introspective habitancy has a high proportion of introverts having sanguine complexions. These are the NP types (20 percent), NPA– types (22 percent) and NA– types (8 percent). Hence, fully 50 percent of this population tends to be motivated introspective introverts.

- The frequencies of PA– and A– types are 5 and 2 percent, respectively. These types may be at higher risk for succumbing to chronic schizophrenia.

Therefore, our model permits the existence of this society comprising primarily ruddy-complexioned, self-sacrificing, dutiful introverts. In comparison with the Punctilious habitancy, they would have a high incidence of rheumatoid arthritis and peptic ulcer. They are known for their characteristic reservedness and may be accused by their detractors of "masochism." They are continually mocked by their neighbors of the Demonstrative habitancy, but show endless patience with them. This unaggressive and somewhat aloof society would not think of attacking its neighbors. However, if they themselves are attacked or their possessions trodden upon, then the members of this habitancy show the highest degrees of resoluteness, even in times of the worst of adversity, and show the world, in their finest hour, that the best offense may simply be a good defense.

TABLE 6

**Theoretical Frequencies of Phenotypes in U.S.A.
and Introspective Habitancies**

(per 100 viable individuals)

| | *Habitancy* | |
Phenotype	*U.S.A.*	*Introspective*
N	8	8
A	4	2
A–	--	2
NA	16	8
NA–	--	8
NP	20	20
PA	10	5
PA–	--	5
NPA	42	20
NPA–	--	22
Gene frequencies	$n = 0.90$ $p = 0.47$ $a\text{-} = 0$ $a = 0.82$ $a + a\text{-} = 0.82$	$n = 0.90$ $p = 0.47$ $a\text{-} = 0.25$ $a = 0.57$ $a + a\text{-} = 0.82$

Q.   What might we expect for the case of a mutation at the n locus?

A.   We denote by **n*** the mutation at the **n** locus and the phenotypic expression of the trait by N*. In the case of a mutation at the site of the gene of aggression **a** we noted that we obtained a state of non-compliant submission, or a profound, but not complete, inhibition of the behavioral complex of aggression and of the aggressive-vindictive rage. By analogy, for a mutation at the site of the gene of narcissism **n** we would expect to find an inhibition of the narcissistic rage and of its prodromal response of blushing, flushing, and lacrimation. Thus, types having the trait N* would tend to be non-sanguine and would lack the ability to display the gingival smile of recognition. They would, however, in some incomplete way be able to exhibit the behavioral complex of narcissism.

If we denote by n_o, $n*$ and n the gene frequencies for the three alleles, then we obtain:

Phenotype	*Genotype*	*Phenotype frequency*
Narcissism (N)	**(nn)**	n^2
Inhibited Narcissism (N*)	**(nn*)**, **(n*n*)**	$n*^2 + 2nn*$
Non-narcissism	**(n₀n)**, **(n₀n*)**, **(n₀n₀)**	$1 - (n + n*)^2$

Without making any quantitative estimations, if we add the **n*** allele to those of the Introspective habitancy we obtain the following additional character types:

Notation	*Character type*
N*, N*P	Non-aggressive asthenic types
N*A, N*PA	Pseudonarcissistic extrovert types
N*A–, N*PA–	Pseudonarcissistic introvert types

Our model now provides us with a basis for the pseudonarcissistic types presented in previous chapters. The pseudonarcissistic extroverts and introverts would resemble the non-sanguine A, PA and A–, PA– types, respectively. In particular, they would lack the capacity to express the narcissistic smile.

In the pseudonarcissistic introvert types (N*A– and N*PA–) neither component of ambition is fully expressed. Hence, these types might have a fairly high risk of succumbing to schizophrenia, but the risk would be less than that of the A– and PA– types. In our present model, therefore, pseudonarcissism appears as a partial defense against schizophrenia.

Q.   Since many of the behavioral disorders that you have discussed previously are confined to definite character types, can we say anything about their relative prevalences in various socio-cultural settings?

A.   Not with any certainty. Consider *schizophrenia* for example. We noted in the last chapter (Table 5) that the prevalence of PA types in the Punctilious habitancy was likely to be at least 5 times lower than that in the U.S.A. habitancy. This is such a large difference in prevalence that one is tempted to predict categorically that the prevalences of schizophrenias of the PA types in the Punctilious habitancy will be much lower than those of the U.S.A. habitancy. Similarly, one might suppose that the prevalences of schizophrenias of the NP types (catatonias and hebephrenias, in particular) would be much higher in the Punctilious habitancy. However, we must keep in mind that both the genetic and environmental risks leading to schizophrenic states may be quite different in the two cultural settings.

How do such predictions agree with the available data on the occurrence of schizophrenia in different cultures? Crider (1979) states:

> Even if schizophrenia were to be found in all cultures and epochs, there is no a priori reason to expect similar *frequencies* of occurrence across cultures. Unfortunately, the available epidemiological data confound any true cultural differences with methodological variation among studies, so the issue is at present irresolvable.

Turning to *manic-depressive disease,* we note that the prevalence of NA individuals is extremely low in the Punctilious habitancy but quite high in the Demonstrative one. Again, one is tempted to predict categorically very low and very high prevalences of *bipolar* NA manic-depressive disease, respectively, for the two habitancies. However, not knowing the frequencies of the (presumed X-linked) gene for the disease in the two habitancies, we cannot really make any safe predictions with regard to the occurrence either of bipolar manic-depressive disease or of other periodic psychoses in the two habitancies.

Q.   Can we learn anything about disorders of human behavior from the study of other primates?

A.   Unquestionably so!

Orangutan. Let us consider first the orangutan, referring to the studies of Galdikas and Brindamour, which have been published in the *National Geographic Magazine.* With the aid of their studies we are drawn to the conclusion that the orangutan is a monomorphic species of the NP type, hence an ape that has lost the ancestral aggressive gene.

We offer the following evidence for this conclusion:

- The orangutan in his natural habitat is a gentle, unaggressive, non-carnivorous ape. Orangutans are described as being rather asocial and phlegmatic, expressing relatively little emotion. Their solitary nature is stressed, with strong tendencies to aloofness especially in the mature males.

- A healthy orangutan will not attack a human, unless provoked into retaliation.

- In many photographs, orangutans display a countenance that can only be described as "melancholic."

- Although orangutans respect an order of priority in their society, they do not "play the game" of dominance and submission. Fighting is extremely rare and is usually limited to males in the presence of an estrous female.

- Orangutans are described as exhibiting limitless patience as they wait their turn to eat from a pile of rice.

- The orangutan shows unmistakable perfectionistic traits as he eats termites, or rice one grain at a time. One individual is described as working "as meticulously as a dentist."

- At times, the orangutan shamelessly shows narcissistic traits. He may exhibit a radiant smile when in the limelight.

Some behavioral activities of orangutans may be interpreted in the context of the narcissistic rage. For example, Galdikas describes a "youngster's rage":

> Sometimes a mother would refuse to share, causing the youngster to throw a violent tantrum. I thought one squealing juvenile might just hurl himself out of the tree headfirst in his fury...

Also, Galdikas describes the behavior of Princess, a young orangutan, "a little terror," that she had hand-raised herself:

> Now I surprised myself by saying things like "Get that rampaging ragamuffin out of here, or I'll put her in a cage!" as Gary fondly gazed at Princess ripping pillows to shreds or splattering milk over the walls.

We cannot resist the comment that the pampered Princess apparently had very permissive "parents" and in being allowed such narcissistic tantrums, she may have unwittingly been raised to become a spoiled child!

Elsewhere, Galdikas describes an incident in which a young female's narcissistic vanity is wounded: when Mute, a young male, flirts with another female in the presence of his regular girlfriend Noisy, his girlfriend attacked her rival! Galdikas thought that this might have been outright "jealousy."

In short, it would come as no surprise to us if someone were to show us a photograph of an orangutan sitting at the side of a pool, smiling coquettishly at his own image.

• Finally, Galdikas' case studies of formerly captive orangutans, which she attempted to rehabilitate, describe grossly abnormal behavior that we can interpret in the context of schizophrenic behavior in an NP type. One female, Sis, had been caged for six years and may have exhibited signs of catatonic stupor:

> While we carried on the wild orangutan research, we continued rehabilitating ex-captives, releasing many of them across the Sikunir Kanan River from our study area. Some adjusted quickly to their new lives. Others, such as Sis, were not so fortunate. We had to bring her back to camp when we found her half-submerged in the swamp, thin, eyes glazed, and unable to raise herself off her haunches.

A photograph of Sis shows the blank visage and half-open mouth of a mentally ill individual.

Another male rehabilitant orangutan, Sugito, also showed distinctly abnormal behavior. He displayed the harried "aggressive" behavior of a paranoid NP schizophrenic and became unapproachable. Finally, in acts most uncharacteristic of a well-adjusted orangutan in the wild, Sugito became a killer of infant orangutans. Galdikas in a striking account describes Sugito's killing of the infant, Doe. We see in her account of a megalomanic urge gone far astray, a variation of the expansive gesture of "the narcissistic arms":

> After some time Sugito slowly turned and came down to the ground. His reaction to Doe's body was totally unlike the others. I watched in fascination as he slowly approached. Then, standing on two legs, he raised both arms over his head and brought them down, fluttering, in front of him. The gesture reminded me of nothing less than a shaman from some obscure tribe performing rituals of obsequiousness to his god.

Q.   How about other primates?

A.   *Gorillas* are gentle, non-carnivorous, do not "play the game" of dominance and submission, smile, and enjoy the limelight in captivity. We suspect that they, too, are a species of NP types. We believe that the gorilla's "aggressive" behavioral display, in which he adorns himself by placing a branch of leaves in his mouth, beats on his chest with his palms and charges forward only to stop and thump the ground with one fist, is essentially a narcissistic display. Gorillas are, in fact, very unaggressive creatures.

Baboons are carnivores, play an aggressive "game" of dominance and submission, glare at each other, groom each other in a perfectionistic manner, and are not known for their smile. We suspect that they are mainly PA types.

Chimpanzees can show perfectionistic and aggressive traits, are omnivorous, smile in the limelight and can be rather overbearing. We suspect that they are mainly NPA and NA types.

If we were forced to assign these various primates to the habitancies presented in Chapter 10, we would (somewhat tongue-in-cheek) assign the orangutans and gorillas to the Punctilious habitancy, the baboons to the Authoritarian habitancy, and the chimpanzees to the Demonstrative habitancy.

Q.   **How about porpoises and whales?**

A.   *Porpoises* are gentle, playful, unaggressive, are known for their persistence and precision, and adore performing in the limelight. Does this sound familiar?

Killer whales (the *orca*) have a bad reputation as a predator. However, Griffin (1966) states that

> During his first months in captivity, I constantly worked and played with Namu, ever alert for some aggressive sign on his part. I never saw one. He was always affable.

Elsewhere he states:

> Though I try not to let the whale know it, I have discovered that he has a mean side. Beware, I have taught myself, when he swings his head sharply or jerks it upward. Especially take care when he bobs his head up and down swiftly. This marks his most severe displeasure and peevishness. "Leave me alone now," it says, and I take the hint.

Could this be narcissistic negativism, with a narcissistic tantrum perhaps not far behind?

Q.   **What would be the implications of identifying the narcissistic trait in lower animals?**

A.   If a genetic or biochemical marker for the behavioral complex of narcissism were found in man, it would be of interest to trace its origin in the animal kingdom. If the trait N is indeed found in the porpoises and whales, then we would conclude that this trait would date back to a common mammalian ancestor living over fifty million years ago. This would follow from our knowledge that these Cetacea are related to the mammalian order of Ungulata, whose common ancestor with the Primates probably dates back to fifty to one hundred million years ago. If the biochemical marker were to be found in birds (the posturing peacock comes to mind), then the common ancestor would be found among the class of Reptilia, or in excess of two hundred million years ago.

Q.   If we accept the fact that many mental illnesses may now be delineated according to character type, what eugenic measures should we take to reduce the incidences of these illnesses?

A.   H.L. Mencken once said, "There is always a well-known solution to every human problem — neat, plausible and wrong."

Those who have thoughtfully pondered the implications of eugenic proposals have come away with the feeling that there are no easy answers. McKusick (1969) states:

> Unfortunately the enthusiasm of the eugenicists has often run ahead of the elucidation of principles on which a sound program must be based. Eugenics can be arbitrarily classified as negative or positive. Negative eugenics is concerned with preventing reproduction of persons carrying undesirable genes... Negative eugenics is inefficient and often the means by which it is practiced are not ethically acceptable to most people... Positive eugenics, encouragement of reproduction by persons of presumably superior genotype, has as many problems as negative eugenics... The difficulties in identifying superior genotypes, as well as detecting deleterious recessives which may be lurking in individuals who have some superior qualities, make [Hermann Muller's] "germinal choice" impractical. Furthermore, the undesirability of doing anything that would encourage genetic uniformity rules out the approach.

Similarly, Dobzhansky (1970) states:

> Selection for one trait may unintentionally induce alterations elsewhere in the genetic system. Such alterations may be expected to conflict with the exigencies of general adaptedness to the environment.

Obviously, the statements above provide enough fodder to feed all kinds of polemics for a least several decades. Our own view is that we should do nothing active in the realm of eugenics until we really know what we are doing — and that time may never come! A main thrust of this book is the gaining of one's free will as an individual. It would be the height of irony if, in trying to gain our free will, we find ourselves inextricably bound in a morass of misguided attempts at improving ourselves in the manner described above by H.L. Mencken.

Q.   I gather that you are presenting your model as a "unified theory of human behavior."

A.   Not quite! Using the language of epidemiology, what we have done is to go out into the field to conduct a *survey* of the activities of mature individuals interacting in a competitive society. We have, so to speak, taken a snapshot (or short film strip) of human behavior from the point of view of personality characteristics, which are assumed to have both genetic and environmental origins. Such an approach appears to have been fruitful. For example, it has allowed us to make various hypotheses and to arrive at quantitative estimations of the relative prevalences of certain character traits.

But a field survey, although generating many hypotheses, is essentially a snapshot in time, and it does not focus on the developmental aspects of an individual as he matures from infancy, through childhood, to adolescence, and finally to adulthood and old age. One can see that a more complete description of an individual's behavior over time (analogous to what the epidemiologist would find in a "prospective study") would require a model of much greater complexity than the one presented here.

Thus, we have not yet arrived at a "unified theory of human behavior." We surmise that such a theory will evolve as gaps are gradually filled in the overlapping areas of genetics (especially identifying other genes important in behavior), anthropology, learning and conditioning, biophysics, biochemistry, neurology, pathology, and psychiatry.

Q.   Could you kindly summarize the main points of this chapter?

A.   Our summary:

- So-called "disorders of human behavior" are, loosely speaking, those conditions which would show a "deviation from the norm" in a given population.

- Many conditions labeled as "disorders" are part of a continuum of behavioral states which merge with those that would be regarded as "normal."

- Disorders of human behavior can have as their bases either primarily endogenous (genetic) or exogenous (environmental) factors.

- The basis for disorders of human behavior, whether rooted primarily in endogenous or exogenous factors, is thought to rest on well-defined biophysical and biochemical mechanisms.

- The occurrence of chronic schizophrenia in the various character types may be interpreted in terms of the inability to express the components of ambition, that is, of the behavioral complexes of narcissism or aggression. Susceptibility to schizophrenia may be of primarily endogenous or exogenous origins.

- Infantile autism is rooted in the NP character type. The reason for the association may be that perfectionistic NP children with developmental disorders are placed by practitioners into categories of "autism" on the basis of clinical and behavioral criteria.

- Some remitting psychoses, including bipolar manic-depressive disease in the NA type, may be primarily endogenous disorders rooted in an X-linked dominant gene.

- Character types having only one component of ambition (either narcissism or aggression) may have an increased susceptibility to psychoses, including depression and schizophrenia.

- Character types having unbridled ambition (lacking the trait of perfectionism) may have an increased susceptibility to psychotic manias and euphorias.

- Alcoholism is not confined to any particular character type. Some types of alcoholism may be engendered by an X-linked recessive gene.

- In the so-called "psychosomatic diseases" both the systemic disease and the characteristic personality picture may have their origins in allelic differences at the **n**, **P** and **a** loci.

- Allelic polymorphism at genes **n** and **a** may be the endogenous basis of the connective tissue diseases, psoriasis, Graves' disease, and other well-known diseases of cryptogenic etiology. Similarly, such polymorphism may be the basis of certain categories of schizophrenia and unipolar depression.

- Endogenous conditions residing in allelic polymorphism at the **a** locus may be related to the HLA (histocompatibility) loci on chromosome 6. Thus, gene **a** (or a gene interacting with gene **a**) may lie on the short arm of chromosome 6.

- The character vectors of submission, resignation, non-aggressive withdrawal, as well as the states of pseudoaggression and pseudonarcissism, may be genetically determined.

- The behavioral complexes of narcissism and perfectionism, as well as that of aggression, can be identified in primates other than man.

- The present model is but a limited step toward a unified theory of human behavior. The latter would require descriptions of the developmental aspects of an individual as he matures from the womb to old age.

CHAPTER 12

TOWARD SELF AND SANITY

The source of all evil is in the very substance of man. To extirpate this evil we must neutralize not only the instincts inherited from our animal ancestors, but the superstitions transmitted by our human ancestors, the excrescences of an uncontrolled mental activity, of misguided ambitions, and replace them with the sense of human dignity. This is not easy, for the ordinary man knows well, or guesses, that the flattering title of conscious Man is only acquired at the cost of restrictions in the activities from which he usually derives all his pleasures.

Lecomte du Noüy (1947)

Conventional theologians as such — those with committed views — are not the prophets of tomorrow. They belong with the priests who enjoy the full splendor of popular following and applause. [Rather] it will be those religious philosophers who are true to the philosopher's vision of seeing things broadly and together... who see sanely in man's religious spirit something that is part and parcel of his mental frame...

Vergilius Ferm (1950)

OUTLINE OF CHAPTER TWELVE

HUMAN FOLLY: POWER & GLORY

RELIGIOUS IMPLICATIONS

SCIENCE & INTUITION

ALIENATION OF THE INDIVIDUAL

> **From the human race**
>
> **From himself**

PREDESTINATION

CONCLUSION

Q. What is the challenge that lies before us?

A.   The challenge that lies before us is concisely stated in the preceding quotation from Lecomte du Noüy's book, *Human Destiny*. As we come to understand our instinctual drives rooted in the behavioral complexes of narcissism, perfectionism and aggression, will we be able to profit from this knowledge? Indeed, will we be able to control our often misguided ambitions, rooted in the narcissistic and aggressive drives, and "replace them with the sense of human dignity"?

Q. The above words have a nice ring to them, but the thoughts conveyed are rather vague. Is there a more immediate, definitive problem before us?

A.   The immediate problem before us is the control of man's misguided ambitions in our age of potential nuclear warfare. It is by no means an exaggeration to say that we are at the very brink of a nuclear Armageddon. And we continue to place our fate in the hands of leaders who have little appreciation for the events of past history, and little understanding of their own motivations.

Q. I assume that you are referring to man's innate aggressive-sadistic tendencies.

A.   As we mentioned earlier, much of human evil is rooted in the behavioral complex of aggression. But this is not the only determinant of evil: each character type has his own particular blindnesses and prejudices. For example, we need only to compare the expansive drives of various world powers during major wars throughout history. To be sure, one type of ambition for world domination has been based on the expression of the gene of the behavioral complex of aggression. It has been based on the iron fist, on a ruthless drive for *Power*.

A second type of drive for domination, which can be identified with nations of primarily N or NP types, has been based on the behavioral complex of narcissism. This drive, too, may be of seemingly boundless intensity. It may appear as a religious-like fervor in a self-appointed leader who can stare into the light of the rising sun and feel a wellspring of electric-like energy emanating from his soul, directing him to galvanize his people into action in order to fulfill his vision of a destiny for all mankind. His vision is such a powerful one that its underlying premises cannot be questioned. It is a vision of unity, a vision of a destiny in which every soul on the face of the earth is massed before him, bowing in recognition to the same rising sun from which he himself derives his electrodynamic energy. Thus, when his people do go into action, the drive that emerges is one toward the attainment of unlimited *Glory* rooted in narcissistic ambition and sometimes mediated by a perfectionist sense of duty to the fatherland.

In short, it would be a serious mistake to assume that all of human folly is rooted in the sadistic trends of the aggressive behavioral complex. One can even imagine a war being initiated by a submissive nation that longs only to be conquered and cared for in a state of subjugation. Although we do not know of any such examples having occurred, no doubt some historian will find one.

We believe that in this age of potential nuclear warfare not only the aggressive sadist, but also the expansive narcissist and the masochistic martyr, may be dangerous individuals in the leadership role. For us peons the challenge that lies before us is, at long last, to begin to choose world leaders who are least likely to become out of touch with themselves, least likely to become prisoners of their own misguided ambitions, and least likely to become the subjugated victims of misguided "powers behind the throne."

Q.   What, finally, are the religious implications of the model presented here?

A.   We were afraid that you were going to resurrect this topic.

It is probable that some theological historians are going to reexamine the accounts of the various past religious leaders from the point of view of their possible character types. And, sooner or later someone will hypothesize that Jesus Christ was an N individual, being the product of an N x NP union. We can see the battle lines forming already, between the Creationists and the Unbelievers. But to those who consider themselves firmly ensconced in one camp or the other, we have the following caution: if you believe that you know all that there is to know about the creation, about what really happened many millennia ago, about the purpose of our existence on this planet, and about the workings of our own minds, consider for a moment the possibility that you are, once again, wrong.

Permit us to invoke the wisdom of Aristotle. He states in his *Ethics* that,

> Wisdom therefore will be a union of intuitive
> reason and scientific knowledge...

Here we interrupt to note that while the translator has provided us the phrase "intuitive reason," in modern parlance we could just as well say "faith."

Continuing from Aristotle's *Ethics,* we read:

> The intuitive reason then is at once beginning and end.
> From the truths perceived by intuitive reason demonstration
> starts, and with them it is concerned. We should therefore

pay no less attention to the undemonstrated assertions and opinions of old and experienced or prudent persons than to demonstrations; for their experience gives their eyes the power of correct vision.

Lecomte du Noüy in his *Human Destiny* states essentially the same view:

In the two cases, that of the religious fanatic, and that of the unbeliever, we come across the same human weakness, pride, and the same error of judgment which consists in neglecting or denying one half of the psychical activity of man. The one denies intelligence, the other intuition. The idea that the human personality emerges from the harmonious combination of the two factors does not enter their minds.

Q.   You spoke earlier of "gaining one's free will," and "coming to terms with oneself." These concepts seem rather vague.

A.   And vague they shall remain unless *we* formulate for *ourselves* a plan for coming into equilibrium with our environments, our societies and our own particular life situations. Understand that our approach is only a starting point on the road to oneself and to sanity. We can now begin to answer that age-old question, "Who am I?" And once we have embarked on this journey we will notice that we see a path before us that is only dimly illuminated. But we know that the path leads somewhere, and that there is no turning back.

We suggest the two-pronged approach that we have outlined in the previous chapters. First, we must consider if we have, by refusing to acknowledge the limitations of the human mind, alienated ourselves from the human race. Second, we must ask ourselves if, in being ignorant of our basic character structures, we have been alienated from ourselves.

Q.   So all of our troubles will disappear once we have come to terms with ourselves. Right?

A.   Wrong.

First, understanding certain aspects of our behavior does not necessarily mean that we will be able to change them easily. As we noted earlier, very powerful forces are at work here. But if things are not working satisfactorily for us now, then little by little we may want to alter our life situations.

Second, just because we feel that *we* have come to terms with *ourselves,* this does not necessarily mean that others with whom we deal have done likewise.

Finally, there will always be the more mundane trials, tragedies and trivia of life to keep us from becoming too smug. When the refrigerator breaks down, we still must have it repaired. And when the baby spikes a fever to 105 degrees and the water pipes burst, this will continue to occur on the night preceding our planned departure on vacation.

Q.   Is there anything else that I should ponder about?

A.   Yes. Perhaps man's future on this planet, even down to the finest details of the activities of all individuals, has been predestined.

Q.   Surely you don't believe in predestination!

A.   The age-old concept of predestination is not as silly as it appears at first glance.

On the one hand, if we are to believe in an omnipotent Creator who has granted us a "free will," it is nevertheless difficult for us to believe that He, in His omniscience, does not already know how all this is going to turn out. And if *He* knows, then it is, in the final analysis, all predestined to occur and we are simply actors religiously playing our parts in something of a bizarre tragi-comedy.

If, on the other hand, we believe not in a Creator as such, but in the fidelity of natural law, then we are drawn to believe that the motion of every electron, every atom, every molecule, every chromosome, every particle and every planet is not a random one but one precisely determined by past history. And if we as Scientists believe that living beings are collections of molecules obeying non-mystical natural laws, then we are led to believe that all events, ever our present actions and thoughts, are determined by motions of molecules following their inexorable, predestined paths. Once again we might be led to the conclusion that man's destiny is foreordained — by forces that are awesome, powerful and of a complexity far beyond the limits of our comprehension.

Q.   So where does this lead us?

A.   Nowhere.

Q.   Nowhere?

A.   Even if we were to accept a concept of predestination, we are not going to sit back on our haunches and simply watch the world go by. In fact, the behavioral complexes of narcissism and aggression give us a guarantee that this will not happen.

And even if we were to accept the possibility of predestination, there is nothing to prevent us from climbing to the top of a mountain and shouting that it is not true. In short, there is nothing to prevent us from ignoring this possible ultimate constraint on our future.

Indeed, we have not come this far, ever so painfully, to try to recover our free will, only to sacrifice it at a nebulous altar of predestination. No, no and NO!

Q.   What if we as Scientists believe that our future is not predestined but is a matter of the complex interaction of events that can be expressed only in the probabilistic terms of quantum mechanics?

A.   Now *that* takes faith!

Q.   One last question.

A.   Yes?

Q.   What should we tell the children?

A.   As with anything else, we should tell them, in due course, what *we* believe to be the truth. Anything else, *they* will not believe.

Q.   Any final comment?

A.   Permit us to close with a quotation by Henrik Ibsen, from his speech to students at Oslo:

> "All I have written... I have mentally lived through...
> Partly I have written on that which only by glimpses, and at
> my best moments, I have felt stirring vividly within me as
> something great and beautiful. I have written on that
> which, so to speak, has stood higher than my daily self...
> But I have also written on the opposite, on that which to
> introspective contemplation appears as the dregs and
> sediments of one's own nature... Yes, gentlemen, nobody
> can poetically present that to which he has not to a certain
> degree and at least at times the model within himself..."

POSTSCRIPT

Testing The Model

Q.   I found your intuitive approach to be interesting — and somewhat amusing. However, let's say that I am an applied scientist. I deal with concrete concepts and hard numbers. How can you get me interested in your model?

A.   We would try to arouse your interest in HLA-associated disorders. We present below for your consideration our hypothetical "letter to the editor" of a major medical journal.

HLA ASSOCIATION WITH BEHAVIOR

To the Editor: More and more diseased states are being found to be associated with the histocompatibility (HLA) antigens, which are coded for by a group of genes located on the short arm of chromosome 6. These states include venerable systemic diseases such as juvenile-onset diabetes mellitus [1,2] and autoimmune thyroiditis [3], as well as disorders of behavior such as endogenous depression [4], schizophrenia [5,6], and perhaps even susceptibility to panic attacks [7]. However, it is becoming increasingly evident that the genetic susceptibility to most of these states does not lie in particular antigens coded for by the HLA genes. Rather, the more likely hypothesis being advanced [1-4] is that these conditions are predisposed by genes that lie in the close vicinity of the HLA loci, in linkage disequilibrium with them.

The tentative conclusion that "genes" in the region of the HLA loci may predispose both to systemic disease and to abnormal behavioral states merits our greatest attention. In fact, many of the HLA-associated systemic diseases had, over the years, been noted to be associated with patterns of behavior — even to the point of being classified by psychiatrists as "psychosomatic diseases." We need only mention juvenile-onset diabetes mellitus [8], Graves' disease [9], peptic ulcer [9], inflammatory bowel disease [9], rheumatoid arthritis [9,10], and systemic lupus erythematosus [11]. For some of the conditions (e.g., ulcerative colitis, Graves' disease, rheumatoid arthritis) rather distinct personality patterns have been reported [9,10]. Finally, with regard to rheumatoid arthritis, the striking finding has emerged that the conditions of rheumatoid arthritis and schizophrenia tend to be mutually exclusive [12,13].

What are we to make of these findings? If we keep in mind Occam's Razor — that the explanation most likely to be correct is the one that explains the greatest number of observations with the fewest suppositions — then we are led to the following hypothesis: there exists a major polymorphic gene on chromosome 6 in the neighborhood of the HLA loci that can simultaneously confer susceptibility to systemic disease and behavior disorder. The precise nature of the phenotypic expression of such a pleiotropic gene would be dependent on other genes and on environmental factors. Variable expression of such a single gene could explain: 1) the overlap in the clinical presentations of some of the HLA-associated diseases, 2) the association of a given systemic disease with certain "personality traits," and 3) the mutual exclusiveness between a given systemic disease and a certain abnormal mental condition. Finally, it is possible that the presumed gene in question could be a determinant of recognizable personality traits in the absence of overt systemic disease, i.e., at the limit of incomplete penetrance.

In short, we speculate that various investigators in somewhat different fields are all pointing to the *same gene* on chromosome 6. No doubt some readers will dismiss this as too simple an explanation. Nevertheless, it may be the correct one.

References

1.   Neel J.V. (1977): *loc. cit.*
2.   Cahill G.F., Jr. and McDevitt H.O. (1981): *loc. cit.*
3.   Strakosch C.R., Wenzel B.E., Row V.V. and Volpe R. (1982): Immunology of autoimmune thyroid diseases. *New England J. Medicine 307* 1499-1507.
4.   Weitkamp L.R. *et al.* (1981): *loc. cit.*
5.   Gattaz W.F., Beckmann H. (1980): HLA-antigens and schizophrenia. *Lancet* (I) 98-99.
6.   Mendlewicz J. and Linkowski P. (1980): HLA-antigens and schizophrenia. *Lancet* (I) 765.
7.   Surman O.S., Sheehan D.V., Fuller T.C. and Gallo J. (1983): Panic disorder in genotypic HLA identical sibling pairs. *Am. J. Psychiatry 140* 237-38.
8.   Dunbar F. (1943): *loc. cit.*
9.   Weiner H. (1977): *loc. cit.*
10.   Moos R.H. (1964): *loc. cit.*
11.   McClary A.R. *et al.* (1955): *loc. cit.*
12.   Nissen H.A. and Spencer K.A. (1936): The psychogenesis problem (endocrinal and metabolic) in chronic arthritis. *New England J. Medicine 214* 576-81.
13.   Gattaz W.F., Kasper S., Ewald R.W. and Beckmann H. (1980): *loc. cit.*

Q.   But other investigators have, over the years, advanced a number of various theories of personality. How is your model different?

A.   As opposed to previous theories of personality, our model is quantitative. It is a predictive model. It is falsifiable. It is testable.

Q.   How is it testable?

A.   By use of the scientific method. Hypotheses based on the model can be advanced, and theory compared to experiment. Here, "experiment" refers to data obtained in biophysical and biochemical protocols, in experimentation with animals, in family studies, and in studies of epidemiology.

Q.   If an aspect of the model does not hold, does the model become worthless?

A.   Not necessarily. As more and more objective data are acquired, the model may require modification. This is, in fact, the fate of most quantitative models in the sciences. For example, it may turn out that more than one gene is involved in the determination of the traits N, P or A.

Q.   What are the most important aspects of testing the model?

A.   They are:

- identification of the N, P and A traits in individuals, and

- identification of the presumed genes controlling the N, P and A traits.

Q.   How might this be done?

A.   Considering first the identification of the N, P and A traits in individuals, some possibilities are:

A behavioral test. Under the stress of time, the individual would be asked to run a gauntlet of decisions under simulated circumstances. The individual's decisions would be evaluated in relation to his possible character vector and the traits N, P and A. This could be done in the context of something in the manner of an "electronic game." The element of stress would be provided in order to make the decision-making process less susceptible to intellectualization.

A questionnaire. However, this would involve all the pitfalls of evaluating not only the individual's character vector and intrinsic behavioral traits, but also the effects of his environmental conditioning, his defense mechanisms and inversions, and his processes of intellectualization.

A clinical test. The trait N, P or A may reveal itself in an individual as a response to a test dose of a drug. For example, investigators who in the past have measured the reaction of flushing in response to alcohol ingestion in maturity-onset diabetics may have been measuring, in part, the presence or absence of the N trait.

A biochemical test. This may be long in coming because of the extreme complexity of the biochemical pathways involved in control of the central nervous system.

We consider next the identification of the presumed genes controlling the N, P and A traits. Once these traits can be objectively and unequivocally identified in individuals, the next step would be to conduct family studies to determine if Mendelian mechanisms of transmission of the traits are operative. The evaluation of known genetic markers is indicated.

Q.   Can you be more specific?

A.   As we implied earlier, an obvious first place to start would be to evaluate if the HLA group of genes is a genetic marker for the A trait. Indeed, it is no longer possible to ignore the striking association of certain diseases and behavioral disorders with the incapacity to express fully the trait of aggression. Among others, such conditions include:

- schizophrenia and endogenous depression

- autism and Asperger syndrome

- cancer and leukemia

- psychosomatic diseases

- collagen vascular diseases

- coronary artery disease

Some of these conditions have been found to be associated with the HLA complex of genes on chromosome 6 (collagen vascular diseases, some leukemias), but some as yet have not (coronary artery disease). Clearly, one of our first priorities would be to conduct family studies to see if the A– trait assorts with HLA haplotype. Hence, it would be of special interest to study families in which submissive types and one of the above diseases were present. This would allow one to test for the concurrent association of HLA haplotype, the A– trait, and the presence of the disease entity.

The validity of the above family studies would hinge, of course, on carefully planned procedural protocols so that tests for statistical significance could be conducted in an unbiased manner. In particular, determination of character type should be done in a blinded manner, without knowledge of an individual's HLA haplotype or the presence or absence of the disease state.

We note that if the results of studies as described above are positive, they would provide not only evidence for the presence of an **a** gene on chromosome 6, but in addition they would suggest that the methods used in assessing character type had some validity.

We emphasize the potential importance of these studies. We are not dealing here with only rare, obscure diseases, but rather with ones that represent major health problems in many countries. In addition, in the area of genetic counseling the results of such studies could allow the physician to identify those family members at risk, and those not at risk, for the condition in question.

Finally, we note in passing the total inadequacy of the "Type A & Type B" classification of personality type in coronary artery disease. The reader may have been surprised that we included coronary artery disease among those conditions associated with an incapacity to express fully the trait of aggression. In fact, in our estimation the most frequently encountered individual with the classic "Type A" personality of coronary artery disease is not at all a dominant XA type, but rather a non-compliant submissive (NPA– or NA–) type.

Q.   Are there any other significant predictions of the model that can be tested?

A.   Yes. A completely unexpected result of the model is the prediction of *infertility* in matings between certain character types. There is a need for studies of the occurrence of miscarriages, stillbirths, and infants who fail to thrive in relation to the character types of the parents.

Again, if in a study there is found a correlation between infertility and parental character type, then not only does the study provide some evidence for verification of the model, but in addition it implies that the methods utilized in assessment of character type may have had some validity.

Q.   What other studies are called for?

A.   A fresh look at the *schizophrenias* is needed. These disorders are a major mental health problem world wide. The proposed genetic basis for the schizophrenias was presented in Chapter 11, and we shall not repeat it here. Clearly, family studies are needed.

Q.   How about animal studies?

A.   Studies with other primates — for example in the orangutan and in the chimpanzee — may be fruitful. It is of interest that our model predicts that schizophrenic-like behavior should be not unknown in the orangutan and gorilla, but should be much less likely to occur in the chimpanzee society.

With regard to personality types in the chimpanzee, it is interesting that outgoing, authoritarian and submissive chimps have been described. A genetic basis for the submissive state could be looked for in "family" studies. However, in view of the chimpanzee's well-known promiscuity, it may be necessary to determine paternity by blood testing — as is sometimes done in our human society.

Q.   You are saying — are you not — that there is a need for a pluridisciplinary approach to the evaluation of these questions of human behavior.

A.   Yes. It is, as they say, an idea whose time has come.

APPENDIX A

The sadist's choice: A vulnerable victim

The following account, "Case Number 27," taken from Stekel's treatise *Sadism and Masochism,* is an extreme example, to be sure, of a PA individual's having gone far astray. But it is precisely in the study of the blatant manifestations of sadistic trends that we may obtain insight into the more subtle ones.

We note in the account the woman's conscious desire to achieve a *master-slave relationship.* In fact, she uses exactly these terms. We note her coldness and austerity, her carefully planned choice of victim. We note the ritualization of the affairs and the repetitive actions involved.

We note her desire to exploit her victim for her own sexual gratification, and her desire to punish in the context of violence. We note how she frustrates him and sends him away in triumphant vindictiveness. We note how she controls the emotions of her victim, while she herself has a total lack of feelings, an emotional barrenness for him. We note, though, her pitiable state of total confusion with regard to the demons that possess her. She wants to be freed of her perversity at any price. She wants to be a normal, healthy woman.

Finally, we note that inevitable "exception that proves the rule": she lives a "blameless life at the side of her husband."

> *Case Number 27.* Mrs. N. L., a beautiful slender woman, twenty-eight years of age, with sharply chiseled features, put herself into my care because of a difficulty which had brought her into the severest conflicts. She is the wife of a high aristocrat and lives a great part of the year upon an estate in the neighborhood of a small town. There she leads a blameless life at the side of her husband, and no one who visits her or knows her at home would suspect to what fearful passions she has yielded. She can stand it at home and with her husband, who idolizes her and grants her every wish, six months at the most, then she has to go to a sanatorium or to a large hotel on the Riviera or to Switzerland, to some house where a large company are gathered together in a free and easy manner. On the very first day of her arrival she makes a survey at table of all those present and…
>
> But I should like to have her speak for herself and of herself: "And I observe at once which of the gentlemen present will be my victim."

"You mean, will fall in love with you?"

"No, I do not mean that. Fall in love... that is a mere trifle. No... I am a sadist. I seek the victim whom I am going to whip..."

"And you always find him and at the first glance? And have you never been in a sanatorium where you did not find a victim?"

"I always find one. I usually find several victims. I have never yet been in a sanatorium where I have not secured a partner. But listen further: I exchange but one look with the gentlemen. I look at them all. My first glance is serious, cruel, and stern. Then I notice in many a man a sudden cringing and I know it at once! The man will be my slave..."

"How does the acquaintance then come about?"

"Oh, that soon happens. Sadists and masochists have a secret language. I might say a secret alliance with secret customs and secret agreement. We speak to each other after the meal, and the first rendezvous is usually arranged then. We say not a word of what we are going to do. I say to him harshly: 'Come this evening at nine or ten o'clock to my room!' I give him a masterful look and go my own way, with no further word to him. If he wants to talk, I shake him off like a dog that would make itself a nuisance."

"And he comes?"

"He is sure to come at the appointed hour. He certainly comes!"

"Then I have him disrobe completely, while I remain dressed. I permit him neither to approach me, nor do I allow any caresses. He has to obey; he undresses and throws himself at my feet. Now I strike him with the riding whip as hard as I can and as long as it is possible without being heard. He moans with pain or pleasure and writhes at my feet."

"Oh, I have seen at my feet proud, renowned men, shining lights in their profession, grateful and kissing my shoes with gratitude for the blows. I strike until the wales on their flesh are bleeding."

"Do you experience great pleasure in this?"

"Of course. Yet not so much as later when the man throws himself upon me and wants to possess me. Then I look at him scornfully, ridicule him. I notice how he squirms with passion, how desire robs him of the last vestige of masculine dignity. He begins to beg, to whimper, to twist and turn before me, pleading for the chief gratification. I remain cold, and this triumph affords me a voluptuous pleasure such as I can never experience in normal intercourse with my husband."

"And he goes away without having possessed you?"

"Always!"

"Have you never found any one who will not do so?"

"Oh, yes, many a one has first writhed in pain and then wanted to be strong and use force. Such a one succeeded beautifully. I was stronger than he and threatened if necessary to scream and cause a scandal. Many a one I have driven out of the door with my whip and forbidden him ever to come again. Believe me, these men then run after me like dogs. I could have asked anything of them; they were utterly in my power."

"And have you never succumbed to your own desire?"

"No; this desire — you mean I suppose for sexual union — no longer exists then for me. I have it with my husband and now and then also with other men. Yet what is this feeble pleasure which I feel in normal intercourse compared with the indescribable bliss of such a sadistic act? The orgasm reaches such a height that I could not feel a stronger one. My whole personality seems exalted; I seem to be proudly uplifted, at one with all my forces."

"And I have found that one has power only over the men to whom one does not completely surrender. All these men are my slaves even today and run after me. Because they have never possessed me, because they are always consumed in ardent desire for me, because they have learned to know my severity and longed afterward to enjoy also my tenderness."

"And what do you expect from me and my medical art?"

"I want to be delivered from my illness. I have heard that people can be changed through hypnosis. You will 'talk out' of me this crazy sadism. I am falling now into wretched embarrassment. In the sanatorium where I was last, a woman next door heard the cracking of the whip and seems to have observed something through the keyhole. I was ordered by the director to leave the house immediately. Think if that comes to the ears of my husband! My reputation and my position in society are endangered!"

"I want to be freed from my perversity — at any price whatsoever. I want to be a normal, healthy woman…"

APPENDIX B

Somerset Maugham, William James and Karen Horney

It is evident, from their writings, that both Maugham and Horney were fascinated by the dynamics of the morbid dependency. To verify this, the reader needs only to read Maugham's *Of Human Bondage* and Horney's *Neurosis and Human Growth.*

In reading Horney's work, one notes that she cites on several occasions William James' works, in particular Volume I of his *Principles of Psychology* published in 1890. However, none of the citations refer to the morbid dependency.

Furthermore, in Maugham's novel *The Razor's Edge* we come upon the following account:

> I looked at it and saw it was William James's *Principles of Psychology.* It is, of course, a standard work and important in the history of the science with which it deals; it is moreover exceedingly readable; but it is not the sort of book I should have expected to see in the hands of a very young man, an aviator, who had been dancing till five in the morning.
>
> I got up. When I left the library Larry was still absorbed in William James's book. I lunched by myself at the club and since it was quiet in the library went back there to smoke my cigar and idle an hour or two away, reading and writing letters. I was surprised to see Larry still immersed in his book. He looked as if he hadn't moved since I left him. He was still there when about four I went away. I was struck by his evident power of concentration. He had neither noticed me go nor come.
>
> I had various things to do during the afternoon and did not go back to the Blackstone till it was time to change for the dinner party I was going to. On my way I was seized with an impulse of curiosity. I dropped into the club once more and went into the library. There were quite a number of people there then, reading the papers and what not. Larry was still sitting in the same chair, intent on the same book. Odd!

This leads us to the question: why was Larry, and by extension Maugham himself, so intrigued by James' work? The answer, we believe, is to be found in the following remarkable account, taken from Volume II of *Principles of Psychology:*

> The passion of love may be called a monomania to which all of us are subject, however otherwise sane. It can coexist with contempt and even hatred for the 'object' which inspires it, and whilst it lasts the whole life of the man [*sic*] is altered by its presence. Alfieri thus describes the struggles of his unusually powerful inhibitive power with his abnormally excited impulses toward a certain lady:

> "Contemptible in my own eyes, I fell into such a state of melancholy as would, if long continued, inevitably have led to insanity or death. I continued to wear my disgraceful fetters till towards the end of January, 1775, when my rage, which had hitherto so often been restrained within bounds, broke forth with the greatest violence. On returning one evening from the opera (the most insipid and tiresome amusement in Italy), where I had passed several hours in the box of the woman who was by turns the object of my antipathy and my love, I took the firm determination of emancipating myself forever from her yoke...

> I was determined never to leave the house, which, as I have already said, was exactly opposite that of the lady; to gaze at her windows, to see her go in and out every day, to listen to the sound of her voice, though firmly resolved that no advances on her part, either direct or indirect, no tender remembrances, nor in short any other means which might be employed, should ever again tempt me to a revival of our friendship. I was determined to die or liberate myself from my disgraceful thraldom... Isolated in this manner in my own house, I prohibited all species of intercourse, and passed the first fifteen days in uttering the most frightful lamentations and groans. Some of my friends came to visit me, and appeared to commiserate my situation, perhaps because I did not myself complain; but my figure and whole appearance bespoke my sufferings. Wishing to read something I had recourse to the gazettes, whole pages of which I frequently ran over without understanding a single word... I passed more than two months till the end of March 1775, in a state bordering on frenzy...

The only good that occurred to me... was that of gradually detaching me from love, and of awakening my reason which had so long lain dormant. I no longer found it necessary to cause myself to be tied with cords to a chair, in order to prevent me from leaving my house and returning to that of my lady. This had been one of the expedients I devised to render myself wise by force. The cords were concealed under a large mantle in which I was enveloped, and only one hand remained at liberty.

Of all those who came to see me, not one suspected I was bound down in this manner. I remained in this situation for whole hours; Elias, who was my jailer, was alone intrusted with the secret. He always liberated me, as he had been enjoined, whenever the paroxysms of rage subsided.

Of all the whimsical methods which I employed, however, the most curious was that of appearing in masquerade at the theatre towards the end of the carnival. Habited as Apollo, I ventured to present myself with a lyre, on which I played as well as I was able and sang some bad verses of my own composing. Such effrontery was diametrically opposite to my natural character. The only excuse I can offer for such scenes was my inability to resist an imperious passion. I felt it was necessary to place an insuperable barrier between its object and me; and I saw that the strongest of all was the shame to which I should expose myself by renewing an attachment which I had so publicly turned into ridicule."

Now we know that Horney read the first volume of James' work, but did she read also the above narrative, which appears in Volume II? Although Horney did not cite the above account, it is probable that her perfectionistic nature would have induced her to peruse both volumes of James' work, especially a chapter alluringly entitled "Will." In addition, we have another clue in the account of Alfieri, in the use of the infrequently employed literary terms "thraldom" and "imperious." These two words would send most individuals to the dictionary, especially someone like Horney whose native language was not English. It seems that Horney did indeed consult the dictionary for the meanings of these two words, and in doing so, associated them together in her mind. This is indicated from one of her journal articles, "The problem of the monogamous ideal," where she is drawn to use the words "imperious" and "imperiousness" in no obvious relation to, but separated only a few paragraphs from, the word "thraldom" (used in a Freudian context).

We therefore conclude that, although both Maugham and Horney refrained from directly citing Alfieri's account as given in James' work, they both read the account and were transfixed by it. It is possible that this passage set both Maugham and Horney on their own separate paths of thought and, in so doing, perhaps in some small measure, affected their subsequent lives.

APPENDIX C

Is the trait N dominant or recessive?

When the frequency of a recessive gene is high, say greater than 0.9, then almost all individuals of the population either exhibit the trait or else carry the gene in the heterozygous state. In such circumstances it may be difficult to differentiate between dominant and recessive inheritance.

Referring to trait N, in the case of *dominant* inheritance one would expect to find progeny of the type NPA × NA → PA, but not of the type PA × PA → NPA. We have not been able to examine enough of such possible cases (which in our habitancy would occur infrequently) to decide whether N could possibly be a dominant trait.

On the assumption of N being *dominant,* we present Table 7 (overleaf), which shows possible phenotypes of the progeny according to the phenotypes of the parents. The table is somewhat different from that presented in Chapter 10 (Table 2). A notable consequence of dominant inheritance of trait N would be that PA types would no longer follow quasi-dominant inheritance (i.e., with N being recessive a PA type must have at least one parent who is either a PA or A type). Obviously, the mode of inheritance of the character traits N, P and A must be clearly delineated before this information can be of use to the genetic counselor or psychiatrist.

Finally, if N were transmitted according to dominant inheritance, one might expect that the calculated frequency distributions of character types for the various habitancies would be different from those given in Chapter 10. However, as it turns out, this is not so. In fact, the numbers of individuals of each character type given in Table 5 would be exactly the same had we made the assumption that trait N is dominant rather than recessive. The gene frequencies of the presumed dominant gene **N'** would, however, be lower. In fact, the gene frequency n' that would give the same distribution of character types as the recessive gene **n** is given by the relation:

$$n = \text{square rt } [n'\,(2 - n')]$$

TABLE 7

**Possible Phenotypes of Children According to Phenotypes of Parents
if Mode of Inheritance of Trait N is Dominant (P Dominant, A Recessive)**

MOTHER \ FATHER	N	A	NP	NA	PA	NPA
N	N — — NA — — 0 A					
A	N — — NA — — 0 A	— — — — — — — A				
NP	N NP P NA NPA PA 0 A	N NP P NA NPA PA 0 A	N NP P NA NPA PA 0 A			
NA	N — — NA — — 0 A	— — — NA — — — A	N NP P NA NPA PA 0 A	— — — NA — — — A		
PA	N NP P NA NPA PA 0 A	— — — — — PA — A	N NP P NA NPA PA 0 A	— — — NA NPA PA — A	— — — — — PA — A	
NPA	N NP P NA NPA PA 0 A	— — — NA NPA PA — A	N NP P NA NPA PA 0 A	— — — NA NPA PA — A	— — — NA NPA PA — A	— — — NA NPA PA — A

GLOSSARY

abject state See *exogenous depression.*

acetylcholine A biochemical that serves to relay nerve signals in both the voluntary and autonomic nervous systems.

acupuncture The originally Chinese medical practice of puncturing parts of the body with needles in order to relieve pain or cure disease. The efficacy of anesthesia by acupuncture is sometimes said to be dependent on the personality of the patient.

acute Occurring over a short period of time. Does not refer to the severity of a condition. See *chronic.*

adornment One of the behavioral characteristics of the unbridled narcissistic trait. Other characteristics of this trait include *expansiveness, exhibitionism,* the desire for *recognition* in a sexual context, and the desire to stand in *honor,* with one's arms extended, above admiring others. There is an instinctual urge for the individual to adorn himself in finery, whether he be a glittering opera singer at a public reception, or an Indian chief adorned in his feathered headdress standing before his tribe. According to this view, much of *art* is an extension of personal adornment, hence is an instinctual urge rooted in the behavioral complex of narcissism.

adrenal medulla The inner portion of the adrenal gland, situated above the kidney, which secretes catecholamines into the blood stream on stimulation by the sympathetic nervous system.

adrenaline A biochemical that is released into the blood stream from the adrenal medulla during a mass discharge of the sympathetic nervous system. Also known as epinephrine.

adrenergic Tending to activate the sympathetic portion of the autonomic nervous system.

affective disorder A disorder of the "emotional tone" of the individual, most commonly in the spectrum of elation to depression.

aggression, behavioral complex of The genetically determined set of behavioral attributes that form the basis of man's desire to survive by maintaining a position of power over his environment.

aggressive-vindictive rage   A mass discharge of the sympathetic nervous system. According to the present model, this can only occur if the trait A is expressed.

alcoholism   Addiction to the drug, alcohol. Since predisposition to such addiction appears to be in part inherited, alcoholism should be regarded as a complex disease.

allele   An alternative form of a gene at a given locus. Genetic variability, or *phenotypic polymorphism,* may be based on the occurrence of multiple alleles at a single locus, or on several genes acting together.

ambition   According to the present model, a functioning individual must have either the trait of *narcissism,* or of *aggression,* or both. If neither the trait N nor A is expressed because of genetic or environmental factors, then the individual is left "ambitionless" and would be incapable of functioning in competitive society. See *schizophrenia.*

amino acids   Organic acids that act as the building blocks of proteins.

anaclitic depression   An acute and striking deterioration in an infant's physical, social and intellectual development during the first year of life, following prolonged separation from the mothering person.

anaphylactic reaction   A generalized allergic reaction that can lead to shock (low blood pressure) and death.

anorexia nervosa   A syndrome marked by aversion to eating, associated with marked weight loss as well as with a wide range of metabolic imbalances. Most frequently found in young females.

antibody   A molecule, produced by the immune system that binds and inactivates a foreign substance (antigen).

antigen   A substance, often a protein, that may stimulate an immune response.

anxiety   Apprehension or tension, the sources of which may be largely unknown or unrecognized.

association   The occurrence together, in a family or population, of two characteristics in a frequency greater than that predicted by chance. The association may or may not reside in the expression of linked genes.

association areas   Areas of the brain which receive information from different sensory inputs and from other association areas. It is thought that information from many different sources "interacts" in these areas.

assortative mating   Non-random or preferential mating among individuals of a population.

asthenic personality   The classic asthenic (or "neurasthenic neurotic") personality is one characterized by weakness, fatigability, poor concentration, lack of ambition, feelings of inadequacy, low sexual drive, and exaggerated attention to bodily organs and functions. It has sometimes been related to *epilepsy.*

autism   Developmental disorders characterized by restricted and repetitive behavior that impair social interaction and communication.

autoimmune disease   A disorder of the immune response system in which substances intrinsic to the body appear to be "mistakenly" perceived as being foreign antigens.

automatisms   Involuntary, mechanical repetitious acts.

autonomic nervous system   The portion of the central nervous system governing many activities that are usually not under conscious control. It is composed of two parts: the *sympathetic* and *parasympathetic* nervous systems.

autosomal   Pertaining to a non-sex chromosome. Autosomes occur singly in gametes and in pairs in somatic (body) cells.

baboon   Large carnivorous Old World monkey, inhabiting mainly Africa and Asia.

basic character vector   One of the three basic approaches to social interaction: *dominance, submission* or *resignation.*

behavioral complex   A genetically determined set of behavioral attributes determined by a pleiotropic gene. The present model identifies three behavioral complexes: those of *narcissism, perfectionism* and *aggression.*

behavioral syndrome   A set of behavioral attributes determined by the synergistic or antagonistic actions of several behavioral complexes, as well as by modifier genes and environmental effects.

bipolar affect disorder   Manic-depressive disease in which mood swings toward either mania or depression may occur in a given individual.

Bleuler, Eugen (1857–1939)   Swiss psychiatrist whose studies of *dementia praecox* led him to propose the term *schizophrenia,* which is presently in common use.

blushing   A little-studied response of vasodilation in an emotional context, in the skin of the face, neck and upper chest. The response is mediated by the autonomic nervous system, and aspects of active vasodilation, as well as the release of vasoconstrictor tone, may be involved. According to the present model, individuals having the N trait may have an increased predisposition to *blushing, flushing,* and some of the so-called *psychosomatic skin disorders* such as *rosacea.*

Brel, Jacques (1929–1978)   Belgian poet and chansonnier.

bridled ambition   In the NPA model character types having the trait of perfectionism, P, are said to have bridled ambition. Character types lacking the trait P (e.g., N, A, NA, NA– and NA =) are said to have unbridled ambition.

carrier   An individual who carries a gene that is not expressed. For example, according to one aspect of the model, individuals of the N and NP types can be carriers of an **a-** gene for non-compliant submission.

catalepsy   Assuming a rigid, trance-like position, with an inhibition of motor (muscular) responsiveness.

catatonia   A category of schizophrenia characterized by either excessive or inhibited muscular activity. In generalized inhibition, one finds negativism, mutism, stupor, or "waxy flexibility" (immobility of the body after having been placed in a given position).

catecholamines   A family of biochemical agents that act as transmitters of nerve signals and as circulating mediators of the sympathetic nervous system. Examples are adrenaline and noradrenaline (or epinephrine and norepinephrine).

Cetacea   The order of whales and porpoises, which have a common ancestor with the ungulates.

chimpanzee   Omnivorous African anthropoid ape.

chromosomes   The cell structures containing the genetic material DNA. The human genome is comprised of 46 chromosomes: 22 pairs of autosomes and 2 sex chromosomes.

chronic   Occurring over a long period of time. Does not refer to the severity of a condition. See *acute.*

cognition   The acts of thinking, feeling, knowing, reasoning and learning, including both awareness and judgment.

color blindness   The inability of an individual to perceive normally the light of certain frequencies (colors). Most genes controlling the ability to perceive colors are located on the X chromosome, hence a defect in one of these genes may be the basis of a form of X-linked color blindness.

compensation neurosis   In psychiatric terms, a mental state in which there is a desire to profit from an adversity in life (secondary gain), e.g., in the context of compensation for an accident or malpractice.

complementary genes   Genes that produce different phenotypic effects depending on whether they are present separately or together.

compliance (*C*)   The state of the basic character vector of profound submission (XA=) in which the individual is not motivated to undergo personality splits to energetic states.

compulsion   An insistent urge to act in a certain manner that is not amenable to logic or reason.

concentration camp syndrome   A set of syndromes resulting from the chronic stress of incarceration, involving both anxiety and depression.

concordance   Identity of matched pairs of individuals for a given trait.

congenital   Present from birth. May or may not be *hereditary* or *familial.*

connective tissue disorder   Descriptive term for a group of diseases whose basic etiology appears to lie in disorders of the immune process.

consanguinity   Referring to individuals related by descent from a common ancestor.

coronary artery disease   Atherosclerotic disease of the coronary arteries of the heart predisposes an individual to myocardial infarction (heart attack). Risk factors include diabetes, hypertension (high blood pressure) and smoking. It has been suggested that the existence of a "Type A" personality (highly stressful, competitive, hostile) is also a separate risk factor.

crossing over   The physical process by which homologous chromosomes interchange DNA during the process of meiosis. See *recombination*. In "intragenic crossing over" there occurs crossing over at points within the gene in question. This could lead to the formation of one "intact" gene from two different mutant ones. If the allele codes for a dominant trait, then the trait could appear abruptly in the next generation, as if by mutation.

cutaneous   Pertaining to the skin.

cyclothymic   Characteristic of alternating periods of elation and depression.

Darwin, Charles (1809–1892)   English biologist and evolutionist. Published *Origin of Species by Natural Selection* in 1859.

delirium   A marked state of confusion characterized by excitement, disorientation and incoherence. Not to be confused with *dementia*. Examples are *delirium tremens* (associated with alcohol withdrawal) and *delirium mortis* (a confusional state of the dying individual). According to the model, character types having only one component of ambition (either N or A) will be more prone to states of delirium.

delusion   A false belief.

dementia   Loss of intellectual function.

dementia praecox   Obsolescent term for schizophrenia, introduced by Morel in 1860 and popularized by Kraepelin.

detachment   In the basic character vector of *resignation* the individual detaches himself from close interactions with others and assumes a lifestyle of quiet independence.

deterministic   Referring to a mathematical model, one in which the mechanisms of causality are specified in part or in whole.

diabetes mellitus   A set of complex diseases characterized by defects in glucose metabolism.

- In *Type I* (juvenile-onset, insulin-dependent) there may be a predisposition to the disease due to a multiallelic gene in linkage disequilibrium with genes of the HLA complex on chromosome 6.

- In *Type II* (adult-onset, non-insulin-dependent) there may be a predisposition to the disease due to an autosomal dominant gene not linked to the HLA complex. According to some reports, this type of diabetes may be associated with facial flushing after the ingestion of alcohol.

diaphoresis Sweating, in certain skin areas, by way of glands mediated by the sympathetic nervous system.

diploid The paired state of chromosomes in somatic cells. The diploid number is 46 in man.

discordance Twins are said to be discordant for a trait if one has the trait and the other does not.

DNA Deoxyribonucleic acid, the primary constituent of the chromosomes. DNA consists of building blocks called nucleotides, the sequences of which code for the production of RNA, as well as for various enzymes and other polypeptides.

dominance The basic character vector whose traits of *narcissism* and/or *aggression* are fully expressed.

dominant trait Referring to a trait that is expressed whether the gene in question is present in either the homozygous or heterozygous state. The phenotypes corresponding to the homozygous and heterozygous states may or may not be identical. See *recessive*. When *multiple alleles* are possible at a given locus, or if a given allele acts as a *pleiotropic gene,* then the concept of dominance and recessiveness is less straightforward. For example, in the present model, if **a-** is an allele for the trait of non-compliance (A–), then it is possible for **a-** to be expressed in the dominant mode with respect to **a** (aggression) but in the recessive mode with respect to a_0 (non-aggression). Furthermore, if the same allele **a-** is a pleiotropic gene determining also the somatic signs of a psycho-somatic disease *(D),* then it is possible for **a-** to be expressed in the recessive mode with respect to both a_0 and **a** in the determination of *D*.

Down's syndrome The condition formerly known as "mongolism" caused by a failure of the number 21 chromosomal pair to segregate normally. This ultimately leads to "trisomy 21," or an extra number 21 chromosome in the zygote.

echolalia The stereotypic echo-like mimicry of the speech of another person.

echopraxia Automatic and repetitive imitation of another person's movements or gestures.

ectomorphic   Having a slender (leptosomatic) somatotype.

embarrassment   An acute subdued state that ensues when an individual's state of vulnerability is revealed before others. According to the model, the character types most prone to embarrassment, hence to *blushing,* are those which have full expression of the N trait but which lack full expression the A trait. Most vulnerable would be the N, NP and submissive types. Least likely to blush would be the A and PA types.

empirical   Based on observation or experience, sometimes without due regard for theory.

endogenous   Without external cause. Assumed to have a cause inherent in the genetic constitution of the individual. The degree to which a so-called *primary* or *endogenous* disease is expressed may, however, depend strongly on environmental factors.

endogenous depression   Depressive disease having a distinct genetic basis, as opposed to *exogenous depression.*

endogenous euphoria   A mental disease, characterized by pathologic states of elation, having a distinct genetic basis.

endomorphic   Having a heavy, rounded (pyknic) somatotype, i.e., with a tendency toward obesity.

energetic state   An individual having a basic character vector of submission or resignation can, on occasion, assume a transient energetic state resembling dominance by undergoing a *personality split.*

environment   The external conditions and influences affecting the development and life of an organism.

enzyme   A molecule, ordinarily a protein, which affects biochemical processes by enhancing or suppressing a chemical reaction.

epidemiology   The study of the geographical distribution and statistical aspects of a pathologic condition.

epilepsy   Common name for *seizure disorder.*

epistasis   The condition in which a gene at one locus suppresses the expression of a gene at another locus.

erythrophohia   Fear of blushing. According to the model, anxiety with respect to blushing would be most often encountered in N, NP and submissive individuals (e.g., Queen Anne of England). It is sometimes mentioned in the context of an acute paranoid reaction.

estrous   Relating to an animal in the state of heat.

etiology   Pertaining to the causation of a pathologic condition.

eugenics   The "science" of the improvement of the human race by selection of individuals having putative advantageous qualities.

evolution   The theory promulgated by the careful investigations of Charles Darwin and others, by which plants and animals have their origins in other preexisting types. The development of different species is due to modifications having occurred over many generations by mutations, natural selection and genetic drift.

exhibitionism   Tendency   toward   display   or   extravagant   behavior. Exhibitionism is a primary manifestation of the trait of *narcissism.* Nevertheless, an individual not having the trait of narcissism can, by a mechanism of inversion, develop intense qualities of exhibitionism, and in particular, of self-adornment (pseudonarcissism).

exhilarated state   See *exogenous euphoria.*

exogenous   Having an external, or environmental cause. However, an individual may have a strong genetic predisposition to a so-called secondary or exogenous disease.

exogenous depression   Mental depression due to an adverse alteration in the individual's life situation. In the present model, a reactive depression, or *abject state,* may occur when the individual's life situation is incompatible with the needs of his basic character structure.

exogenous euphoria   Elation due to favorable alterations in the individual's life situation. In the present model, an *exhilarated state* may occur when the individual's life situation is wonderfully compatible with the needs of his basic character structure.

explosive personality   A state characterized by volcanic outbursts of rage, or of verbal or physical aggressiveness.

expressivity   The degree to which a genetic trait is observed in the phenotype. Variable expressivity may be caused by modifier genes or by environmental effects (including prenatal ones). At the limit of variable expressivity, a trait will not be detectable at all in some individuals and would thus be *incompletely penetrant.*

extrapunitive   Pertaining to a character type whose aggression is directed toward others. According to the model, this applies best to types having the A trait. *Intrapunitive,* thus, would pertain to character types in which the A trait is not expressed, as if the individual's aggressive tendencies were directed inwardly.

extrovert   An individual whose attention and interests are directed primarily toward others. See *introvert.*

failure to thrive   In pediatric terminology, referring to an infant who for unexplainable reasons does not develop normally and eventually succumbs. According to the model, certain parental combinations of character types will be prone to miscarriages, stillbirths, and infants who fail to thrive. This principle may explain, in part, the *relative infertility* of many past monarchs, who have often desperately attempted to produce legitimate heirs to the throne.

familial   Tending to recur in a family pedigree. May or may not be hereditary.

"flight or fight" reaction   Behavioral response associated with mass discharge of the sympathetic nervous system, as described by the American physiologist W.B. Cannon.

flushing   A response of vasodilation of the skin of the face, neck and upper chest. See *blushing.*

folie à deux   A condition in which two persons, usually closely related, share similar paranoid delusions.

Freud, Sigmund (1856–1939)   Austrian neurologist and psychiatrist; founder of psychoanalysis.

frontal lobes   The more rostral (forward) areas of the frontal lobes of the brain are thought to be *association areas* involved in the elaboration of thought. Destruction or removal of these prefrontal areas causes marked or subtle defects in personality.

"game"   Individuals having the aggressive (A) trait constantly *"play the game"* of dominance and submission. In so doing they may frequently undergo stressful *personality splits*.

gamete   A germ cell, containing the haploid number of chromosomes: i.e., an *ovum* in the female, or a *spermatozoon* in the male.

gene   A fundamental unit of heredity, composed mainly of DNA. Genes are arranged in linear order on the chromosomes and determine the genotype of the individual.

gene frequency   The numerical probability of an allele's existing at a given chromosomal locus in an individual of a given population.

genetic marker   A trait that can be traced in a family pedigree.

genotype   The genetic constitution of the individual with respect to a gene locus or loci.

geographic trail   In population genetics, a set of connected geographic points corresponding to the location of the birth of parents in successive preceding generations where a particular *dominant trait* is passed from parent to child. In NPA theory a non-sanguine individual (lacking the trait N) must have a non-sanguine parent, so non-sanguine types have ancestral geographic trails.

In NPA theory the three dominant traits relevant to geographic trails are: 1) absence of narcissism, 2) absence of aggression, and 3) perfectionism.

gingival smile   A smile revealing the gums of the upper teeth. Seen especially in N and NP individuals. Not usually seen in individuals lacking the N trait.

globulins   Proteins found in blood. Alpha, beta and gamma globulins are distinguished. Antibodies are gamma globulins.

gout   A spectrum of disorders characterized by an increased secretion of uric acid in the urine. "The king of diseases and the disease of kings." The caricatured patient afflicted with endogenous gout is an endomorphic male having a flushed face, suggesting an enhanced association with the N trait. Endogenous gout is reminiscent of non-insulin-dependent (Type II) diabetes in having a polygenic inheritance, and, apparently, no close association with the histo-compatibility antigens.

habitancy   In NPA population genetics, the inhabitants of a region, taken collectively. A subpopulation. For ease of communication we define the following habitancies:

Polymorphic — a mixture of all NPA character types (mixed)
Sublime — mainly N types (sanguine)
Punctilious — mainly NP types (sanguine)
Corybantic — mainly NA types (sanguine)
Demonstrative — mainly NPA types (sanguine)
Authoritarian — mainly PA types (non-sanguine)
Militant — mainly A types (non-sanguine)
Introspective — mainly NPA– types (sanguine)
Acquiescent — mainly NA– types (sanguine)

The Authoritarian and Militant habitancies have mainly non-sanguine types; the other habitancies have a preponderance of sanguine types. Examples — *Polymorphic*: USA; *Sublime*: Polynesian Islands, Eastern Africa; *Punctilious*: Switzerland, areas of China, Korea, indigenous Yucatan; *Corybantic*: Brazil, indigenous New Guinea, Bougainville; *Demonstrative*: Southern Italy, Colombia; *Authoritarian*: Western Russia, Serbia; *Militant*: areas of Mideast; *Introspective/ Acquiescent*: New Zealand, Nova Scotia.

hallucination   A perception that has no physical basis.

haploid   The single (non-paired) state of chromosomes in gametes (ova and spermatozoa). The haploid number in man is 23.

haplotype   A set of alleles of the HLA (histocompatibility) gene complex on chromosome 6. Two haplotypes, one from each parent, constitute an individual's histocompatibility genotype.

haptoglobin   A blood plasma protein whose structure is coded for, in part, by a polymorphic gene on chromosome 16. Some haptoglobin types are reported to be associated with subgroups of depressive disease. Not to be confused with the *haplotypes* of the *histocompatibility loci* on chromosome 6.

Hardy-Weinberg principle   The principle states that if two different alleles occur in a randomly mating population, the expected proportions of the different genotypes will remain constant from one generation to the next. In the present model infertile parental pairs would mate, producing non-viable P and null types. Hence, in order for the proportions of the genotypes of the population to remain constant, the occurrence of assortative mating or of reproductive advantages in certain NPA types would have to be invoked.

hebephrenia   A category of schizophrenia characterized by shallow and inappropriate emotional tone, smiling and giggling, silly and regressive behavior, and frequent hypochondriacal complaints.

heterozygous   Having non-identical alleles at a locus of a homologous pair of chromosomes.

Hippocrates (460-377 BCE)   Greek physician, called "Father of Medicine." Some ascribe to Hippocrates the theory of the existence of four basic human temperaments, corresponding to four body humors:

choleric (yellow bile)
melancholic (black bile)
sanguine (blood)
phlegmatic (phlegm)

Interpreted in the light of the present model, the ruddy-complexioned *sanguineous* is an N type, while the sallow *cholericus* is an A type. Thus, *phlegmaticus* is a sanguine perfectionist (NP type), while the austere *melancholicus* is a perfectionist choleric (PA type).

histocompatibility   The major histocompatibility complex is the group of HLA genes on chromosome 6 controlling the synthesis of proteins (antigens) that have fundamental roles in the immune process.

history   It repeats itself, *ad nauseum.* The reason why it does so is rooted in the fact that the cast of characters, with little variation, is always the same.

HLA loci   The genes of the major histocompatibility complex, located on the short arm of chromosome 6. The loci are highly polymorphic (multiallelic).

HLA association   There are several possible reasons why a genetic trait may show association with particular antigens of the HLA complex on chromosome 6:

1) *Non-homogeneous sampling,* i.e., the subpopulation sampled may not be representative of the entire population or may even be an isolate.

2) *Causality,* i.e., there may be a biochemical mechanism underlying the association between the trait and particular HLA antigens. The mechanism could be a direct one, or one in which the genes act indirectly as modifier genes, for example by a process of epistasis.

3) *Linkage disequilibrium,* i.e., a gene predisposing to the trait may also be on chromosome 6, closely linked to the HLA complex.

homologous chromosomes   A matched pair of chromosomes of the diploid state. One of the chromosomes derives from the male parent and the other from the female. Homologous chromosomes *segregate* during the process of meiosis. *Crossing over* may have occurred before segregation.

homozygous   Having identical alleles at a locus of a homologous pair of chromosomes.

Horney, Karen (1885–1952)   German-American psychiatrist of Dutch and Norwegian heritage who dared to counter the tenets of Freudian psychoanalysis. Her "neo-Freudian" views were largely ignored by others in her field.

hybrid   An individual whose parents belong to two different varieties of a species, or to two different species.

hypochondria   The persistent preoccupation with the body and with presumed diseases of various organs. It may appear in a poorly adjusted individual of any character type, in particular in a melancholic NP type or a hysterical NA type.

hypomanic   In psychiatric terms, an individual who has a heightened emotional tone. He is said to be active, socially aggressive, talkative, flippant, distractible, intolerant of criticism, and may be openly erotic in his speech or behavior. In *mania* the individual is psychotic and usually requires hospitalization.

hypothalamus   The portion of the limbic system of the brain that acts as the principal locus of integration of the autonomic nervous system. It is involved in the regulation of body temperature, blood pressure, emotions, sleep, sexual reflexes and complex processes of metabolism.

hysterical   Tending to bizarre, volatile behavior, sometimes accompanied by delusions, hallucinations or hypochondria

iatrogenic   Induced inadvertently by the physician during the course of treatment.

ICU psychosis   Intensive care unit psychosis. An acute syndrome of psychosis associated with sensory and sleep deprivation during the confinement of a patient in an intensive care unit. May be associated with *toxic psychosis*. According to the present model, individuals having only one component of ambition (N or A) would be more susceptible to acute psychotic episodes.

idiot savant   A person of low general intelligence who can perform mental tasks, especially ones involving numbers, which are completely beyond the capability of most persons.

illusion   A false perception or misinterpretation of a real sensory experience.

immune system   The body's system of mobilization against agents perceived as foreign to it, involving the production of both molecular antibodies and specific cells.

imprinting   The association of a behavioral response with a specific stimulus pattern in early maturation.

inborn error of metabolism   Concept developed by the English pediatrician A.E. Garrod by which a disease may be caused by a *genetic mutation*. Thus, a mutation may produce a "defective" enzyme and lead to an alteration in the normal biochemical pathways on which human metabolic processes depend.

independent assortment   The random separation of chromosomes during meiosis in the reduction of the diploid number to the haploid number.

infertility   The inability of a mated couple to produce viable offspring. Thus, the female will be prone to natural abortions (miscarriages) and may deliver stillborn offspring. According to the model, the parental combinations (of dominant types) that will lead to *relative* or *complete infertility* are:

- N × A
- N × PA
- NP × A
- NP × PA

In addition, parental combinations conducive to producing, on an endogenous basis, the phenotypes A-, PA-, N- and N-P in the offspring may also be relatively infertile. This would follow from a hypothesis that the latter phenotypes are associated with a reduced viability during fetal development and maturation.

instinct   An inherited disposition to a primitive behavioral pattern

intelligence   The faculty of understanding and reasoning, especially with respect to coping with new situations.

intra-psychic conflict   A conflict in which the demands of the individual's life situation cannot all be satisfied with respect to the demands of his character structure.

intrapunitive   See *extrapunitive.*

introvert   An individual whose interests are predominantly concerned with his own mental life.

inversion   A mode of behavior in which an individual consciously or unconsciously "bends over backwards" to do the opposite of what is demanded of him by his basic character structure. Under stress the inversion may dissipate, the individual showing his true colors. In historical accounts individuals are sometimes described by terms or activities that seem to be "out of character." The explanation for such behavior may lie in inversions, but other possibilities must be considered, namely:

- the individual's activities may have been influenced by dominant individuals in his entourage, especially by "powers behind the throne,"

- his activities may have been constrained by the cultural mores of the time,

- he may have become demented or frankly schizophrenic, or

- the historical record may have been inaccurate.

involutional melancholia   Older term for a unipolar depression occurring in the middle life period characterized by anxiety, guilt, hypochondria and insomnia.

isolate   A genetically separate subpopulation.

James, William (1842–1910)   American psychologist and philosopher.

jealousy   An emotion felt by an individual when the needs of his character structure are threatened by another person. In a *dominant* individual this appears as intolerance to rivalry. In a *submissive* or *subjugated* individual it appears as an intolerance to infidelity.

Kanner's syndrome   The syndrome of *infantile autism,* described by the Austrian-American psychiatrist Leo Kanner.

Kraepelin, Emil (1856-1926)   German psychiatrist who developed an extensive system of classification of psychoses, especially *dementia praecox* (schizophrenia).

la belle indifférence   The display of a lack of concern that seems to be inappropriate to the real life circumstances.

lacrimation   Secretion of tears into the eyes by glands controlled by the parasympathetic nervous system.

laughter   A complex behavioral response that may occur when circumstances are perceived in the context of real or vicarious amusement, i.e., in harmony with the individual's character structure. The initial and final stages of laughter may mimic the narcissistic smile of recognition.

Lecomte du Noüy, Pierre (1883-1947)   French biophysicist and philosopher.

lethal gene   A gene that renders non-viable an organism or cell possessing it. According to the present model, the genes corresponding to the *absence* of traits N and A, when present together, act *as complementary genes* to produce a lethal effect (a nonviable zygote).

leukemia   A debilitating disease in which there occurs a progressive proliferation of abnormal white blood cells (leukocytes). According to popular lore, "Only nice people get leukemia." The predisposition to leukemia may be partially rooted in NPA character type.

limbic system   A group of cortical (external) and subcortical (internal) structures of the brain, centered around the hypothalamus, which are intimately involved in the expression of emotions.

limelight   The reward of fulfilled narcissistic ambition is, literally or figuratively, honor and applause in the spotlight of recognition.

linkage   The location of genes on the same chromosome. In *close linkage* the genes are close to one another on the same chromosome.

linkage disequilibrium   The tendency in a population for certain alleles at closely linked loci to occur together more often than expected by chance.

lupus erythematosus   A systemic disorder of the immune response system associated with the HLA gene complex. The disease is known for its characteristic facial rash and its associated psychotic depressions. In the *lupus syndrome* due to a drug reaction (e.g., hydralazine or procaineamide) there is also, interestingly enough, an association with facial rashes and with psychotic depressions or personality changes.

malar flush   A reddening of the cheeks, as sometimes seen in various diseased states, such as *post-rheumatic mitral stenosis* and some of the autoimmune disorders. Most commonly seen in N, NP and submissive individuals.

mania   A mood disorder characterized by excessive elation, hyperactivity, agitation, accelerated thinking and speaking, and sometimes hypersexuality. Such a psychotic state often requires hospitalization.

manic-depressive disease   A major disorder of the emotional tone of the individual. It is characterized by severe mood swings toward mania, depression, or both.

Marfan's syndrome   A hereditary condition, transmitted by a pleiotropic autosomal dominant gene, causing vascular defects and changes in skeletal and soft tissues. The individual's body habitus is often ectomorphic and the personality type unaggressive.

masochism   Satisfaction derived from being dominated or abused by others.

Maugham, W. Somerset (1874–1965)   English novelist, playwright and world traveler.

meiosis   The process by which gametes are formed. Half of the chromosomes (the haploid number) normally segregate, with independent assortment, to form the chromosomal complement of an ovum or spermatozoon.

Mendel, Gregor (1822–1884)   Austrian priest and botanist who derived quantitative laws of inheritance based on the binary nature of inherited characteristics.

mesomorphic   Of a muscular somatotype.

metabolic disorder   An imbalance in the normal biochemical functions on which life processes depend. Its cause may be primarily *endogenous* (of genetic origin) or *exogenous* (of environmental origin).

MMPI, Minnesota Multiphasic Personality Inventory   An empirically designed, self-administered personality questionnaire intended to facilitate psychopathologic diagnosis.

modifier genes   Genes which modify an observed physical or behavioral trait.

monogamous ideal   The idea that we should all fall in love, marry and have children with a single mate. And if we do not succeed, we are failures.

morbid dependency   A symbiotic relationship, essentially sadomasochistic in nature, between individuals assuming dominant and submissive roles.

morphology   Pertaining to physical structure or form.

Muller, H. J. (1890–1967)   American geneticist and proponent of eugenics.

multiple personality   A personality state in which the individual may alternately adopt several widely different character states, with partial or total amnesia for the ones not in awareness.

Munchausen syndrome   A psychiatric syndrome in which an individual having but a minor ailment, and traveling as a vagabond, repeatedly gains admission to one hospital after another. Named after the Baron von Munchhausen, the legendary teller of tall tales.

mutation   A change in the DNA structure of a gene. If the change occurs in a gamete, then an alteration in gene function may be perpetuated in subsequent generations.

narcissism   From Narcissus, the figure in Greek mythology who fell in love with his own reflected image. In psychiatric terms, it has been considered to be "self-love," as opposed to "object-love" (love of another person). In the present model, narcissism is based on a genetically determined behavioral complex and is the basis of man's non-aggressive ambition to achieve success and recognition.

narcissistic arms gesture   A gesture of recognition in which the arms are extended to the front or sides, with the fingers slightly spread apart. According to the model, this gesture, like the smile, is often instinctual and has its genetic basis in the narcissistic behavioral complex.

narcissistic behavioral complex   The genetically determined set of behavioral attributes that form the basis of man's non-aggressive drive of ambition. The reward of fulfilled narcissistic ambition is recognition, honor, and attention in a sexual context.

narcissistic personality disorder (NPD)   In the NPA model patients diagnosed with NPD will tend to be poorly-adjusted individuals having the unbridled N trait, being mostly NA and N types. The DSM-IV diagnostic criteria for NPD are poor, since they do not differentiate the trait of narcissism from that of aggression.

narcissistic rage   A mass discharge of the parasympathetic nervous system. According to the present model, this can only occur if the trait N is expressed.

Narcissus   The figure of mythology who became enraptured with his own reflected image. The modern-day Narcissus still lives, as shown by the following quotation (G. Tully, 1949):

> ... One morning, about the middle of October, I became curious about this man Roosevelt and went to a beautiful old mirror of the early Federal period and took a careful look at him in the glass. He smiled. I smiled back. And after a careful examination I decided that all that this villain looked like to me was a man who wanted to be re-elected President of the United States.

narcolepsy   A sudden disposition to unconsciousness occurring at irregular intervals and without obvious predisposing cause; "sleep epilepsy."

natural selection   The process allowing the survival of those species that are best adapted to their environment. Thus, according to the *theory of evolution* there is a "survival of the fittest" if the individual's adaptation confers to him a reproductive advantage.

neocortex   The phylogenetically youngest, layered exterior portion of the mammalian brain.

neoplastic disease   Medical term for cancer. Over the years a personality type prone to cancer has been described, being characterized by dependency, dutifulness and a tendency to suppress emotions, especially anger. Thus, our model furnishes us the hypothesis that, as in some *psychosomatic diseases,* predisposition to some types of cancer may be partially rooted in allelic polymorphism at the **a** and **n** loci. See *oncogene.*

neurotic   According to classical psychiatric theory, a neurosis is an emotional maladaptation arising from unresolved, unconscious conflicts. In psychoanalysis the patient speaks freely with the analyst, especially about his childhood experiences and dreams, in an effort to understand the underlying psychic phenomena. In a *neurosis* the individual's perception of "reality" is grossly intact. See *psychosis.*

neurotransmitter   A biochemical, for example noradrenaline, dopamine or acetylcholine, that governs neural activity.

non-compliance (*n-C*)   The state of the basic character vector of submission (XA–) in which the individual is motivated to undergo frequent *personality splits* to energetic states.

non-sanguine   Referring to individuals who lack the trait N. The dominant non-sanguine NPA types are A (aggressive) and PA (perfectionist-aggressive). The facial complexion in non-sanguine individuals having light skin color tends to be dull, dusky, sallow or milky white. See *sanguine.*

nucleotides   Chemical compounds comprising the "building blocks" of DNA. Each nucleotide is composed of a five-carbon sugar, a phosphate group and an organic nitrogen-containing base. The nucleotides are linked together by bonds between their sugar and phosphate groups to form spiral (helical) chains. Information is coded in these chains by the precise sequence of four different kinds of nitrogenous bases.

null zygote   According to the present model, a zygote in which neither the traits of narcissism, perfectionism nor aggression are expressed. Null zygotes do not develop into viable individuals.

obsession   A persistent idea or impulse that is not amenable to logic or reason.

Occam's razor (principle)   The explanation most likely to be correct is the one that explains the greatest number of observations with the fewest number of assumptions. Also called the principle of "parsimony."

oncogene   A gene involved in predisposition to cancer. See *neoplastic disease.*

orangutan   Non-carnivorous, arboreal, ruddy anthropoid ape inhabiting Borneo and Sumatra.

orca   The killer whale, a cetacean.

ovum   A female gamete, or egg cell.

panphobia   Fear of everything. Sometimes called *pantophobia.*

paranoia   A behavioral pattern characterized by hypersensitivity, suspicion, jealousy, envy and a tendency to blame others and ascribe evil motives to them.

paraphrenia   A paranoid type of schizophrenia.

paraphrenia systematica   A type of paranoid schizophrenia, described by Kraepelin, having an insidious onset and poor prognosis for recovery.

parasympathetic nervous system   A division of the autonomic nervous system that controls, among other functions, pupillary constriction, lacrimation, and slowing of the heart rate. The NPA model suggests that the trait of *narcissism* is mediated by the parasympathetic nervous system.

parkinsonism   Parkinson's disease, or *paralysis agitans.* The syndrome is characterized by muscular rigidity, loss of postural reflexes, and a characteristic tremor. The personality of the patient is phlegmatic and sullen, accompanied by a monotonous voice and an immobile facial expression somewhat reminiscent of a depressed submissive phenotype. An association with paranoid psychoses is sometimes reported. Thus, our model provides us the hypothesis that susceptibility to parkinsonism may be related, in part, to allelic polymorphism at the **n** and **a** loci.

passive-aggressive   A behavioral pattern characterized by obstructionism, procrastination, intentional inefficiency and stubbornness.

pathology   The study of impaired structure or function.

pedigree   A register recording the ancestral line of an individual.

penetrance   The appearance of the trait when the genotype is present. Thus, in "incomplete penetrance" a certain proportion of individuals will not exhibit the trait although the appropriate genotype is present. Incomplete penetrance may be caused by modifier genes or by environmental influences.

Peoples Temple   A religious sect, led by the Reverend Jim Jones, which met its demise by mass suicide in Guyana in 1978.

perfectionism, behavioral complex of   The genetically determined set of behavioral attributes that form the basis of man's achievement by order, persistence and repetition. In the present model perfectionism mediates the two components of unbridled ambition: narcissism and aggression.

periodic psychosis   Psychotic states that show periodicity, with intervals of "normal" behavior between attacks.

peripheral vasodilatation   Physiologic widening of small blood vessels, for example in cutaneous flushing and blushing.

perseveration   An automatism in which a behavioral response is continued once initiated.

personality   A collection of behavioral patterns unique to an individual that is consistent over time. In the NPA model personality has two major genetic bases:

- *NPA character type* (the three NPA genetic loci), and
- *temperament*, or the general level of activity or excitability of an individual in the Pavlovian sense (probably many genes).

personality split   According to the model, an individual having the A trait may alter his personality state in stressful situations. In "playing the game" dominant and submissive types can assume *subdued* and *energetic* states, respectively. Resigned types, although having renounced the "game," may occasionally be activated to the energetic state of dominance.

phenocopy   The term *genetic heterogeneity* refers to a situation in which similar phenotypes have more than one endogenous cause. The heterogeneity may be due to allelic polymorphism at a single gene locus or due to one or several genes at other loci. Thus, the terms *genocopy* or *genetic mimic* describe genetic traits that are superficially similar but fundamentally distinct from one another. Finally, the term *phenocopy* refers to a situation in which environmental (exogenous) factors result in a trait similar to one caused by a genetic mechanism. For example, according to the model the phenotypic trait of non-compliant submission XA– may have primarily either genetic or environmental bases. Specifically, the XA– trait may be caused by an **a-** gene, but there may also occur phenocopies of this trait based on adverse factors operative during the period of maturation.

phenotype   The observable trait or traits in an individual.

phobia   An obsessive, intense but illogical fear of some object or situation.

phrenology   The study of the conformation of the skull, in the belief that traits of character can be deduced there from.

physiognomy   The age-old art of determining temperament and character from outward appearance. For example, Plutarch in his *Lives of the Noble Romans* gives us this account:

> Nor was Caesar without suspicions of him, and said once to
> his friends, "What do you think Cassius is aiming at? I
> don't like him, he looks so pale." And when it was told him

> that Antony and Dolabella were in a plot against him, he
> said he did not fear such fat, luxurious men, but rather the
> pale, lean fellows, meaning Cassius and Brutus.

piloerection   Erection of body hair as a result of the stimulation of the sympathetic nervous system.

pleiotropism   The determination of many different somatic and/or behavioral characteristics by a single gene.

poisoning delusion   A delusion ("My food is poisoned!") that often appears in the agitated phase of paranoid schizophrenia.   Although it may occur in schizophrenias of both the NP and PA types, it appears to be especially common as part of a hypochondriacal syndrome in the agitated schizophrenic break of an NP individual.   See, for example, accounts of the psychotic illnesses of Robert Schumann, Vincent van Gogh and the Empress Carlota of Mexico.

poker face   A countenance of suppressed emotions.   It can be melancholic (NP type), dour (PA type), or apprehensive (submissive types).

polygenic   Referring to the influence of several genes determining the expression of a trait or syndrome.

polymorphism, allelic   The occurrence, in a population, of multiple alleles at a given genetic locus.   According to the model, allelic polymorphism at the **n, P,** and **a** loci may play a role in:

- subtypes of the major character types,

- states of submission and resignation,

- states of non-aggressive withdrawal,

- states of pseudoaggression and pseudonarcissism,

- states of schizophrenia and other psychoses,

- cancers and leukemias, and

- many of the "psychosomatic diseases."

polymorphism, phenotypic   The definition, according to C.E. Ford, is "The occurrence together of two or more discontinuous forms of a species in such proportions that the rarest of them cannot be maintained merely by recurrent mutation." In a *balanced polymorphism* where certain NPA phenotypes are at a reproductive disadvantage because of relative infertility, the gene frequencies in question may nevertheless be maintained through other phenotypic forms, for example by assortative mating and by compensatory reproductive advantages.

polypeptide   A large organic molecule composed of a chain of amino acids.

porphyria   A group of inherited metabolic disorders characterized by abdominal pains, skin rashes, and dark urine; sometimes associated with intermittent episodes of acute psychosis. According to the model, the character types most susceptible to psychosis would be those having only one functioning component of ambition (N or A). An historical example may have been George III of England.

positive feedback   When the result of a process causes the process to be accentuated even more.

power behind throne   A symbiotic relationship between a perfectionist-aggressive individual (the Power) and a figurehead individual on whom he depends (the Throne). The Throne may be the Power's mate, and he maybe subjugated in the context of a morbid dependency.

Prader-Willi syndrome   A syndrome originating in infancy, characterized by obesity, short stature and mental retardation. The affable nature of these children is sometimes stressed.

predestination   The philosophy that all events follow inflexible, predetermined pathways to their ultimate destinies.

premorbid personality   The personality of the individual before the onset of psychopathology.

pride   Self-esteem, in a non-pejorative sense, based on one's innate character structure as well as on positive attributes that one has gained on the basis of environmental factors. *Happiness* is determined by the degree to which one's accomplishments (dependent on one's pride) measure up to one's expectations in life. Thus, pride is the primary psychic attribute that permits one the justification of one's very existence.

Primates   The order of mammals that includes man, apes, monkeys, and related forms such as lemurs and tarsiers.

proband   The individual, having the trait in question, who first comes to the attention of the investigator and leads to the study of his family. Sometimes called the *propositus.*

prodromal   Pertaining to early signs of a developing physiological or psychopathological process.

prognosis   Pertaining to the course and outcome of a pathologic condition.

proptosis   Bulging of the eyeballs. This physical sign is often seen in the patient with an overactive thyroid gland (hyperthyroidism), especially in Graves' disease. However, the eyes of a dominant NPA type may sometimes have a proptotic appearance as he makes intense eye contact, with a reduced blink rate, during conversation with others.

proteins   Complex organic molecules formed by the linkage of polypeptides.

pseudoaggression   An individual not having the trait of aggression may rebel against his society, inverting his basic character structure. He, thus, may exhibit profound anti-social or sociopathic behavior. However, when stressed such an individual does not manifest the aggressive-vindictive rage.

pseudonarcissism   In an individual lacking the narcissistic trait N, an inversion of his basic quality of aggression to a character structure having a superficial resemblance to a narcissistic type. If the inversion is a profound one, the individual can be drawn to extreme self-adornment, exhibitionism, transvestism, or trans-sexualism. When stressed such an individual may manifest an aggressive-vindictive rage, rather than the narcissistic rage of withdrawal.

psoriasis   A hereditary skin disease sometimes complicated by arthritis. Transmission of the condition appears to be related to the HLA (histocompatibility) loci on chromosome 6.

psychiatry   The branch of medicine concerned with the diagnosis and treatment of psychopathologic conditions.

psychogenic   Originating in the mind, as opposed to *somatogenic.*

psychosis   A major mental disorder in which the individual's ability to interpret "reality" is grossly impaired. See *neurotic.*

psychosomatic disease   A disease whose origins, onset or course is said to be intimately related to emotional factors or to the individual's personality. According to the present model, many of the psychosomatic diseases are endogenous to allelic polymorphism at the **n, P** and **a** loci (other genes and exogenous factors may also be involved).   Such polymorphism may be responsible for many diseases that have been associated with the HLA histocompatibility locus on chromosome 6.   Of particular interest is the model's prediction of only certain mental disorders in association with various psychosomatic diseases.

pupillary constriction   Narrowing of the pupils of the eyes, controlled by activation of pathways of the parasympathetic nervous system.

rage   A mass discharge of a portion of the autonomic (sympathetic-parasympathetic) nervous system.   Subjectively, it is perceived as a type of anger. The present model proposes that the N and A traits allow the release of inhibition of two different types of mass discharges, allowing the occurrence of *narcissistic* and *aggressive* rages, respectively.

rationalization   In psychiatric terms, an intra-psychic mechanism in which logical explanations are consciously provided in order to account for unconsciously unacceptable thoughts, feelings and actions.

reactive depression   Synonym for exogenous depression.

recessive   Referring to a trait that is expressed only when the gene in question is present in the homozygous state, i.e., on both chromosomes of an autosomal pair. In X-linkage, the trait is ordinarily expressed if the relevant allele is present in a single dose in the male (one X chromosome) but only in the homozygous state in the female (two X chromosomes).   See *dominant trait.*

recombination   The consequence of *crossing over* in which DNA is exchanged between homologous chromosomes in the process of meiosis.   Thus, genes formerly on separate chromosomes of a homologous pair can come to lie on the same chromosome, and vice versa.

repression   In psychiatric terms, an intra-psychic mechanism by which unacceptable thoughts and impulses are kept from consciousness.

Reptilia   The class of "cold-blooded" air-breathing vertebrates that, according to evolutionary theory, gave rise to birds and mammals.

resignation   The basic character vector in which an individual at maturity renounces the "playing of the game" of dominance and submission and adopts a philosophy of independence.

RNA   Ribonucleic acid. Various forms of these nucleic acids, present in both the nuclei and cytoplasm of cells, aid in the production of enzymes and other polypeptides. RNA is produced from chromosomal DNA "templates."

rosacea   A skin disorder, sometimes said to be "psychogenic" in origin, which in its early stages resembles permanent blushing.

sadism   Satisfaction derived from aggressively dominating or abusing others.

sadomasochism   A symbiotic relationship between two individuals having sadistic and masochistic trends. According to the present model, only individuals having a measure of the trait A play an active "game" of sadomasochism, or dominance and submission. This view is consistent with that of classic psychiatry, where it is often stated that "sadism and masochism often occur in the same individual." In the context of mating, it is consistent with the popular phrase that "you only hurt the one you love."

sanguine   Blood-red, or flushed. According to ancient physiology, belonging to one of the four temperaments in which blood predominates over the other three humors, leading to a ruddy countenance. In the NPA model a sanguine character type is any type having the N trait, for example the dominant types N, NP, NA and NPA. See *non-sanguine*.

savannah   A tropical or sub-tropical grassland containing shrubbery and scattered trees.

schizo-affective disorders   A poorly defined set of psychiatric illnesses in which aspects of both schizophrenia and affect disorder (elation-depression) are present. These probably have included endogenous periodic psychoses, as well as exogenous psychoses, in individuals of various character types.

schizoid   Withdrawn; tending to avoid close relationships with others.

schizophrenia   A group of psychotic disorders characterized by disturbances in thought, mood and behavior.   In the present model schizophrenia appears as a *loss of ambition* in which the individual is reduced to an aimless (null) or perfectionistic state because of a lack of normal expression of both the N and A traits.

schizophreniform psychosis   A psychiatric illness of short duration resembling schizophrenia (lasting months rather than years). According to the model, such episodes of psychosis, rooted in either endogenous or exogenous causes, could occur in certain vulnerable character types.

scoliosis   Curvature of the spine. May be of primarily endogenous or exogenous origins.

seal   An aquatic mammal of order Carnivora, known for its affability and indefatigability when performing in the limelight.

segregation   The separation of homologous chromosomes during the process of meiosis.

seizure disorder   A disorder characterized by convulsions, associated with abnormal neural discharges of the central nervous system. Commonly called *epilepsy.*

self-effacing personality   A personality state characterized by emotional dependency on others, feelings of inferiority, and a lack of self-confidence. See *submission.*

senile dementia   A chronic condition of loss of intellectual function associated with aging.

septal rage   A rage produced in a laboratory animal by stimulation of the "septal" region of the limbic system of the brain. In some reports it is said to cause mainly fear and anxiety, associated with a tendency for the animal to withdraw.

sex chromosome   An X or Y chromosome. In the human XX determines the female gender, while XY determines the male.

sham rage   A rage produced in a laboratory animal by stimulation of centers within the hypothalamus of the brain. The animal's behavior may be superficially similar to the "flight or fight" response.

smile   An instinctual, complex facial response of recognition. According to the present model, the warm smile of recognition is based on the narcissistic behavioral complex. Not to be confused with a *grin* or with *laughter.* (In many languages, however, the words for *smile* and *laugh* stem from the same root). See also *gingival smile.*

somatic   Affecting the body, as opposed to the mind.

somatotype   The body type, with respect to physique, as quantified by the American psychologist W.H. Sheldon.

spermatozoon   A male gamete.

stochastic   Referring to a mathematical model, one in which the basic mechanisms of the process are not specified. Thus, the process is described by empirical, probabilistic methods, but the mechanisms of operation remain unknown, as in a "black box."

stress   According to the present model, an individual may be subject to psychic stress emanating, in part, from the following interrelated sources:

1) if his accomplishments do not measure up to the demands of his environment or of his own expectations,

2) if two or more behavioral complexes pull him in conflicting directions, and

3) if he plays a stressful "game" of dominance and submission in the context of a competitive society or of mating.

subdued state   An individual having a basic character vector of dominance (XA) can, on occasion, be reduced to a transient subdued state resembling submission (XA–) by undergoing a *personality split.*

subjugated state   A chronic state resembling submission of an individual with respect to his companion or mate.

submission   The basic character vector in which an individual's aggressive trait has been partially suppressed by endogenous or exogenous factors before maturity.

submissive type   An NPA character type in which the trait of aggression, A, is not fully expressed. See *compliance* and *non-compliance.*

symbiosis   A relationship between two individuals that has elements of mutual advantage.

sympathetic nervous system   A division of the autonomic nervous system that controls, among other functions, peripheral vasoactivity, pupillary dilation, cardioacceleration, piloerection, diaphoresis and secretion of catecholamines into the blood stream from the adrenal medulla.

systemic disease   A disease affecting the body generally, i.e., not just localized to a single area or to a single organ.

tantrum   A defensive, diffusely directed *narcissistic rage* of agitation and sometimes violence. Not to be confused with the aggressive-vindictive rage.

taxonomy   The classification of plants and animals according to their presumed natural relationships. The hierarchal classification is as follows: kingdom, phylum, class, order, family, genus, species, variety.

temperament   The general level of activity or excitability of an individual in the Pavlovian sense (probably dependent on many genes).

toxic psychosis   An acute syndrome of psychosis developing as the result of the ingestion of a drug or other foreign substance.

trait (genetic)   Any genetically determined characteristic or set of characteristics.

Type A, Type B   A binary classification of personality type used by some physicians, especially with reference to the risk of myocardial infarction (heart attack). The *Type A* individual is an active, achievement-oriented person. The *Type B* is said to be a placid, more phlegmatic individual. According to the model, the Type A group would include not only dominant types having the aggressive trait but also non-compliant submissive types.

type casting   In the performing arts, the little-used principle of matching the character type of the actor with that of the individual he is to portray.

unbridled ambition   In the NPA model character types having the trait of perfectionism, P, are said to have bridled ambition. Character types lacking the trait P (e.g., N, A, NA, NA–, NA=) are said to have unbridled ambition.

unconscious   In psychiatric terms, the division of the mind encompassing repressed emotions and memories that normally remain out of awareness.

ungulates   Hoofed mammals.

unipolar depressive disease   Depressive disease in which mood swings toward depression, but not mania, occur in the individual.

vanity   Self-esteem rooted in narcissistic ambition. An individual's vanity is wounded if his narcissistic drives are frustrated. The reaction to wounded vanity may be the *narcissistic rage.* Similarly, if an individual's aggressive drives are frustrated, then his pride may be wounded and he may respond in the *aggressive-vindictive rage.* If both are wounded, he may respond by an NA or NPA rage.

vasoconstriction   Physiologic narrowing of blood vessels in response to a neural or biochemical stimulus.   Peripheral vasoconstriction, causing blanching of the skin, is associated with the "flight or fight" response mediated by the sympathetic nervous system.

vasodilatation   Physiologic widening of blood vessels in response to a neural or biochemical stimulus. According to the present model, peripheral vasodilatation, causing flushing of the skin, is associated with blushing and with the narcissistic rage response mediated by the sympathetic-parasympathetic nervous system.

Venn diagram   A graph that employs circles to represent logical relations between sets. After the English logician, John Venn.

verbigeration   The meaningless utterance of words or phrases.

vindictive   Disposed to seeking revenge. According to the model, personages having the A (aggressive) trait have the greatest tendency to be *actively vindictive.* However, those lacking the A trait may be *passively vindictive,* for example by withholding information from others, or by recalcitrance.

X   The symbol X can take the values of N, or P, or both, or neither, as follows:

    XA   Character type having the trait A, i.e., A, PA, NA or NPA

    XA–   Non-compliant submissive type, i.e., A–, PA–, NA– or NPA–

    XA=   Compliant submissive type, i.e., A=, PA=, NA= or NPA=

    X–A   Resigned type, i.e., –A, P–A, N–A or NP–A

X-linkage   The location of genes on the **X** chromosome.

zygote   A fertilized ovum, whether or not destined to become a viable individual.

CHAPTER REFERENCES

PREFACE

Max Planck quote: Planck M. (1949).
Christina of Sweden quote: Stolpe S. (1960).
Somerset Maugham: Maugham W.S. (1915, 1949): *Of Human Bondage, A Writer's Notebook.*
Albert Einstein: Einstein A. (1949).
Lecomte du Noüy: Lecomte du Noüy P. (1947, 1948).
Sigmund Freud: Freud S. (1896, 1924).
Karen Horney: Horney K. (1950).
Desmond Morris: Morris D. (1967, 1969).

CHAPTER 1

B. Rimland quote: Rimland B. (1964).
Obsessions and compulsions: Horney K. (1950).
Anxiety: Horney K. (1950).
Psychiatry: Arnheim R. (1972); Kolb L.C. (1973); Torrey E.F., in Brady J.P. and Brodie H.K.H. (1980): "Does psychiatry have a future?"

CHAPTER 2

Peer Gynt: Mencken H.L. (1935): *Eleven Plays of Henrik Ibsen.*
J.A. Thompson quote: Simpson G.G. (1949).

CHAPTER 3

G.G. Simpson quote: Simpson G.G. (1949).
Origin of universe: Hunt P. (1960); True W.P. (1960); Sagan C. (1980); Parker B. (1981).
Evolution: Simpson G.G. (1949); Moore R. (1953); Morris D. (1967); Dobzhansky T. (1970); Jerison H.J. (1973); Leakey R.E. and Lewin R. (1977); Bonner J.T. (1980); Lovejoy C.O. (1981).

CHAPTER 4

Desmond Morris quote: Morris D. (1967).
Obsessions and compulsions: Horney K. (1950).
Animal behavior: Morris D. (1967, 1969); Keeton W.T. (1967); Reynolds V. (1967); Bonner J.T. (1980).
Dominance, submission and resignation: Horney K. (1945, 1950).
Personality splits: Horney K. (1950).

CHAPTER 5

Darwin quote: Darwin C., *Origin of Species*, in Hunt P. (1960).
Intelligence: Keeton W.T. (1967); Munn L. (1946).
Narcissism: Horney K. (1939, 1950); Aristotle's *Nicomachean Ethics*, in Loomis L.R. (1943).
Narcissistic rage: Moore B.E. and Fine B.D. (1968); Horney K. (1950).
Personal adornment: Morris D. (1977).
Perfectionism: Horney K. (1950).
Grooming in apes: Morris D. (1967).
Chimpanzee smile: Darwin C. (1890); Hunt P. (1960).
Blushing: Darwin C. (1890).
Aggression: Morris D. (1967, 1969); Sandler M. (1979).
Playing the "game" of dominance and submission: Morris D. (1967).
"Flight or fight" response: Keeton W.T. (1967).
Phobias: Kolb L.C. (1973).
Arrogant-vindictive personality: Horney K. (1950).
Moving "toward," "against," "away from": Horney K. (1945).
Aggressive-vindictive rage: Horney K. (1950).
Vindictive triumph: Horney K. (1950).
Sadism: Horney K. (1945, 1950); Stekel W. (1929).
NA type: Horney K. (1967): "The overvaluation of love: A study of a common present-day feminine type"; "Personality changes in female adolescents."
Hypomanic-depressive or cyclothymic personality: *See* Chapter 11.
Hysterical or histrionic personality: Weintraub M.I. (1977).
Hypersexuality: Horney K. (1950); "Nymphomania," in Vetter H.J. (1972).
Obsessive-compulsive personality: Rowe C.J. (1954).
Perfectionist "deal with life": Horney K. (1950).
Explosive personality: Rowe C.J. (1954); Quick-tempered people of Aristotle's *Nicomachean Ethics,* in Loomis L.R. (1943).
Autistic and schizophrenic child: *See* Chapter 11.
Schizoid personality: Rowe C.J. (1954); Kolb L.C. (1973); Bleuler M. (1978).
Passive-aggressive personality: Rowe C.J. (1954).
Paranoid states: Rowe C.J. (1954); Kolb L.C. (1973).
Folie à deux: Rowe C.J. (1954); Kolb L.C. (1973).
Compensation neuroses: Rowe C.J. (1954).
Power behind the throne: Horney K. (1950); *See* Chapter 9.
Compliant type: Self-effacing personality of Horney K. (1950).
Non-compliant type: Inverted sadist of Horney K. (1945).
Masochistic trends: Horney K. (1939, 1950); Horney K. (1967): "The problem of feminine masochism."

CHAPTER 5 (ctd.)

The Peter Principle: Peter L.J. and Hull R. (1969).
Morbid dependency: *See* Chapter 9.
Resigned type: Horney K. (1945, 1950).
Rebellious rage: Horney K. (1950).
Hypersexual resigned type: Horney K. (1950).
Asthenic personality: Rowe C.J. (1954); Kolb L.C. (1973).
Abject states, depressions: *See* Chapter 11.
Exhilarated states, euphorias: Leonhard K. (1979); *See* Chapter 11.

CHAPTER 6

Somerset Maugham quote: Maugham W.S. (1938).
Karen Horney quote: Horney K. (1950).
Inverted behavior: e.g., inverted sadism, in Horney K. (1945).
Autism, schizophrenia, involutional melancholia: *See* Chapter 11.
Dieter's father (of autistic child): Bosch G. (1970).
Sexuality of NA type: Horney K. (1967): "The overvaluation of love."
Nymphomania: Vetter H.J. (1972).
Vampire: Webster's New Collegiate Dictionary (1980): 2b. a woman who
    exploits and ruins her lover.
Morbid dependency: *See* Chapter 9.
"The Ant and the Grasshopper": Maugham W.S. (1951).
"The Master Builder," Ibsen and Emilie Bardach: Le Gallienne E. (1955); Fjelde
    R. (1965); Meyer M. (1971).
"Rain": Maugham W.S. (1951).
Identical twins reared apart: Bouchard T.J., Jr. and coworkers: University of
    Minnesota studies (publication pending); Browne M.W. (1980); Farber S.L.
    (1981).
Imprinting: Keeton W.T. (1967); Morris D. (1967).
Gestures: Morris D. *et al.* (1977, 1979).

CHAPTER 7

W.E. Henley quote: Hayward J. (1964); Browne M.W. (1980).
Albert Einstein quote: Einstein A. (1949).
"The Master Builder": *See* Chapter 6.
"Close Encounters of the Third Kind" (Second version): Spielberg S. (1980).
Biblical quotes from Revelation: Whealon J.F. (1978).
Maugham quotation: Maugham W.S. (1949).

CHAPTER 7 (ctd.)

Pride: Horney K. (1950).
The psychotic: Horney K. (1950).
"The search for glory": Kolb L.C. (1973).
The ideal image: Horney K. (1950).
Anxiety: Horney K. (1950); Kolb L.C. (1973).
Rationalization and externalization: Horney K. (1950).
Munchausen syndrome: Spiro H.R. (1968); Lanners E. (1973).
Hedonism: Maugham W.S. (1949).
Self-hate and self-destruction: Horney K. (1950).
Winces: Calder K.T. (1980).

CHAPTER 8

Albert Schweitzer quote: Anderson E. (1965)
Joseph Goebbels quotes: Lochner L.P. (1948).
Friendship: Aristotle's *Nicomachean Ethics*, in Loomis L.R. (1943).
Sadistic trends: Horney K. (1945, 1950).
Inverted sadism: Horney K. (1945).
Sadistic envy: Horney K. (1945, 1950).
Remotivating activities: Morris D. (1967, 1977).
Masochistic trends: Horney K. (1937, 1939, 1950).
Masochistic suffering: Horney K. (1950).
Santayana: Runes D.D. (1959).
Arrogance in medicine: Ingelfinger F.J. (1980).

CHAPTER 9

Dr. Fell: Chapter 27 in Arnheim R. (1972).
Confined monkeys: Morris D. (1967).
Folie à deux: *See* Chapter 11.
Power behind throne: Horney K. (1950).
Ibsen's "A Doll's House": Fjelde R. (1965).
Morbid dependency: Horney K. (1942, 1950); Poem "Lines," in Maugham W.S.
    (1949).
William James: Runes D.D. (1959); James W. (1890, 1908).
Bisexuality: Horney K. (1967): "The neurotic need for love."
The Reverend Jones: Krause C.A. (1978).
Kierkegaard's "Diary of a Seducer": Horney K. (1945); Gillhoff G. (1966).
Morbid dependency anecdote: James W. (1908).
The monogamous ideal: Belloc H. (1938); Horney K. (1967).

CHAPTER 10

William James quote: James W. (1908).
Genetics: Srb A.M. et al. (1965); McKusick V.A. (1969); Watson J.D. (1970);
Fuller J. L. and Thompson W.R. (1978).
Population genetics: Srb A.M. et al. (1965); McKusick V.A. (1969); Dobzhansky
T. (1970).
Dilger's hybridization of parrots: Keeton W.T. (1967).
Aggression in strains of rats: Mandel P., Mack G. and Kempt E., in Sandler M.
(1979).
Stress and human behavior: Engel G.L. (1978); Tache J., Selye H. and Day S.B.
(1979).
"Flight or fight" reaction: Keeton W.T. (1967); Morris D. (1977).
Blushing and flushing: Darwin C. (1890); McClary A.R. *et al.* (1962);
Appenzeller O. (1970); Whitlock F.A. (1976); Mulley G.P. (1978).
Flushing in relation to alcohol consumption: Wolff P.H., cited in Roebuck J.B.
and Kessler R.G. (1972); Leslie R.D.G. and Pyke D.A. (1978, 1979).
Autonomic nervous system: Appenzeller O. (1970).
Human gene mapping: McKusick V.A. (1969); Sparkes R.S. (1980).
Down's syndrome: Penrose L.S. and Smith G.F. (1966); Lilienfeld A.M. and
Benesch C.H. (1969).
Miscarriages and stillbirths: McKusick V.A. (1969); Biggers J.D. (1981).
U.S.A. Habitancy: Healy P. (1956).
Balanced polymorphism: McKusick V.A. (1969); Dobzhansky T. (1970).
Geographical distribution of character types according to Ambroise Paré: "On the
practice of aforesaid rules of temperaments," in Johnson T. (1634).

CHAPTER 11

Sigmund Freud quote: Blumenthal R. (1981).
Sandor Rado quote: J.D. Rainer, in Mendlewicz J. and Shopsin B. (1979).
Freud: Freud S. (1896, 1924).
Murray's hierarchy of needs: Arnheim R. (1972).
Horney's theory of neurosis: Horney K. (1950).
Hippocrates: Paré A., "Of humors," in Johnson T. (1634); Adams F. (1891).
Minnesota Multiphasic Personality Inventory (MMPI): Kolb L.C. (1973).
Sheldon's somatotypes: Munn L. (1946).
Intelligence: Munn L. (1946); Keeton W.T. (1967).
Neurosis: Horney K. (1950); Rowe C.J. (1954); Kolb L.C. (1973).
Learning and conditioning: Keeton W.T. (1967).
Personality disorders: Rowe C.J. (1954); Kolb L.C. (1973).
A multiple personality disorder: Thigpen C.H. and Cleckley H.M. (1957)
Anorexia nervosa: Kolb L.C. (1973); Schwabe A.D. (1981).

CHAPTER 11 (ctd.)

Self-effacing, detached, inverted sadistic personalities: Horney K. (1945, 1950).

Infantile autism: Rimland B. (1964); Churchill D.W. *et al.* (1971); Ritvo E.R. (1976); Rutter M. and Schopler E. (1978).

Infantile autism: "Personality characteristics of parents," McAdoo W.G. and DeMyer M.K., in Rutter M. and Schopler E. (1978).

Infantile autism: "Family factors," Cantwell D.P., Baker L. and Rutter M., in Rutter M. and Schopler E. (1978).

Infantile autism, genetic studies: Spence M.A., in Ritvo E.R. (1976).

Infantile autism, lack of smile: Nelson W.E. *et al.* (1969); Ruttenberg B.A., in Churchill D.W. *et al.* (1971).

Richard, an autistic adolescent: Bosch G. (1970).

Idiot savant: Rimland B. (1964).

Schizophrenia: Bellak L. (1958); Crider A. (1979); Leonhard K. (1979).

Schizophrenia and genetics: Karisson J.L. (1966), and in Kaplan A.R. (1972); Gottesman I.I. and Shields J. (1972); Kaplan A.R. (1972); Heston L.L. (1977); Rutter M., cited in Heston L.L. (1977); Baron M. (1980).

Schizophrenia, family studies: Reed S.C. *et al.* (1973); Bleuler M. (1978).

Schizophrenia, occurrence in U.S.A.: Dohrenwend B.P. *et al.* (1980).

Folie à deux: Rowe C.J. (1954); Gottesman I.I. and Shields J. (1972); Kolb L.C. (1973).

Depression: Beck A.T. (1967); Burrows G.D. (1977); Usdin G. (1977); Leonhard K. (1979); Perris C. and Strandman E.(1980).

Manic-depressive disease: Winokur G. *et al.* (1969); Depue R.A. and Monroe S.M. (1978); Mendlewicz J. and Shopsin B. (1979); Leonhard K. (1979).

Affective illnesses, X-linkage and color blindness: Winokur G. *et al.* (1969); Fieve R.R. *et al.* (1973).

Affective illness in relatives of unipolar and bipolar patients: Winokur G.: "A family history (genetic) study of pure depressive disease," in Mendlewicz J. and Shopsin B. (1979).

Periodic psychoses: Beck A.T. (1967); Leonhard K. (1979).

Paranoid psychoses: Rowe C.J. (1954); Kolb L.E. (1973).

Paranoid involutional states: Rowe C.J. (1954).

Schizo-affective psychosis: Beck A.T. (1967); Mendlewicz J., in Mendlewicz J. and Shopsin B. (1979); Pope H.G., Jr. and Lipinsky J.F. (1980).

Alcoholism: Shields J., in Edwards G. and Grant M. (1976); Israel Y. and Mardones J. (1971); Roebuck J.B. and Kessler R.G. (1972).

Alcoholism, color blindness and genetic polymorphism: Cruz-Coke R., in Israel Y. and Mardones J. (1971).

Anaclitic depression: Rowe C.J. (1954); Bellak L. (1958).

Concentration camp syndrome: Kolb L.C. (1973).

Hyperactive child: Kolb L.C. (1973).

CHAPTER 11 (ctd.)

Stuttering: Barbara D.A. (1962).
Psychosomatic disease: Dunbar F. (1943); Whitlock F.A. (1976); Weiner H. (1977); Tache J., Selye H. and Day S.B. (1979).
HLA type and autoimmune disease: Rosenberg L.E. and Kidd K.K. (1977); Talal N. (1977); Rose N.R. and Friedman H. (1980); Rotter J.I. and Rimoin D.L., in Brownlee M. (1981).
HLA type and mental disorder: James N.M. *et al.* and Perris C., in Mendlewicz J. and Shopsin B. (1979); Oattaz W.F. *et al.* (1980); Smeraldi E. and Bellodi L. (1981); Weitkamp L.R. *et al.* (1981).
Rheumatoid arthritis, schizophrenia and HLA association: Moos R.H. (1964);
Weiner H. (1977); Lawrence J.S. (1977); Gattaz W.F. *et al.* (1980).
Diabetes mellitus: Neel J.V. (1977); Talal N. (1977); Leslie R.D.G. and Pyke D.A. (1978, 1979); West K.M. (1978); Keys A. (1980); Brownlee M. (1981); Cahill G.F., Jr. and McDevitt H.O. (1981); Wilkinson D.G. (1981).
Socio-cultural factors in schizophrenia: Benedict P.K., Chapter 17 in Bellak L. (1958); Crider A. (1979).
Eugenics: McKusick V.A. (1969); Dobzhansky T. (1970).
The orangutan: Galdikas-Brindamour B. and Brindamour R. (1975); Galdikas B.M.F. and Brindamour R. (1980).
The gorilla: Schaller G.B. (1964).
The baboon: Strum S.C. and Ransom T.W. (1975).
The chimpanzee: Van Lawick-Goodall J. (1971).
The porpoise: Conly R.L. and Nebbia T. (1966); Lockley R.M. (1970).
The Orca: Griffin E.J. (1966).

CHAPTER 12

Lecomte du Noüy quotes: Lecomte du Noüy P. (1947).
Vergilius Ferm quote: "Philosophies of religion," in Ferm V. (1950).
Aristotle's Nicomachean Ethics: Loomis L.R. (1943).
Ibsen's speech to Oslo students: Le Gallienne E. (1961).

APPENDIXES A & B

Case study: Stekel W. (1929).
Somerset Maugham: Maugham W.S. (1915, 1944).
William James: James W. (1890, 1908).
Karen Horney: Horney K. (1942, 1950). Horney K. (1967): "The problem of the monogamous ideal."

BIBLIOGRAPHY

Adams F. (1981): *The Genuine Works of Hippocrates.* William Wood, New York.

Anderson E. (1965): *The Schweitzer Album.* Harper and Row, New York

Appenzeller O. (1970): *The Autonomic Nervous System.* North-Holland, Amsterdam.

Arehart-Treichel J. (1980): *Biotypes: The Critical Link Between Your Personality and Your Health.* Times Books, New York.

Arnheim R. (Ed.) (1972): *Psychology Today.* CRM Books, Del Mar, California.

Barbara D.A. (1962): *The Psychotherapy of Stuttering.* Thomas, Springfield.

Baron M. (1980): Schizophrenia on paternal and maternal sides: An analysis of familial factors. *Brit. J. Psychiatry 137* 505-09.

Basch V. (1931): *Schumann: A Life of Suffering.* Trans. by C.A. Phillips. Tudor, New York.

Beck A.T. (1967): *Depression: Clinical, Experimental and Theoretical Aspects.* Harper and Row, New York.

Bellak L. (Ed.) (1958): *Schizophrenia: A Review of the Syndrome.* Grune and Stratton, New York.

Belloc H. (1938): *Louis XIV.* Harper, New York.

Biggers J.D. (1981): In vitro fertilization and embryo transfer in human beings. *New Engl. J. Medicine 304* 336-42.

Bleuler M. (1978): *The Schizophrenic Disorders: Long-term Patient and Family Studies.* Yale University Press, New Haven.

Blumenthal R. (1981): Scholars Seek the Hidden Freud in Newly-emerging Letters. *N.Y. Times,* Aug.18, 1981, p. C-2.

Bonaparte M. (1949): *The Life and Works of Edgar Allan Poe: A Psychoanalytic Interpretation.* Imago, London.

Bonner J.T. (1980): *The Evolution of Culture in Animals.* Princeton University Press, Princeton, New Jersey.

Bosch G. (1970): *Infantile Autism.* Springer-Verlag, New York.

Brady J.P. and Brodie H.K.H. (Eds.) (1980): *Psychiatry at the Crossroads.* W.B. Saunders, Philadelphia.

Browne M.W. (1980): And What if the Findings Are Unpleasant? *N.Y. Times,* Nov. 18, 1980, p. C-3.

Brownlee M. (Ed.) (1981): *Handbook of Diabetes Mellitus.* Garland STPM Press, New York.

Burrows G.D. (Ed.) (1977): *Handbook of Studies on Depression.* Excerpta Medica, New York.

Cahill G.F., Jr. and McDevitt H.O. (1981): Insulin-dependent diabetes mellitus: The initial lesion. *New Engl. J. Medicine 304* 1454-65.

Calder K.I. (1980): An analyst's self-analysis. *J. Amer. Psychoanal. Assn. 28* 5-20.

Castelot A. (1965): *Josephine.* Trans. by D. Folliot. Harper and Row, New York.

Churchill D.W., Alpern G.D. and DeMyer M.K. (Eds.) (1971): *Infantile Autism.* Thomas, Springfield.

Conly R.L. and Nebbia T. (1966): Porpoises: Our Friends in the Sea. *Natl. Geographic Magazine 130* 396-425.

Cooper T. and panel (1981): Coronary-prone behavior and coronary heart disease: A critical review. *Circulation 63* 1199-1215.

Crider A. (1979): *Schizophrenia: A Biopsychological Perspective.* Lawrence Erlbaum Associates, Hillsdale, New Jersey.

Darwin C. (1890): The *Expression of the Emotions in Man and Animals.* Greenwood Press, New York (1969).

Depue R.A. and Monroe S.M. (1978): The unipolar-bipolar distinction in the depressive disorders. *Psychological Bulletin 85* 1001-29.

Dobzhansky T. (1970): *Genetics of the Evolutionary Process.* Columbia University Press, New York.

Dohrenwend B.P., Dohrenwend B.S., Gould M.S., Link B., Neugebaner R. and Wunsch-Hitzig R. (1980): *Mental Illness in the United States.* Praeger, New York.

Dunbar F. (1943): *Psychosomatic Diagnosis.* Paul B. Hoeber, New York.

Edwards G. and Grant M. (Eds.) (1976): *Alcoholism: New Knowledge and New Responses.* University Park Press, Baltimore.

Einstein A. (1949): *The World As I See It.* Philosophical Library, New York.

Engel G.L. (1978): Psychologic stress, vasodepressor (vasovagal) syncope, and sudden death. *Annals Internal Medicine 89* 403-12.

Eysenck H.J. (1980): *A Model for Personality.* Springer-Verlag, New York.

Farber S.L. (1981): *Identical Twins Reared Apart: A Reanalysis.* Basic Books, New York.

Ferm V. (Ed.) (1950): *A History of Philosophical Systems.* Philosophical Library, New York.

Fieve R.R., Mendlewicz J. and Fleiss J.L. (1973): Manic-depressive illness: Linkage with the Xg blood group. *Amer. J. Psychiatry 130* 1355-59.

Fjelde R. (1965): *Henrik Ibsen: Four Major Plays.* Signet Classics, New York.

Fox R.H., Goldsmith R. and Kidd D.J. (1962): Cutaneous vasomotor control in the human head, neck and upper chest. *J. Physiol. (London) 161* 298-312.

Frances A. (1980): The DSM-III personality disorders section: A commentary. *Amer. J. Psychiatry 137* 1050-54.

Freud S. (1896): *Heredity and the Aetiology of the Neuroses*, in Early Psychoanalytic Publications. Hogarth Press, London (1962).

__________ (1924): *A General Introduction to Psychoanalysis.* Permabooks, New York.

Fuller J.L. and Thompson W.R. (1978): *Foundations of Behavioral Genetics.* C.V. Mosby, St. Louis.

Galdikas B.M.F. and Brindamour R. (1980): Living with the great orange apes. *Natl. Geographic Magazine 157* 830-53.

Galdikas-Brindamour B. and Brindamour R. (1975): Orangutans, Indonesia's "people of the forest." *Natl. Geographic Magazine 148* 444-72.

Gattaz W.F., Kasper S., Ewald R.W. and Beckmann H. (1980): Arthropathies and schizophrenia. *Lancet (II) No. 8193, 536-37.*

Gilihoff G. (Trans.) (1966): *Soren Kierkegaard: Diary of a Seducer.* Frederick Ungar, New York.

Gottesman I.I. and Shields J. (1972): *Schizophrenia and Genetics.* Academic Press, New York.

Griffin E.I. (1966): Making friends with a killer whale. *Natl. Geographic Magazine 129* 418-46.

Guilleminault C., Dement W.C. and Passouant P. (Eds.) (1976): *Narcolepsy.* Spectrum Publications, New York.

Hall R.C.W. (Ed.) (1980): *Psychiatric Presentations of Medical Illness: Somatopsychic Disorders.* SP Medical and Scientific Books, New York.

Harris J.E. and Weeks K.R. (1973): *X-raying the Pharaohs.* Scribner's, New York.

Hayward J. (Ed.) (1964): *The Oxford Book of Nineteenth Century English Verse.* Oxford University Press, Oxford.

Healy P. (Ed.) (1956): *Stagecoach '56.* O'Toole and Sons, Stamford, Connecticut.

Herndon W.H. and Weik J.W. (1888): *Life of Lincoln.* Fine Editions Press, Cleveland.

Heston L.L. (1977): Schizophrenia: Genetic Factors. *Hospital Practice, No. 6,* 43-49.

Horney K. (1937): *The Neurotic Personality of Our Time.* Norton, New York.

___________ (1939): *New Ways in Psychoanalysis.* Norton, New York.

___________ (1942): *Self-analysis.* Norton, New York.

___________ (1945): *Our Inner Conflicts.* Norton, New York.

___________ (1950): *Neurosis and Human Growth.* Norton, New York.

___________ (1967): *Feminine Psychology.* Norton, New York.

___________ (1980): *The Adolescent Diaries of Karen Horney.* Basic Books, New York.

Hunt P. (Ed.) (1960): *The Wonders of Life on Earth.* Time Inc., New York.

Ingelfinger F.J. (1980): Arrogance. *New Engl. J. Medicine 303* 507-11.

Israel Y. and Maroons J. (Ed.) (1971): *Biological Basis of Alcoholism.* Wiley-Interscience, New York.

James W. (1890): *Principles of Psychology* (2 Vols.). Holt, New York.

___________ (1908): *The Varieties of Religious Experience.* Longmans, Green and Co., London.

Jerison H.J. (1973): *Evolution of the Brain and Intelligence.* Academic Press, New York.

Johnson T. (1634): *The Collected Works of Ambroise Paré.* Milford House, Pound Ridge, New York (1968).

Kaplan A.R. (Ed.) (1972): *Genetic Factors in Schizophrenia.* Thomas, Springfield.

Karlsson J.L. (1966): *The Biologic Basis of Schizophrenia.* Thomas, Springfield.

Keeton W.T. (1967): *Biological Science.* Norton, New York.

Keys A. (1980): *Seven Countries: A Multivariate Analysis of Death and Coronary Heart Disease.* Harvard University Press, Cambridge.

Kolb L.C. (1973): *Modern Clinical Psychiatry.* W.B. Saunders, Philadelphia.

Kraepelin E. (1919): *Dementia Praecox and Paraphrenia.* R.E. Krieger, Huntington, New York (1971).

Krause C.A. (1978): *Guyana Massacre: The Eyewitness Account.* Berkley, California.

Kuhn T.S. (1957): *The Copernican Revolution.* Random House, New York.

Ladee G.A. (1966): *Hypochondriacal Syndromes.* Elsevier, New York.

Lavater J.K. (1774): *Essays on Physiognomy.* Trans. by T. Holcroft. Ward, Lock and Co., London.

Lawrence J.S. (1977): *Rheumatism in Populations.* Heinemann, London.

Leakey R.E. and Lewin R. (1977): *Origins.* Dutton, New York.

Lecomte du Noüy P. (1947): *Human Destiny.* McKay, New York.

___________ (1948): *The Road to Reason.* Longmans, Green and Co., New York.

Ledbetter D.H., Riccardi V.M., Airhart S.D., Strobel R.J., Keenan B.S. and Crawford J.D. (1981): Deletions of chromosome 15 as a cause of the Prader-Willi syndrome. *New Engl. J. Medicine 304* 325-29.

Le Gallienne E. (1955): *Henrik Ibsen: The Master Builder.* N.Y. University Press, New York.

___________ (1961): *The Wild Duck and Other Plays of Henrik Ibsen.* Modern Library, New York.

Leonhard K. (1979): *The Classification of the Endogenous Psychoses* (5th Ed.) Irvington Publishers, New York.

Leslie R.D.G. and Pyke D.A. (1978): Chlorpropamide-alcohol flushing: A dominantly inherited trait associated with diabetes. *British Medical Journal 2* 1519-21.

___________ (1979): Sensitivity to enkephalin as a cause of non-insulin-dependent diabetes. *The Lancet,* 17 Feb. 1979, 341-43.

Lilienfeld A.M. and Benesch C.H. (1969): Johns Hopkins Press, Baltimore.

Lochner L.P. (1948): *The Goebbels Diaries.* Doubleday, New York.

Lockley R.M. (1970): *Whales, Dolphins and Porpoises.* Norton, New York.

Loomis L.R. (Ed.) (1943): *Aristotle: On Man in the Universe.* W.J. Black, New York.

Lovejoy C.O. (1981): The origin of man. *Science 211* 341-50.

Lynch H.T. (Ed.) (1981): *Genetics and Breast Cancer.* Van Nostrand, Reinhold, New York.

McClary A.R., Meyer E. and Weitzman E.L. (1955): Observations on the role of the mechanism of depression in some patients with disseminated lupus erythematosus. *Psychosomatic Medicine 17* 311-21.

McKusick V.A. (1969): *Human Genetics.* Prentice-Hall, Englewood Cliffs, New Jersey.

______________ (1978): *Mendelian Inheritance in Man: Catalogs of Autosomal Dominant, Autosomal Recessive and X-linked Phenotypes* (Fifth Ed.). Johns Hopkins University Press, Baltimore.

Maugham W.S. (1915): *Of Human Bondage.* Random House, New York.

______________ (1925): *The Painted Veil.* Doubleday, New York.

______________ (1930): *The Gentleman in the Parlour.* Doubleday, New York.

______________ (1930): *Cakes and Ale.* Doubleday, New York.

______________ (1932): *The Narrow Corner.* Doubleday, New York.

______________ (1938): *The Summing Up.* Doubleday, New York.

______________ (1939): *Christmas Holiday.* Bantam Books, New York.

______________ (1944): *The Razor's Edge.* Doubleday, New York

______________ (1949): *A Writer's Notebook.* Doubleday, New York.

______________ (1951): *The Complete Short Stories.* Heinemann, London

Mencken H.L. (Ed.) (1935): *Eleven Plays of Henrik Ibsen.* The Modern Library, New York.

Mendlewicz J. and Shopsin B. (Ed.) (1979): *Genetic Aspects of Affective Illness.* SP Medical and Scientific Books, New York.

Meyer B.C. (1967): *Joseph Conrad: A Psychoanalytic Biography.* Princeton University Press, Princeton, New Jersey.

Meyer M. (1971): *Ibsen: A Biography.* Doubleday, New York.

Moore B.E. and Fine B.D. (1968): *A Glossary of Psychoanalytic Terms and Concepts.* The American Psychoanalytic Assn., New York.

Moore R. (1953): *Man, Time and Fossils: The Story of Evolution.* Knopf, New York.

Moos R.H. (1964): Personality factors associated with rheumatoid arthritis: A review. *J. Chronic Dis. 17* 41-55.

Morgan T. (1980): *Maugham.* Simon and Schuster, New York.

Morris D. (1967): *The Naked Ape.* McGraw-Hill, New York.

______________ (1969): *The Human Zoo.* McGraw-Hill, New York..

______________ (1977): *Manwatching: A Field Guide to Human Behavior.* Abrams, New York.

______________, Collett P., Marsh P. and O'Shaughnessy M. (1979): *Gestures.* Stein and Day, New York.

Mulley G.P. (1978): *Flushing. J. Royal College London 12* 359-64.

Munn L. (1946): *Psychology.* Houghton Mifflin, Boston.

Neel J.V. (1977): The genetics of juvenile-onset type diabetes mellitus. *New Engl. J. Medicine 297* 1062-63.

Nelson W.E., Vaughan III V.C., and McKay R.J. (1969): *Textbook of Pediatrics.* Saunders, Philadelphia.

Parker B. (1981): The age of the universe. *Astronomy, No. 9,* 66-71.

Penrose L.S. and Smith G.F. (1966): *Down's Anomaly.* Little, Brown and Co., Boston.

Perris C. and Strandman E. (1980): Genetic identification of a subgroup of depressed patients. *Psychiatria Clinica 13* 13-24.

Peter L.J. and Hull R. (1969): *The Peter Principle.* William Morrow, New York.

Planck M. (1949): *Scientific Autobiography and Other Papers.* Trans. by F. Gaynor. Philosophical Library, New York.

Plutarch (ca. 100 CE): *Lives of the Noble Romans.* Trans. by J. Dryden. Dell, New York (1969).

Pope H.G., Jr., Lipinski J.F., Cohen B.M. and Axekod D.T. (1980): "Schizo-affective disorder": An invalid diagnosis? A comparison of schizo-affective disorder, schizophrenia and affective disorder. *Am. J. Psychiatry 137* 921-27.

Raphael F. (1974): *Somerset Maugham and His World.* Scribner's, New York.

Reed S.C., Hartley C., Anderson V. E., Phillips V.P. and Johnson N.A. (1973): *The Psychoses: Family Studies.* Saunders, Philadelphia.

Reichenbach H. (1942): *From Copernicus to Einstein.* Philosophical Library, New York.

Reynolds V. (1967): *The Apes.* Dutton, New York.

Rimland B. (1964): *Infantile Autism.* Appleton-Century-Crofts, New York.

_______________(1969): Psychogenesis versus biogenesis: The issues and the evidence, in Plog S.C. and Edgerton R.B.: *Changing Perspectives in Mental Illness.* Holt, Rinehart and Winston, New York, pp. 702-35.

Ritvo E.R. (Ed.) (1976): *Autism.* Spectrum Publications, New York.

Roebuck J.B. and Kessler R.G. (1972): *The Etiology of Alcoholism.* Thomas, Springfield.

Rohrlich F. (1983): Facing quantum mechanical reality. *Science 221* 1251-55.

Rose N.R. and Friedman H. (1980): *Manual of Clinical Immunology.* American Society for Microbiology, Washington D.C.

Rosenberg L.E. and Kidd K.K. (1977): HLA and disease susceptibility: A primer. *New Engl. J. Medicine 297* 1060-62.

Rowe C.J. (1954): *An Outline of Psychiatry.* William C. Brown, Dubuque, Iowa.

Rubins J.L. (1978): *Karen Horney: Gentle Rebel of Psychoanalysis.* Dial, New York.

Runes D.D. (1959): *Pictorial History of Philosophy.* Philosophical Library, New York.

Rutter M. and Schopler E. (Eds.) (1978): *Autism.* Plenum, New York.

Sagan C. (1980): *Cosmos.* Random House, New York.

Sandler M. (Ed.) (1979): *Psychopharmacology of Aggression.* Raven Press, New York.

Schaller G.B. (1964): *The Year of the Gorilla.* University of Chicago Press, Chicago.

Schwabe A.D. (Moderator) (1981): Anorexia nervosa. *Annals Internal Medicine 94* 371-78.

Siegler M. and Osmond H. (1974): *Models of Madness, Models of Medicine.* Harper and Row, New York.

Simpson G.G. (1949): *The Meaning of Evolution.* Yale University Press, New Haven.

Skinner B.F. (1961): *The Behavior of Organisms.* Appleton-Century-Crofts, New York.

Smeraldi E. and Bellodi L. (1981): Possible linkage between primary affective disorder susceptibility locus and HLA haplotypes. *Am. J. Psych. 138* 1232-34.

Sparkes R.S. (Moderator) (1980): Human gene mapping, genetic linkage, and clinical applications. *Annals Internal Medicine 93* 469-79.

Spielberg S. (1980): *Close Encounters of the Third Kind* (2nd Version). Columbia Pictures, Hollywood. Written version (1978): Dell Books, New York.

Spiro H.R. (1968): Chronic factitious illness: Munchausen's syndrome. *Arch. Gen. Psychiatry 18* 569-79.

Srb A.M., Owen R.D. and Edgar R.S. (1965): *General Genetics.* W.H. Freeman, San Francisco.

Stekel W. (1929): *Sadism and Masochism.* Liveright, New York.

Stolpe S. (1960): *Christina of Sweden.* Macmillan, New York.

Stone M.H. (1980): *The Borderline Syndromes.* McGraw-Hill, New York.

Strum S.C. and Ransom T.W. (1975): Life with the pumphouse gang. *Natl. Geographic Magazine 147* 672-91.

Tache J., Selye H. and Day S.B. (1979): *Cancer, Stress and Disease.* Plenum, New York.

Talal N. (Ed.) (1977): *Autoimmunity: Genetic, Immunologic, Virologic and Clinical Aspects.* Academic Press, New York.

Talbott J.H. and Yu T.-F. (1976): *Gout and Uric Acid Metabolism.* Stratton, New York.

Thigpen C.H. and Cleckley H.M. (1957): *The Three Faces of Eve.* Kingsport Press, Kingsport, Tenn.

True W.P. (1960): *The Structure of the Universe.* Simon and Schuster, New York.

Tully G. (1949): *F.D.R.: My Boss.* Scribner, New York.

Usdin G. (Ed.) (1977): *Depression: Clinical, Biological and Psychological Perspectives.* Brunner/Mazel, New York.

388   BIBLIOGRAPHY

Valzelli L. (1981): *Psychobiology of Aggression and Violence.* Raven Press, New York.

Van Lawick-Goodall J. and Van Lawick H. (1971): *In the Shadow of Man.* Houghton Mifflin, Boston.

Vetter H.J. (1972): *Psychology of Abnormal Behavior.* Ronald Press, New York.

Watson J.D. (1970): *Molecular Biology of the Gene.* W.A. Benjamin, New York.

Weiner H. (1977): *Psychobiology and Human Disease.* Elsevier, New York.

Weintraub M.I. (1977): *Hysteria: A clinical guide to diagnosis. Ciba Clinical Symposia 29 No .6,* Summit, New Jersey.

Weitkamp L.R., Stancer H.C., Persad E., Flood C. and Guttormsen S. (1981): Depressive disorders and HLA: A gene on chromosome 6 that can affect behavior. *New Engl. J. Medicine 305* 1301-06.

West K.M. (1978): *Epidemiology of Diabetes and Its Vascular Lesions.* Elsevier, New York.

Whealon J.F. (Imprimatur) (1978): *Good News New Testament.* American Bible Society, New York.

Whitlock F.A. (1976): *Psychophysiological Aspects of Skin Disease.* Saunders, London.

Wilkinson D.G. (1981): Psychiatric aspects of diabetes mellitus. *Brit. J. Psychiatry 138* 1-9.

Winokur G., Clayton P.J. and Reich T. (1969): *Manic-Depressive illness.* C.V. Mosby, St. Louis.

SOURCES OF ILLUSTRATIONS

PLATE

1 By permission of E. Schweizerbart'sche Verlagsbuchhandlung, Stuttgart. From Adolf H. Schultz (1933): Die Korperproportionen der erwachsenen catarrhinen Primaten. *Anthropologischer Anzeiger.* Jahrg. X, Heft 2/3, pp. 154-185. Reproduction copy courtesy of Dr. Vernon Reynolds, Oxford.

2 Photo by Julius Kirschner. ©American Museum of Natural History.

3 Male baboon with young, Gilgil, Kenya. By T.W. Ransom. ©1975 The National Geographic Society.

4 Black-faced chimpanzee. ©New York Zoological Society.

5 Snowflake, the albino gorilla of the Barcelona Zoo. ©David Hosking, London.

6 Illustration ©1980 by Margaret LaFarge after photographs of J. Goodall. From John Tyler Bonner: *The Evolution of Culture in Animals.* Reprinted by permission of the Princeton University Press.

7 Lill's Performing Chimps, St. Louis Zoo, St. Louis, Missouri. ©Hallmark Cards, Inc., 1979.

8 Photo by F.D. Schmidt. ©The Zoological Society of San Diego.

9 Sea lion at Berlin Zoo. ©Rami Harcsztark, Berlin.

10 Aborigines of the Txukuhamae tribe, Upper Xingu River, Brazil. Photo: Jesco, Brasilia, and the Summer Linguistic Institute. Reprinted from R.L. Beals and H. Hoijer (1971): *An Introduction to Anthropology,* by courtesy of the Macmillan Publishing Co., New York.

11 Collection of author.

12 Mrs. J. F. Kennedy with Queen Elizabeth after having dinner at Buckingham Palace on 5th June 1961. Mrs. Kennedy chose a gown by a New York designer. It was a heavy pale blue sleeveless dress with a bateau neckline in front and a plunging V-neckline in back. ©United Press International.

13 Photo by Christian Jam, Nancy. Collection of author.

14 Prince Andrew on return from the Falkland Islands, September 1982. By Associated Press. ©Wide World Photos, New York.

15 Still from *Encore* (1951). Courtesy of the Rank Organisation, London.

16 From painting by Giovanni Bellini: *St. Francis in Ecstasy.* ©The Frick Collection, New York.

17 Leonard Bernstein rehearses the N.Y. Philharmonic Orchestra. Photo by Edward Hausner. ©The New York Times.

18 Culver Pictures, New York.

19 ©1979 F.J. Maroon, Washington D.C. Courtesy of the National Geographic Society.

20 Alfred Hitchcock with ravens. *Life Magazine* cover of 1 February 1963. ©Philippe Halsman, *Life.*

21 Drawings by Anthony Moore, from Desmond Morris (1977): *Manwatching.* Courtesy of Elsevier/Equinox Publishing Projects, Oxford.

22 Drawn from photographs by Patricia Moss-Vreeland. Montage by author.

23 Photo ©Association for the Advancement of Psychoanalysis of the Karen Horney Psychoanalytic Institute, New York.

24 The four characters of man, from Johann Kaspar Lavater's *Physiognomics* (ca. 1775).

PLATE

25 Opening stanzas of "Annabel Lee," reproduced from G.R. Woodberry (1909): *The Life of Edgar Allan Poe, with his chief correspondence with men of letters.* By courtesy of the Houghton Mifflin Co., Boston.

26 Portrait by K. Pochwalski, 1914. The Heeresgeschichtliches Museum, Militar-wissenschaftliches Institut, Vienna.

27 Peter the Great interrogating his son, Alexis. Nineteenth century painting by Nikolai Gyé at the Tretyakov Gallery, Moscow. Photo: Culver Pictures.

28 Winston Churchill flanked by the Generals Sikorski and de Gaulle. ©Imperial War Museum, London.

29 Still from *Quartet* (1948). Courtesy of the Rank Organisation, London.

30 Charles I: Triple portrait by Van Dyck at Windsor Castle. By courtesy of H.M. the Queen. Richelieu: Triple portrait by Champaigne. By permission of the Trustees, The National Gallery, London.

31 Daguerreotype by Mathew Brady. Courtesy of the Library of Congress.

32 Portrait by J. van Meytens. Kunsthistorisches Museum, Vienna.

33 Portrait by the studio of T. Lawrence. The National Portrait Gallery, London.

34 Portrait by M. Sittow, catalogued as "Isabella von Aragon." Kunsthistorisches Museum, Vienna.

35 Photograph by Alexander Hesler, June 3, 1860. Courtesy of the Chicago Historical Society.

36 Daguerreotype by N.H. Shepherd, the Library of Congress.

37 From portrait by Max Koner. The Bettmann Archive.

38 Portrait by Pesne. Courtesy of Verwaltung der Staatlichen Schlosser und Garten, Schloss Charlottenburg, Berlin.

39 Portrait by Hans Holbein, the younger. Courtesy of the Thyssen-Bornemisza Collection, Lugano.

40 Portrait of Napoleon in his study by Jacques-Louis David. Courtesy of National Gallery of Art, Washington D.C.

41 Portrait of the French School, 16th century. Musée du Louvre. ©Réunion des Musées Nationaux, Paris.

42 Lithograph by Franz Eybl after portrait of Anton Einsle (ca. 1872). Bild-archiv der Osterreichischen Nationalbibliothek, Vienna.

43 Count Wenzel Anton Kaunitz-Rietberg, minister under Maria Theresa. Engraving by J. Schmutzer from painting by J. Steiner (1765). Archiv fur Kunst und Geschichte, Berlin.

44 Nineteenth century lithograph after contemporary portrait. Courtesy of New York Public Library Picture Archive.

45 Courtesy of Ny Carlsberg Glyptotek, Copenhagen.

46 Culver Pictures, New York.

47 Self-portrait 1889-90. Musée du Jeu de Paume. ©Réunion des Musées Nation-aux, Paris.

48 Portrait by J. van Meytens. Akademie der bildenden Kunste, Vienna.

INDEX

Boldface type denotes major entries.

Abject states,47, **66-67**, 161. *See also* Depression.
    depressed ambition &, 134, **295-296**
    environmental bases of, 47
    in hospitalized patients, 181
    in NA type, 113-114
    in NPA type, 119
    psychotic, 276-77
Aborigines, American, 352, 480, Pl. 10
Adolescence
    NA–type &, 125
    personality development, 314
    resignation &, 47
Adornment, **39**
    in Alpine dancers, Pl. 18
    Aristotle on, 49
    in identical twins, 130
    in Elizabeth II & Mrs. Kennedy, Pl. 12
    in gorilla, 311
    in N type, 49, 90, 92 ff.
    in N type schizophrenic, 283
    in NA type, 111-112, 114
    in NA– type, 125
    in NA= type, 121
    object between teeth, 311, Pl. 14
    in pseudonarcissism, 132
    in S. American aborigines, Pl. 10
    subdued, 99, 104, 116, 120-121
    in Sublime habitancy, 259
Affective disorder, 70. *See also* Manic-depressive illness.
Aggression. **457-459**. *See also* Character types (of model); Character vectors; War.
    in animal model, 23-25, 36
    in childhood, 134-135, 244, 296
    eye contact in, 38. *See also* Eye contact.
    in habitancies, 259-261
    in human model, **36-38**
    inhibition or lack of, 279-280, 290-291, 295, 297, 303, 312
    mannerisms of, 37 ff., 142
    physiology of, 243 ff.
    pseudoaggression, 131
    psychosomatic disease &, 301
    schizophrenia &, 283
    survival &, 253
Akbar of Hindustan, 119

Alcohol, and flushing, 243, 328
Alcoholism, 161, 243, **292-294**
Ambition, **36 ff.**
    loss of & schizophrenia, 43, 45, 132 ff., **273 ff.**
    misguided, 317, **319-320**
    religion &, 149 ff.
    suppression of, 45, 64, 118, **134-135, 294-296**
Ambition, unbridled
    in Militant habitancy, 261
    in NA type, 50, 109, 148, 237
    in NA– type, 125, 190
    psychotic euphorias &, 292
Amenhotep II (Egypt), 303
Andrew, Prince (England), Pl. 14
Animal behavior, 21 ff., Pls. 1-9. *See also* Primates; Porpoise; Whale.
    aggression in rats, 239
    birds, 135, 239, 313
    complexity of, 31, 70
    intelligence, 35
    link with human behavior, 17, 32, 35 ff., 41, 190, 209, 317, 330
Anne, Queen (England), 349
Anorexia nervosa, 297
Antony, Marc, 94, 363-364
Anxiety, **148 ff.**
    exchanged for abject state, 209
    free-floating, 154, 195
    in NA type, 113
    in "power behind throne," 191 ff.
    in resigned type, 61 ff.
    in subjugated individual, 173, 204 ff.
    not understood, 47
    pervasiveness of, 89 ff.
Aristotle
    on anger, 54
    on science & intuition, 320-321
    on true friendship, 165
    on narcissistic personality, 49
Arrogance, 89, 158, 185
    in A type, 50, 97
    in NA type, 109
    in NPA type, 114 ff.
    in medicine, **179 ff.**
Arthritis
    rare in schizophrenia, 303, 325
Arthritis, rheumatoid, **303-304**. *See also* Diseases & syndromes.
    personality types in, 303, 325
    in Introspective habitancy, 306

Art
    character type &, 101, 129, 176, 254
    self-adornment &, 341
Arts, literary
    poetry & character type, 101, Pl. 25
    N type as author, 93
    submissive types, 123
Arts, performing
    comedians, 158, 186
    mime, 92
    motion pictures, 126, 148, Pl. 20
    N type &, 92
Asperger syndrome, 328
Attila the Hun, 98
Autism, infantile, 45, 101-102, 274, **275 ff.**
    environmental factors in, 279, 299
    hereditary psychosis &, 282
    hormonal factors in, 279
    in identical twins, 290
    incidence of, 278
    lack of aggression &, 279-280
    narcissistic rages &, 279
    perfectionism in, 40 ff., 51, 278
    personality of parents, 101-102, 278
    schizophrenia &, 275 ff.
    sex ratio in, 279
    speech disorders in, 297
Autocrat, NPA type, 54, 119, Pl. 38

Baboon. *See* Primates.
Barbarossa, 119
Beck, A.T., 290
Behavioral complex, **237**, 296. *See also* Aggression; Narcissism; Perfectionism.
Behaviorists, 269
Bellini, Giovanni, 142, Pl. 16
Belloc, Hilaire, 196, 208
Bernstein, Leonard, Pl. 17
Biochemical test, 244, 328
Biochemistry
    aggression &, 243 ff.
    basis of behavior states, 266-267, 325, 340, 354
    basis of disease, 297
    neurotransmitters &, 244
Bleuler, Eugen, 344, 433
Bleuler, Manfred, 281, 287-288, 294, 432, 433-437
Blushing, 40
    autonomic nervous system &, 245
    character type &, 243

inhibition of in pseudonarcissism, 307
narcissistic rage &, 245
in N type, 93
in NP type, 93, 100
vasodilatation &, 243
Bonaparte, Napoleon, 95-96, Pl. 40
Borderline states, 45, 134, 295, **485 ff.** *See also* Autism; Character types (of model):
                Submissive, Resigned, Non-aggressive withdrawn; Schizophrenia.
Bosch, G., 101, 279, 280
Bouchard, T.J., Jr., 130
Brel, Jacques, *Pref.*
Brothers, Dr. Joyce, 121
Brown, Thomas, 183
Browne, Malcolm, 130

Caesar (J.), 363-364
Caligula, Emperor (Rome) Pl. 45
Caroline of Brunswick, 187
Catatonic state, 101, 190, 290. *See also* Schizophrenia.
   periodic, 290
   in orangutan, 311
Catherine II, Empress (Russia), 95
Catherine of Aragon, 103, Pl. 34
Catherine de Medicis, Pl. 41
Cetacea, 312
Chagall, Marc, 95
Chaplin, Charlie, mother of, 283
Character structure
   breakdown of, 443 ff., 101, 190, 267, 271 ff.
   ignorance of, 3 ff., 89 ff., 144
   individual's merits &, 151
   instinctual origins of, 3, 21 ff.
   rigidity of, 21, 89 ff., 151, 321
   understanding of, 21
Character traits
   discreteness of, 131, 235
Character types. *See also* Interactions between types; Personality types.
   confusion of, 31, 47, 70, 133 ff.
   discreteness of, 4, 237
   genotypes of, **246 ff.**
   intelligence &, 266-267
   of parents & progeny, 247 ff. *See also* Progeny.
   of "power behind throne," **191-195**
   readily identifiable, 5
   somatotype &, 266
   subtypes of, **130,** 135, 282, **299,** 301 ff.
   value judgments of, 65

Character types (of model), **48 ff., 149 ff.**
    dominant types, **48 ff., 92 ff.**
        A type, **50, 97-98**
        N type, **48-49, 92-96**
        NA type, **50-51, 109-114**
        NP type, **52-53, 98-104**
        NPA type, **53-54, 114-120**
        P type, **51, 97**
        PA type, **54-57, 104-108**
    non-aggressive types, 64 ff., **474 ff.**
    non-aggressive withdrawn types, **64 ff., 129 ff.**
    null type, **247-249**
    submissive types, **57 ff., 120 ff.**
        compliant type (*C*), **57-58, 120-122**
        non-compliant type (*n-C*), **59-61, 122-126**
    resigned types, **61 ff., 126 ff.**
Character vectors, basic, 23 ff. *See also* Dominance; Submission; Resignation; Non-
        aggressive withdrawal.
    in animal model, **23 ff.**
    dominance, **41 ff.**
    non-aggressive withdrawal, **64 ff.**
    prior personality theories &, 265
    resignation, **47 ff.**
    submission, **45 ff.**
Charles I (England), 297, Pl. 301
Chateaubriand, 123
Cheney, Richard B., 195
Childhood
    character traits in, 244
    delayed speech in, 138
    hyperactive children, 296
    narcissistic trait in, Pls. 11 & 13
    personality development in, 65, 134-135, 265, 268, 269, 314
        See also Schizophrenia (juvenile onset).
    sadism in television &, 178
    schizophrenia in, **271 ff.**
    smile in, 279, Pl. 11
    submission in, 45 ff.
    suppression of traits A & N, 45 ff., 65, 257
Chimpanzee. *See* Primates.
Christ, Jesus, 320
Christina of Sweden, *Preface*
Chromosome, 237
Chromosome *6*, 245, 303, 325-326, 352, 353
Chromosome *16*, 352
Chromosome *21*, 246, 298
Chromosome *X*. *See* Gene, X-linkage.
Claudius I, Emperor (Rome), 297

Color blindness, 289, 292, 294
Complexion
    in A type, 97
    in N type, 92, 94
    in NP type, 100, 283
    in N–P type, 129
    in NPA type, 114, 116, Pl. 21
    in NA– type, 125
    in PA type, 104 ff., Pls. 21 & 40
    blanching of skin, 244, 364
    in character types, **139**
    in Introspective habitancy, 306
    in Maugham's Mildred, 202
    non-sanguine, 104 ff., 308, 474 ff., Pl. 39
    pale, 104 ff., 126, 202, 283, 363-364, Pls. 21 & 40
    Plutarch on, 363-364
    in pseudonarcissism, 308
    in rages, 94, 116, Pls. 21 & 38
    in resigned type, 126
    in rheumatoid arthritis, 303
    sallow, 97, 104
    sanguine, 94, 100, 114, 116, 125, 126, 129, 283, 303, 306
    in schizophrenias, 126, 139, 283
    in submissive types, 126, 306
Concentration camp syndrome, 296
Conrad, Joseph, 123
Creativity, 39, 44, 152
Crime. *See also* Psychopathic behavior; Violence.
    in habitancies, 258, 259
    NP type &, 53
    PA type &, 56
    submissive type &, 60
Criticism, reaction to,
    in A type, 97
    in N type, 49
    in NPA type, 116
    violence, 172
Cruz-Coke, R., 292, 294
Cynicism
    and sadism, 171, 174
    in PA type, 55, 106
    misinterpretation, 122

Dance, 125, 132, 259, Pl. 18
Darwin, Charles, 33
de Gaulle, Charles, Pl. 28

"Deal with life"
    and perfectionism, 53, 102
    NPA type &, 119, 160
Death, reaction of NP type, 102
Death wish, in *n-C* type, 59, 61
Defense mechanisms, 3, 89, 132-133
    general, **154 ff.**
    specific, **157 ff.**
Delirium, 296
Delusion, of poisoning, 280-281, 364
Dementia, 296
Denial, 157
Depression
    anaclitic, 296, 299
    endogenous, **291 ff.** *See also* Manic-depressive illness.
    exogenous, **292**. *See also* Abject states; Depression, reactive.
    psychotic, 275 ff.
    reactive, 47
Destiny of man, 91, 319 ff.
    and predestination, 322
    as vision of narcissism, 150, 319
Devil's pact, 150
Dilger, W.C., 239
Disease, 134, 297 ff.
Diseases & syndromes. *See also* Endogenous, Exogenous disorders; Psychosomatic
        disease.
    Addison's disease, 304
    alcoholism, 292, 294
    allergic reactions, 301, 304
    anaphylactic reaction, 243, 301
    anorexia nervosa, 270, 297
    arthritis, 304, Pl. 48
    arthritis, rheumatoid, 297, 300, **303-304**, 306, 325
    asthma, 300
    autoimmune, 303, 325
    bacterial endocarditis, 304
    cancer, 305, 328, Pl. 23
    carcinoma of breast, 305
    carcinoma of lung, 305
    celiac disease, 305
    collagen vascular disease, 304, 328, 329
    connective tissue disease, 303, 305
    dermatitis, atopic, 305
    dermatitis herpetiformis, 300, 305
    dermatologic disorders, 305
    dermatomyositis, 304
    diabetes mellitus, 297, 298, 304, 325, 328

Down's syndrome, **246**, 298
eczema, 300
emphysema, 305
endocrine disorders, 304
epilepsy, 281
glaucoma, 300
Graves' disease, 300, 304, 325
gout, 304
heart disease: congenital, 304
heart disease: coronary, 305, 328-329
heart disease: rheumatic, 304
hemochromatosis, 305
hives, 304
Hodgkin's disease, 305
hypertension, essential, 300
immune system &, 304
infectious disorders, 298, 305
inflammatory bowel disease, 325
intensive care unit psychosis, 305
leprosy, 298, 305
leukemia, 305, 329
lupus erythematosus, 300, 303, 304, 325
Marfan's syndrome, 266, 305
migraine headache, 305
mononucleosis, 305
multiple sclerosis, 305
Munchausen syndrome, 158
myasthenia gravis, 305
neoplastic disease, 305
peptic ulcer, 300, 306, 325
periarteritis nodosa, 303
pernicious anemia, 305
polymyositis, 303
psoriasis, 300, 305
rheumatic fever, 298, 305
rosacea, 300, 304
sarcoidosis, 304
scleroderma, 303
scoliosis, 305
skin disorders, 305
tuberculosis, 298, 305
ulcerative colitis, 300, 305
urticaria, 304
Disorders of behavior, **263 ff.**
Divorce rate
in Western countries, 91, 208
in Punctilious habitancy, 258

DNA, 237-238, 242
Dobzhansky, T., 313
Doctor Fell, 183
Dominance, **48 ff.**
     in animal model, 23 ff.
Dominance & submission. *See also* Interactions between types; Morbid dependency;
               "Playing the game"; Sadomasochism.
     in animal model, 23 ff.
     mannerisms in, 38
     submissive type &, 122 ff.
Dominant trait, 347
Don Quixote, 94
Dreams, narcissistic, 148
Drugs, abuse of, 161
Duty, sense of
     and containment of sadism, 170
     in internally directed aggression 291
     in NP type, 52, 99, 152, 258
     in NPA type, 54, 115 ff., 188, 196
     in PA type, 107
     paucity of: in A, NA types, 97, 110
     to fatherland, 319

Ecstasy, Pls. 16 & 17
     narcissism &, 142,149, 319, Pl. 16
     sadism &, 189, 333
     religious, 149
Edward VIII (Windsor), 123
Egyptian pharaoh, 303, Pl. 19
Einstein, Albert, *Preface*, 145
Elizabeth II (England), Pl. 12
Embarrassment
     blushing &, 40, 100
     as objective in sadism, 174
Embryology, 13, 14
Emotions, and sadism
     barrenness of, 172 ff., 331
     control over, 172 ff.
     in hospital setting, 179 ff.
Empathy *vs.* sympathy, 61
Empiricism, 9
Endogenous disorders, **297 ff.**
Energetic state, 25, 45. *See also* Personality splits.
Environment. *See also* Abject, Exhilarated states; Exogenous disorders.
     behavioral complexes &, 25 ff., 296
     counseling &, 269
     genetic factors &, 267 ff.
     imprinting, 135, 268

individual's merits &, 151 ff.
infantile autism &, 278, 299
isolation, effects of, 299
learning & conditioning, 239, 268, 314
personality development &, 23 ff., 43 ff., 132-133
real life situation &, 89
schizophrenia &, 271 ff.
stress &, 23, 27 ff., 267
training (animal), 25, 134
Envy, 175, 205
Epidemiology, 246, 298, 309, 314. *See also* Genetics, of populations;
     Habitancies.
Epilepsy, 281
Ethnic factors
character type &, 48
Down's syndrome &, 246
population genetics &, 254
Eugenics, 313-314
Eugénie, Empress (French), 187
Euphoria, endogenous, **291-292**
Euphoria, exogenous, **292**. *See also* Exhilarated states.
Evil, **167 ff.,** 319 ff
Lecomte du Noüy on, 317
Evolution
continuing, 11, 33
denial of, 13, 62
implications of, **13 ff.**
"jerky," 14
relative reality &, 15
trait N &, 312-313
"Exception that proves the rule," 91
in covert sadism, 176
in PA type, 90, 107, 331
in "power behind throne," 191
Exhibitionism
narcissism &, 39
pseudonarcissism &, 132
Exhilarated states, **68-70**, 161
happiness &, 156
sadism &, 205
Exogenous disorders, **292 ff.**
Externalization, 157, 173
Extroversion. *See* Character types (of model); Personality types; Temperament.
Eye contact
averted & sadistic behavior, 176
averted in haughty PA type, Pl. 28
averted in physician, 181

during courtship, 38
in A type, 97
in NA type, 109
in NP type, 100
in NPA type, 115
in PA type, 105 ff., 332, Pl. 3
in dominance & submission, 38, Pl. 27
in "power behind throne," 193
in resigned type, 126
in sadomasochism, 38, 110, 196, 200
in submissive type, 38, 120, 122, 124, Pls. 27 & 29
intense, 115, Pls. 3 & 26
intimidating glare, 38, 106, 142, 193, Pl. 3
Eyes
elevated, of narcissism, 143, 149
in N type, 96
in NPA type, 115
in PA type, 106 ff.
in catatonic orangutan, 311
in hyperthyroidism, 303
in rheumatoid arthritis, 303
in submissive type, 122, 303, Pl. 30

"Failure to thrive," 29, 136, 329, 350. *See also* Infertility.
Ferm, Vergilius, 317
Fertility & alcoholism, 294
"Flight or fight" response, 38, 243. *See also* Rages, aggressive.
Flushing
alcohol ingestion &, 243
autonomic nervous system &, 245
character type &, 243
disease &, 304
inhibition of & pseudonarcissism, 307
malar flush, 358
narcissistic rage &, 245
Focus, F-score in NPA test, 462
Folie à deux, 59, 209, **310-311**
Ford, C.E., 286
Fragmentation, 9, 157
Franz Ferdinand, Archduke (Habsburg), Pl. 26
Free will, 431
Einstein on, 145
elusive entity, 299-300
granted by Creator, 322
Henley poem, 145
loss of, 9-10, 21, 150
regaining, 9-10, 314, 321 ff.
Schopenhauer on, 145

Freud, Sigmund, *Preface*
    on heredity, 263
    Horney &, 265
    sexual repression &, 150
    tripartite theory of, 263
Friedrich Wilhelm I (Prussia), 119, Pl. 38
Friendship, Aristotle on, 165
Fugues, 126

Gambling, compulsive, 53
Gender differences
    alcoholism &, 292
    lack of, 48, **113**
    infantile autism &, 279
    manic-depressive illness &, 289
    periodic psychoses &, 289
Gender differences, female
    appearance in N type, 92-93
    and NA type, 111-113
    "brassiness" in NPA type, 115
    handwriting in PA type, 105
    laugh in NA type, 109
    love relations in NA– type, 109
    "masculine," 115
    smile in NA type, 109
    voice in NPA type, 114
Gender differences, male
    aloofness in orangutan, 309
    appearance in N type, 90, 92
    father figure, 112, 147-48
    voice in N type, 93
    voice in NPA type, 114
Gene(s), 238 ff.
    of aggression, 23, 29, 38, 41, 301 ff., 327
    character traits &, 41, 131, **239-242, 244 ff.,** 266, 282, 325-326
    complementary, 238, 345
    dominance & recessiveness, **238, 239 ff.**
    frequencies of, **250 ff.,** 347
    modifier, 238, 268, 272, 282, 284
    of narcissism, 40-41
    of perfectionism, 41
    of submission, **301 ff., 305 ff.**
    X-linkage, 238
        alcoholism &, 292, 294
        color blindness &, 289, 292, 294
        manic-depressive disease &, 289 ff., 309
        periodic psychoses &, 290

Genetic defects, 287
Genetic markers, 245, 303-303, 326
Genetics
    allelic polymorphism, **297 ff.**
        behavior states &, 300 ff.
        endogenous depression &, 291
        psychosomatic disease &, **300 ff.**
        systemic disease &, 300 ff.
    association, 238
    basic principles of, **237 ff.**
    character vector of submission &, **299-300, 305 ff.,** 329
    crossing over, intragenic, 242
    environment &, 3, 268. *See also* Environment.
    epistasis, 246, 304
    expressivity, 243, 298
    genocopies, 299, 363
    genotypes of character types, 246-249
    independent assortment, 238, 245
    linkage, 238, 245
    linkage disequilibrium, 266, 357
    Mendel's laws, 238, 358
    mutations, 29, 242, 300 ff.
    penetrance, 237, 242-243, 298, 362
    phenocopies, 299, 363
    phenotypes of parents & progeny, 246-249
    pleiotropism, 238-239
    polygenic inheritance, 302
    of populations, 239 ff., **473-480**
        alcoholism &, 292, 294
        color blindness &, 292, 294
        hypothetical habitancies, **251 ff.,** 352
        isolates, 254
        mating, assortative, 253-254
        mating, random, 239
        monomorphism of character type, 255
        natural selection, 253
        non-compliant submission &, **305 ff.**
        "Out of Africa," **473 ff.**
        polymorphism of character type, **254 ff.**
        psychosomatic disease &, **300 ff.**
    recombination, 238
Genghis Khan, 98
Geographic trails, 351, 475 ff.
George IV (England), 187, Pl. 33
George VI (England), 297
Gestures. *See also* Aggression, Narcissism, Perfectionism: mannerisms of.
    aggressive, in animals, 23

aggressive, in resigned type, 63
in conductor of music, Pl. 17
in character types, **142**
"narcissistic arms," 59, **142**, 311, Pls. 15,16, 19 & 20
in NP type, 100
in NPA type, 116
in schizophrenics, 142
Glory
aim of narcissism, 36, 39-40, 94, 112, 125
world domination &, 319-320
God. *See* Religion.
Goebbels, Joseph, 163
Gorilla. *See* Primates.
Griffin, E.I., 312
Grin. *See also* Laughter, Smile.
as trait of aggression, 39
contrasted to smile, Pl. 22
in PA type, 105
Guilt
in submissive type, 48, 57
induced in others, 168
of sadistic type, 209

Habitancies, hypothetical, **250-261, 305-307,** 352
and non-human primates, 312
Acquiescent, 352
Authoritarian, 256, **259-260**
Demonstrative, 256, **259**
Introspective, **305-307**
Militant, 256, **260-261**
Punctilious, **255-258**
Sublime, 256, **258-259**
U.S.A., **251-253**, 256
Haig, Jr., Alexander, 195
Handwriting, **143**
of A type, 97
of N type, 94
of NP type, 99, 103
of NPA type, 116
of medical, nursing student, 179, 181
of physician, 181
of PA type, 105
of submissive type, 122, 179
in U.S.A. habitancy, 251
Happiness
exhilarated states &, 156
rarity of, 156, 269
self-esteem &, **156-158**

Hardy-Weinberg approach, 239, **250 ff.**, 305-307. *See also* Genetics, of
          populations.
    validity of, 252-253
Hedonism
    in character types, 98, 110
    Maugham on, 158-159
Henley, W.E., 145
Henry III (France), Pl. 44
Henry IV (Navarre), 187
Henry VIII (England), Pl. 39
Heston, L.L., 284
Hippocrates, 265, 353
Histocompatibility, 245, 325-326, 353. *See also* HLA loci.
Hitchcock, Alfred, Pl. 20
HLA association, 353
HLA loci, **304-305**, 353
Hormonal factors, 279
Horney, Karen, *Preface*, 354, Pl. 23
    break with Freud, 265
    expansive types of, 87
    morbid dependency &, 197, **335-338**
    perfectionistic nature of, 337
    theory of neurosis, 147, 265, 354
Human behavior
    limitations of, 3, 15, 65, 90-91, 130, 157-158, 178
    unified theory of, 314
Humor
    in N type, 92
    in NP type, 100
    in PA type, 104-105
    sardonic wit, 55, 104-105, 158
Humors, ancient theory of, 165, 353, Pl. 24. *See also* Temperaments.
Hypochondria
    in infantile autism, 280-281
    in NA– type, 126
    in N–P type, 129, 190
    in schizophrenia, 364
Hypomanic behavior, 113. *See also* Character types (of model): NA type.
Hysteria, 51, 126. *See also* Personality types: hysterical, histrionic.

Ibsen, Henrik
    NA type &, 112, 147-148
    "power behind throne" &, 192
    speech to students, 323
    as submissive type, 123
    works of, 7, 112, 147-148, 192
Ideal image
    life expectations &, 154

of narcissism, 112, 166
    not measuring up to, 161
Idiot savant, 280
Imprinting, **135**, 268
Independence, 61, 121, 126
Indians. *See* Aborigines.
Individual, uniqueness of, 4, 70
Infancy
    anaclitic depression &, 296, 299
    character traits in newborn, 244
    personality development, 134-135, 299
    smile in, 244, 279
Inferiority complex, 48, 120
Infertility, **247-249,** 355
    in animal model, 29
    character type of parents &, 355
    rates of, 252
    schizophrenic parents &, 285
    in habitancies, **251 ff.**
Inheritance. *See* Genetics.
Insanity, self-questioning of, 3, 147-148, 207, 272
Insight, lack of, 7, 9, **147 ff.**
Instinct
    aggression &, 21 ff.
    character traits &, 3, 21 ff., 36 ff.
    sadism &, 37, 209
Instinctual drives, understanding of, 319
Intellectualization
    about love, 167
    as defense mechanism, 157
    as human quality, 35
Intelligence
    animal, 35
    character type &, 266-267
    human, 15, 35
    idiot savant, 280
    non-discrete nature of, 70
Interactions between types, **185 ff.** *See also* Character types (of model).
Intra-psychic conflicts
    in animal model, 27
    in non-compliant type, 59
    personality splits &, 43
    sources of, **169-160**
Introversion. *See* Character types (of model); Personality types; Temperament.
Inversions, **90**, 356
    of aggression, 132-133
    as "exception that proves the rule," 91
    avoidance of horror movie, 177

God as inversion of human ambition, 149
love as inversion of self-glorification, 166
pseudoaggression, pseudonarcissism &, **131 ff.**
Inverted denial, 90, 148
Inverted modesty, praise
in character types, 90, 92 ff., 116
in religious leader, 49, 142-143
Inverted sadism, 176-177
Inverted sadomasochism, 195-196
I.Q., 35, 130, 266
Isolation. *See* Depression, anaclitic; Environment.
Ivan, the Terrible, 119

Jackson, Andrew, Pl. 31
James II (England), 297
James, William, **355-358**, 384
on inheritance of personality, 235
morbid dependency &, 196, 206, 336-337
Joan of Arc, pose of, 143
Jones, Reverend Jim, 198-199
Josephine, Empress (France), 95-96

Kanner, Leo, 356
Kanner's syndrome, 275. *See also* Autism.
Karlsson, J.L., 284
Kaunitz, Count (Austria), Pl. 43
Kelvin, Lord: admonition of, 243
Kennedy, Mrs. John F., Pl. 12
Kermit the Frog, 188
Kierkegaard, Soren, 201-202
Kleptomania, 53, 176
Kraepelin, Emil, 275, 281, 282, 296, 356

La belle indifférence, 194, 357
Landers, Ann, 113, 272
Laughter, 141, 357
derisive, 106
in exhilarated states, 70, 156
mimics smile, 39, **141**, Pl. 22
in N type, 48
in NA type, 109, 114
in NP type, 100
in NPA type, 115, 117
in PA type, 106, 108
Lavater, Johann Kaspar, 436, Pl. 24
Leadership
and aggressive type, 152
and narcissistic type, 319-320

and NA– type, 125
and NP type, 100
and NPA type, 118-119, 254
and "power behind throne," 191 ff.
and submissive type, 112-113
of cult, 198-199
world leaders, 6, 178, 319-320, Pl. 12
Learning & conditioning, 268. *See also* Environment.
Lecomte du Noüy, Pierre, 317, 319, 321, 357
Lemur. *See* Primates.
Leonhard, Karl, 275, 282
classification of:
endogenous depression & euphoria, 291
manic-depressive illness, 288
periodic psychoses, 288
schizophrenias, 304 ff.
Life, expectations, **152 ff.**
Lincoln, Abraham, 103-104, 142, 197-198, Pl. 35
Lincoln, Mary Todd, 197-198, Pl. 36
Lorenzo the Magnificent (Medici), 95
Louis XIII (France), 297
Love, **165 ff.**
character interactions &, 165-166, **186-191**
devil's pact &, 150
of God & hedonism, 158-159
inverted sadism &, 177
James on, 196, 206
"lightning & thunder" affair, 123
"love-hate" confusion, 202-203
Maugham on, 197, 201, 335
reverence of nature, 61, 129, 163, 166
sadomasochism &, **195 ff.,** 368
Love relations, **186-191**
"power behind throne" &, 191 ff.
not understood, 91, 166-167, 207-209
of A type, 98
of NA type, 110-114
of NA– type, 126
of NP type, 52, 102
of NPA type, 53, 119
of PA type, 107
of submissive type, 57-58, 60, 123
outlandish expectations in, 209
Love triangle, 60
Ludwig II (Bavaria), 126, Pl. 46

McKusick, V.A., 252, 313

Madison, James, 303
"Madness," 287
Manic states, 70, 291-293. *See also* Euphoria.
Manic-depressive illness (bipolar), 70, **288 ff.,** 309. *See also* Depression.
    classification of Leonhard, 288
    misdiagnosis of, 390
    possible genetic heterogeneity, 289
Marguerite de Valois, 187
Maria Theresa, Pls. 43, 48
Marie-Antoinette, 95, Pl. 32
Marriage. *See also* Interactions between types.
    Belloc on, 208
    between character types, 102, 107, 111, 119, 186 ff.
    choice of older mate, 112, 188, 190
    entering into, 166-167, 208
    in *folie à deux,* 286
    of Lincoln, 187-188
    "power behind throne" &, 194-195
    resigned type &, 62
    sadomasochism &, 171
    schizophrenia &, 285
    stability of, 167, 186 ff., 258
    submissive type &, 121-130, 186 ff.
Martial (Rome), 183
Martyrdom
    masochistic, 204, 320
    narcissism &, 49
Masochism, 171 ff.
    hospitalized patient &, 181
    Introspective habitancy &, 306
    love &, 166
    mannerisms of, 331 ff., 335 ff.
    submissive triumph &, 57 ff., 120 ff., 331 ff., 335 ff.
    submissive types &, 120 ff.
    vulnerable victim &, 331-334
Maugham, W. Somerset, 358
    characterization of:
        N type, 95
        NA type, 111, 113-114, 191
        NPA type, 117-118
        PA type, 108
        resigned types, 127-128, 191
        submissive types: *See* Maugham W.S., works of: *Of Human Bondage.*
        end of an affair, 207
    in "narcissistic arms" pose, 142, Pl. 15
    on discrete character types, 87
    on evolution, *Preface*

on hedonism, 158-159
on ignorance of others, 150-151
on vindictiveness, 201-202
morbid dependency &, 197, 201, 206, 335 ff.
religious views of, *Preface*
stuttering in, 297
as submissive type, 123, Pl. 29
works of:
> *The Ant and the Grasshopper,* 111
> *Cakes and Ale,* 95, 118
> *Christmas Holiday,* 108
> *The Gentleman in the Parlour,* 128
> *Of Human Bondage, Pref.,* 123, 165, 197, 201-202, 206, 207, 335
> *The Lion's Skin,* 117
> *The Narrow Corner,* 95, 127
> *Rain,* 113
> *The Razor's Edge,* 127, 191, 207, 335-336
> *A Writer's Notebook,* 151, 158

Medicine, 265 ff. *See also* Diseases & syndromes; Psychosomatic disease.
> disease, concept of, 297
> Hippocrates, 265
> intensive care unit, 181, 243
> pathology, 314
> pediatrician, 244
> physical disability, 158
> physician-patient relationship, 181, 198, 202
> sadism in, **179-182**

Melancholia, involutional, 291. *See also* Depression, endogenous (unipolar).
Mencken, H.L., 313, 314
Mendel, Gregor, 4, 238
Mendlewicz, J., 290
Metabolism, inborn error of, 287, 298
Minnesota, University of
> personality inventory (MMPI), 266
> study of identical twins, 130

Miscarriages, 252, 285, 329. *See also* Infertility.
Miss Piggy, 188
Model, animal, **19 ff.**
Model, human, **35 ff.,**
> genetics of, **237 ff.**
> immutable principle of, 90
> implications of, 5

Models, mathematical, **13-15**
Models, religious, 17
Mona Lisa, & smile, 139
Monkey. *See* Primates.
Monogamous ideal, 208

Monomania, 196, 336
Morbid dependency, **195-209**
    absence in non-aggressive types, 38
    and Horney, 197, 335 ff.
    and Maugham, 197, 335 ff.
    and NA type, 111
    and PA type, 189, 190
    and "power behind the throne," 192
    and submissive type, 58, 60, 111, 189
    as source of stress, 159
    basic dynamics of, **195 ff.**
    in non-love relationships, 198 ff.
    lack of understanding of, 199
    persistence of, 206 ff.
Morris, Desmond, *Preface*, 19, 170
Motivation, 3, **134**. *See also* Ambition; Schizophrenia.
    lack of understanding of, 3, 17, **147 ff.**
Muller, H.J., 313, 359
Munchausen syndrome, 158
Murray, Henry, 265
Music
    narcissism &, 92, 149, 283, Pl. 18
    NP type &, 101
    religion &, 142, 149, 283
    singing, 92, 142, 283

Napoleon I (Bonaparte), 95-96, Pl. 40
Napoleon III, 95, 187
Napoleonic stance, 143
Narcissism, **37-42, 457-459**
    and blushing, 40
    and choice of older mate, 112, 188, 190
    and endogenous euphoria, 292
    and exhibitionism, 39
    and personality splits, 64
    and Sublime habitancy, 258-259
    and survival, 253
    and vision of destiny, 319-320
    in chimpanzee, 41, 312, Pls. 4 & 7
    in gorilla, 311
    in orangutan, 309-311, 337
    lack of, 131-133, 284, 474
    malignant, 179, **432, 460, 471-472**
    mannerisms of, 39, 49, 142, 311
        in Alpine dancers, Pl. 18
        in animals, **309-312**, Pls. 4, 5, 7, 8, 9
        bowing, 143, 257, 283, 319
        in child, Pls. 11 & 13

Joan of Arc pose, 143
"narcissistic arms" pose, 39, 141, **142,** 312, Pls. 15, 16, 19 & 20
object between teeth, 312, Pl. 14
in Prince Andrew, Pl. 14
physiologic bases of, 244
smile of, **39, 139, 369**
symbols of
colored lights, 149
flying, 148, 149,
music, heights, 163
Narcissus
of mythology, 36
modern day, 360
orangutan as, 310
Nervous system, autonomic, **244-246**. *See also* Rages.
parasympathetic, 245
sympathetic, 38, 244, 301
Nervous system, central, 244-246
"Neurotic" behavior, 5, 268
Nicholas II (Tsar), 103
Non-aggressive withdrawal, 64 ff., 129 ff., 295
"Normal" behavior, 5, 65, 177, 267, 272
self-questioning in NA type, 110-111
NPA personality test, 461-469
Nudity, 92

Obesity, 113, 266
Obsessive-compulsive
behavior, 3, 21, 37, 152, 185
personality, 52, 136, 270
Occam's razor, 325, 361
Occupation
and "power behind throne," 191 ff.
as defense mechanisms, 157-158
of N types, 92
of NP types, 101
of resigned types, 62
Opportunism
as defense mechanism, 157-158
sadism &, 171
submissive type &, 59-60, 122-123
Orangutan. *See* Primates.
Oriental societies & groups, 209, 243, 279. *See also* Habitancies: Punctilious.

Panic attacks, 59, 120
Paranoid behavior, 361. *See also* Schizophrenia.
and poisoning delusion, 280-281
and premorbid personality, 281-282

in covert sadism, 169, 174
in endogenous depression, 291
in *folie à deux,* 190, 286
in infantile autism, 280-282
in non-compliant type, 60
in PA type, **55-57**, 106
in periodic psychoses, 290
in schizophrenias, **281 ff.**
Parenting
   "failure" in nurturing, 268
   instinct of, 170
   of orangutan by human, 310
   preconceived ideals of, 299-300
   withdrawn children &, 135, 268, 296, 299
Passive aggressiveness. *See* Character types (of model): PA type.
Passive resistance, 101
Pecking order, 57
Peoples Temple sect, 198-199
Perfectionism, 35-36, **40-41**, 118. *See also* Character types (of model): P, NP, PA, NPA
           types.
   autism &, 40-41, 278
   covert sadism &, 56, 173-174
   "deal with life" &, 53, 102, 119, 160
   Horney &, 337
   lack of & unbridled ambition, 36, 97, 125, 292-293. *See also* Character types (of
        model): N, A, NA, NA–types.
   mannerisms of, 40-41
      grooming in baboon, 41, 312
      head-banging in child, 244
      in chimpanzee, 312, Pl. 6
      in orangutan, 310
      in porpoise & whale, 312
   passive aggressiveness &, 55
   in poet, Pl. 25
   in "power behind throne," 192
   religion &, 150
   schizophrenias &, 273
   stubbornness &, 100
"Personality conflicts," 185
Personality development, 3, 156-157, 314. *See also* Adolescence; Childhood;
        Environment; Infancy; Schizophrenia.
Personality disorders, 269-271. *See also* Character types (of model); Personality types.
Personality splits, **31, 41-47, 68, 294**
   absence in non-aggressive types, 64, 66-67
   as source of stress, 27, 31, 159
   in animal model, 31
   in NA type, 113
   in NPA type, 188, 196

in resigned type, 63
submissive type, 59
Personality theories, **266**, 327
Personality types. *See also* Character types (of model); Disorders of behavior;
          Temperaments.
of alcoholism, 292
"ambitious predator," 136
"anal sadistic," 270
animal-like, 45
anointed visionary, 152
of anorexia nervosa, 270
antisocial, 179, 271
"arrogant dynamo," 136
arrogant-vindictive, 271
of arthritis, 303, 325, Pl. 48
asthenic, 64, 129, 138, 271, 308
"asthenic aesthete," 129, 138
austere melancholic, 136
"bird of prey," 63, 110, 123, 147
"blushing boy," 100
borderline, 45, 134, 295
bovine, 136
"brooding non-sanguine autocrat," 136
in cancer, 305, 328, 360, Pl. 23
choleric, 136, 353, 476-479
"chronic complainer," 55, 104, 136
"chronic criticizer," 55, 104, 136
"clinging vine," 57
in compensation neurosis, 56
in coronary artery disease, 305, 329
cyclothymic, 51, 136, 270
detached neurotic, 271
of Down's syndrome, 246
expansive types of Horney, 87
explosive, 54, 136, 160, 270
in gout, 304
in Graves' disease, 325
"henpecked husband," 187-188
histrionic, 125, 136, 270
hypomanic, 51, 136, 270
hysterical, 51, 136, 270
homo universalls, 90
"humors," theory of, 265, 353, Pl. 24
"jovial endomorph," 266
in infantile autism, 274
"inferiority complex," 120
inverted sadistic, 271
in leukemia, 328

litigious, 56
in manic depression (premorbid), 299
in Marfan's syndrome, 266
"martinet with panache," 119
masochistic, 137
melancholic, 136
in migraine headache, 305
multiple, 270
narcissistic, 270
neurasthenic, 125, 271
"neurotic," 268
"nervous bird," 136
in non-human primates, 309-312, 330
non-sanguine autocrat, 136, Pl. 39
obsessive-compulsive, 52, 136, 270, 291
paranoid, 55, 136, 271
passive-aggressive, 55, 104, 136, 271
passive dependent, 271
perfectionist, 270
phlegmatic, 136
phlegmatic-melancholic, 136
prophet, 152
in pseudoaggression, 132 ff., Pls. 40, 43, 45, 47
in pseudonarcissism, 132 ff.
"quiet achiever," 63, 136
reactive depressive, 136, 137
in religious leaders, 317, 320
in rheumatoid arthritis, 303, 325
sanguine, 136
sanguine autocratic tyrant, 136, Pl. 38
"sardonic wit," 55, 104, 136, 158
schizoid, 136, 138, 271
schizoid-paranoid, 137
in schizophrenias (premorbid), 282-283
"self-anointed glory seeker," 136
self-effacing, 121, 166, 271
"shrinking violet," 45, 57, 120, 137
"Simple Simon," 99
sociopathic, 132, 179, 271
"solid citizen," 115
"suspicious manipulator," 136
type A & B, 266, 329
tyrannical despot, 56
in ulcerative colitis, 325
in urticaria, 304
world leaders, 178, 320
Peter principle, 57
Peter the Great (Tsar), Pl. 27

Phobia, 363
    resigned type &, 62
    submissive type &, 59, 126
Photographs
    of character types 4, 140-141
    of PA type, 105
    of submissive types, 122, 142
    in U.S.A. habitancy, 251
Physiognomy, 363-364. *See also* Character types (of model); Smile.
    according to Lavater, Pl. 24
    poker face, 93, 99, 104, 105
    in rages, Pl. 21
    in submissive types, Pls. 29 & 30
Piniella, Lou, 116-117
Planck, Max, *Preface*
"Playing the game," 351
    absence in non-aggressive types, 43-44
    absence in orangutan & gorilla, 310-311
    in animal model, **25 ff.**
    in baboons, 312, Pl. 3
    mannerisms in, 27 ff., 38
    sadism &, 37
    as source of stress, 27, 47-48, 159-160, 273
    in various types, 50, 97, 107,119, 125, 141
Poe, Edgar Allan, Pl. 25
Porpoise, 35, 312
Power
    trait of aggression &, 36 ff.
    world domination &, 319
"Power behind the throne," 107, 136, **191-195**, 365
    covert sadism &, 176
    historical examples, 195
    source of stress, 160
    subjugated world leader &, 125, 320
    suppression of ambition &, 118, 134
    vulnerability of submissive type to, 123, 125
Predestination, 322-323
Pride, 151-152. *See also* Self-esteem.
Primates
    animal model &, 23 ff.
    apes, 35, Pl. 1
    baboon, 45, 312, 343, Pl. 3
    chimpanzee, 41, 312, 330, Pls. 1, 4, 6 & 7
    comparative anatomy of, Pl. 1
    gorilla, 311, Pl. 5
    lemur, Pl. 8
    monkey
        aggression in, 170

reared in isolation, 19, 190, 299
    submissive, 190
  orangutan, 309-311, 330, Pls. 1 & 2
  relationship to Ungulata, 313
Progeny
  character type of parents &, 247 ff., 339-340
  phenotypes of, **247-248**
  viability of, 239, 247 ff. *See also* Infertility.
Pseudoaggression, **132-133**, 179, 366, Pls. 40, 45, 47
  genetic factors in, 307-308
  nature of, 91
  schizophrenia &, 132-133, 295
Pseudonarcissism, **144-145**, 394
  coercive nature of, 95
  genetic factors in, 324-325, 332-333
  in Maugham novel, 145-146
  schizophrenia &, 146, 147, 320
Psychiatry. *See also* Horney; Freud.
  concepts of:
    anxiety, 154
    paranoia, 190
    personality disorders, 51, 52, 54, 56, 63, 64, 70, 129, **265 ff.**
    psychosomatic disease, 300 ff.
    sadomasochism, 166
  counseling in morbid dependency, 202
  of future, 244, **268-269**, 314
  genetics &, 263
  limitations of, 5, 147, 268-269
  subjugation of patient by physician, 181
Psychic unity, 89, 91, 120, **147 ff.**
Psychoanalysis, 197, 263
"Psychopathic" behavior, 45, 53, 56, 60
Psychoses, 366. *See also* Autism; Disorders of behavior; Depression; Schizophrenia.
  family history of, 272, 278, 282, 284-285, 286, 288 ff.
  in intensive care unit (ICU), 305
  patterns of aggression in, 290-291
  periodic, **288-291**
  poisoning delusion &, 280, 364
  response to stress, 4, 272-273
  schizo-affective, 290
  toxic, 296
Psychosomatic disease, **300-305**, 367
  medical implications of, 304-305, 325 ff.
Psychosomatic symptoms, 113, 126. *See also* Hypochondria.
Public speaking
  aggressive type &, 152
  narcissistic gestures &, 142
  N type &, 92, 94

NP type &, 100, 142
NPA type &, 90, 116
submissive types &, 120, 122, 123-124, 125

Racial factors
   facial complexion &, 100. 139
   flushing &, 243
   population genetics &, 254, 473-480
Racism, & genetics, 130
Rado, Sandor, 263
Rages, **70, 90**, 133, Pl. 21
   infrequency of, 109, 116, 160
   not understood, 47
Rages, aggressive-vindictive, **38, 244-245**, Pl. 21
   in A type, 50, 98
   absence in pseudoaggression, 131
   in animal model, 27, 31
   extensively studied, 243
   in morbid dependency, 202
   in NA type, 109
   in PA type, 56, Pl. 21
   sadism &, 169, 172
   in submissive types, 47, 58, 60, 155, 202
   submissive triumph &, 155
   vindictive triumph &, 155
Rages, combined, **245**
   in Mary Lincoln, 188
   in NA type, 51, 114
   in NPA type, 53-54, 90, 116-117, 188, 196, Pls. 21 & 38
   red-faced, 125
   in sports figure, 116-117
   in submissive types, 60, 125
Rages, narcissistic, **40, 245**. *See also* Tantrum.
   absence in pseudonarcissism, 132, 307
   infrequency in N–P type, 129
   narcissistic triumph &, 169
   in N type, 49
   in NP type, 53, 101, 187
   NP schizophrenias &, 281-282
   in orangutan, 310
   red-faced, 94
   in resigned type, 63, 156
Rages, septal, 244
Rages, sham, 244
Rationalization, 89, 156-157, 168
Real self, finding, 173
Reality
   relative, **15**

Recognition. *See also* Smile.
    trait of aggression &, 154
    trait of narcissism &, 35-36, 39-40, 99, 142-143, 165
Religious views
    Einstein, *Preface*
    Lecomte du Noüy, *Preface*
    Maugham, *Preface*
    Thompson, J.A., 9
    Shaw (J.B.), 208
    rejection of, 9-10
Religion. *See also* Martyrdom.
    "acts of God," 198
    Christianity, 149-150, 208, 320
    conventional dogma, 9-10, 17, 317
    containment of sadism &, 170
    creationism, 320
    as defense mechanism, 158, 167
    devil's pact, 150
    fanatic of, 211
    genetics &, 149-150, 254
    God
        as ultimate image of narcissistic ambition, 149-150
        omnipotent, 322
        sadomasochism &, 198
        visions of, 149-150, 283
    heaven & hell, 149, 198
    hedonism &, 158-159
    immortality, 150
    narcissism &, 142, 149, 152, 165, 258-259, 319-320, Pl. 16
    perfectionism &, 92, 149-150
    philosophers of, 317
    priest & penitent, 198
    prophesy & evangelism, 49, 92, 152, 156, 168, 317
    reincarnation, 150
    religious leaders, 317, 319-320
    St. Francis of Assisi, 142, Pl. 16
    science &, 9-10, 15, 17, 70, 320-324
    unbelievers, 320
    universal nature of, **149-150**
    various character types &, 62, 99, 115, 129, 152, 166
Renoir, Pierre-Auguste, 303
Reptilia, 313
Resignation, **47, 61-63, 126,** 367
    in animal model, 27-29
    causes of, 61, 244
    as a healthy realization, 209
Richelieu, Cardinal de, Pl. 30
Rimland, Bernard, 1

420   INDEX

Ritualism
    arts of self-defense &, 279
    covert sadism &, 173-174, 209, 331 ff.
    infantile autism &, 40-41, 278, 279
    NP type &, 53
    PA type &, 56, 331 ff.
    perfectionism &, 41-42, 173-174
    some schizophrenias &, 136
Rome, ancient, 94, 183, 297, 364, Pl. 45
Roosevelt, F.D., 360
Roosevelt, Theodore, 95
Rudolf of Mayerling (Habsburg), 129
Rutledge, Ann, 104
Rutter, M., 282

Sadism, 4, 36-37, **167 ff.**, 368. *See also* Aggression; Morbid dependency;
        "Playing the game"; Sadomasochism.
    A type &, 50, 98, 168
    absence in non-aggressive types, 37, 101, **177 ff.**
    American television &, 178
    choice of vulnerable victim, 171, 331-334
    containment of, 119, 169-170, 177-178
    covert, 56, **168-169, 173-176**, 200-201
    dissimulation &, 176
    general manifestations of, **171-172**, 197-198
    inverted, 169, **176-177,** 196
    laughter &, 106, 141
    in medicine, **181-182**
    NA type &, 51, 110, 112, 147-148
    NPA type &, 119, 196
    not recognized, 37, 173, 176-177, 199
    opportunistic, 171
    overt, **169 ff.**
    "power behind throne" &, 193
    PA type &, 56, 107, 331-334
    PA– type &, 45
    resigned type &, 63
    sexual behavior &, 36-37, 170, 198-199
    situational, 60, 119, 171, 188
    subliminal, 177-178
    submissive type &, **60-61**
    teasing &, 101, 110, 176
    world domination &, 319-320
Sadomasochism, 110 ff., 123, 165, 189-190, 286, **331-334**. *See also* Morbid
    dependency; "Playing the game."
St. Francis of Assisi, 142, Pl. 16
Santayana, admonition of, 178
Schizo-affective illness, 290, 368

Schizoid behavior, 271
  covert sadism &, 173-174
  folie à deux &, 286
  PA type &, 56-57, 107, 189, 283
  perfectionism &, 41, 43, 173-174
  "power behind throne" &, 194
  predicted by model, 4-5
  resigned type &, 45
  schizophrenia &, 274-275, 284
  sociopathic behavior &, 45, 173 ff.
  submissive types &, 45, 47, 60, 190
Schizophrenia, 47, 58, 68, 129, 136 ff., 190, **271-288**, 330, **368**. *See also* Autism;
        Endogenous, Exogenous disorders; Psychosomatic disease.
  alcoholism &, 292
  arthritis &, 303, 325-326
  Bleuler (M.) on, 287-288
  brain morphology &, 299
  classes of, **374 ff.**
  classification of Kraepelin, 275, 281, 282
  classification of Leonhard, 275, 281, 282, 284-285
  complexions in, 126, 139, 283
  damnosa hereditas, 287. *See also* Psychoses, family history of.
  HLA loci &, 303-304, 325-326, 329
  imprinting &, 135
  infectious agent &, 305
  lower risk to, 126, 286, 303, 325
  NA, NPA types &, 286
  in orangutan, 311, 330
  paranoid forms, **281-282**
  perfectionism &, 40-41, 53
  poisoning delusion &, 280-281, 364
  pseudoaggression, pseudonarcissism &, 132-133, 179, 295, 308
  quantitative aspects of, **284-286**
  risk in Introspective habitancy, 306
  smiles in, 279, 283
  socio-cultural factors &, 308-309
Schopenhauer, 145
Schumann, Robert, 129, 364
Schweitzer, Albert, 163
Seal (sea lion), 369, Pl. 9
Self-consciousness, & submissive type, 122, 123-124, 142
Self-destruction, 102-103, 129, 147
  self-hate &, **161-162**
Self-esteem, **151 ff.** *See also* Pride.
  of narcissism, 35-36, 48, 94, Pl. 18
Self-hate, **161-162**
Selfishness, 89, 158, 167-168, 176
Self-justification, 156-157, 169, 185

Self-preservation, 170, 253
Senility, 296
Sexual behavior
    of aggression & sadomasochism, 23, 36, 38, 39, 58, 170, 177, 198, 199, 253, 331-334
    amorality of innocence, 93, 95
    in character types, 254, 257
        in N type, 49, 92-93, 95, 186
        in NA type, 51, 110-111, 147-148, 188
        in NP type, 52, 102, 187-188
        in N–P type, 129, 190
        in NPA type, 53
        in resigned types, 62, 63, 166
        in submissive types, 60-61, 189-190
    as defense mechanism, 158
    in Demonstrative habitancy, 259
    eye contact in courting, 38
    Freudian view & psychosexual factors, 150, 269
    masturbatory activity, substitute, 122
    in monkeys & apes, 170, 309 ff.
    narcissism &, 253, 309 ff.
    promiscuity & "hypersexuality"
        as counterfeit lovability, 167
        and population genetics, 253 ff.
        in various character types, 50 ff., 92-93, 110-111, 119, 186, 188, 190
    relationships not understood, 91, 166, 173
    seduction, Kierkegaard on, 200-201
Sexual exploitation, 172, 181, 198-199
Sexual preference
    bisexuality, 198, Pl. 44
    character type &, 48
    morbid dependency &, 198
    N type &, 92-93, 95
    pseudonarcissism &, 132
    resigned type &, 126
    sadistic subjugation &, 198
    submissive type &, 126, 198
Sexual repression, 122, 150, 177
Shame
    in submissive type, 48, 57, 120-121
    objective in submissive triumph, 202
Shaw, G.B., 204, 208
Sheldon, W.H., 266
Simpson, G.G., 11
Smile, 369, Pls. 4, 5, 7, 10, 11, 12
    in character types, 139-140
        in N type, 92, 95
        in NA type, 109
        in NP type, 99

     in NPA type, 115
     in PA type, 105, 108
     in submissive type, 142
  gingival, 99, 257, 351, 369, Pls. 5, 7, 10, 11
  in hebephrenias, 283
  in infant, 244, 279
  lack or infrequency of, 98, 105, 107, 122, 129, 279, 283, 308, 312, Pls. 3, 22
  laughter &., 139, 140, 141, Pl. 22
  of narcissism, 35-36, **39, 99-100,** 351, 369
  poses and gestures with, 142, Pls. 12, 18
  Primates &., 41, 310, 311, 312, Pls. 4, 5, 7
  sardonic, 105, 108, 139, 140
  in sexual context, 253
  sincerity of, 99-100,155
Smoking (tobacco), 161
Social class, & character type, 48
Sociobiology. *See* Socio-cultural factors.
Socio-cultural factors. *See also* Genetics, of populations; Habitancies; Religion.
  behavior disorders &., 308-309
  character traits &., 91, 135, 254-255
  lovability &., 166-167
  sadomasochism &., 169-170, 209
Sociopathic behavior, 53, 56, 132-133, 179
Somatotype, 266,
Sophie, Archduchess (Bavaria), Pl. 42
Speech disorders, 122, 138, 281, 297
Spielberg, Steven, 148
Sports
  character type &., 62, 94, 98, 100, 116-117, 130, 177
  of high risk, 161
Stage fright, & submissive type, 59
Stekel, W., 331
Stillbirth, 136, 252, 285, 329. *See also* Infertility.
Stress, 23 ff., 64, **174 ff.** *See also* Anxiety; Environment; Intra-psychic conflicts.
Stubbornness, 100, Pl. 34
Stuttering, 297. *See also* Speech disorders.
Subdued states, 23-25, 42-45
Subjugation, 23-25, 31, 43-45, 134, 294. *See also* Morbid dependency;
       "Playing the game"; Sadomasochism.
Submission, **44-46**, 133, 370
  in animal model, 25 ff.
  diseases &., 328
  genetic & environmental factors in, 25-27, 45, 134-135, 244, **301-304**
  imprinting &., 135
  in monkeys & apes, 19, 355
  schizophrenia &., 273, 285
Submissiveness, S-score on NPA test, 461

Success, fear of, 57, 62, 120 ff.
Suffering
    masochism &, 155
    morbid dependency &, 202
    submissive triumph &, 155
    submissive type &, 60, 158
Suicide. *See also* Self-destruction.
    lack of insight in, 162
    prevention of, 162
    vindictive triumph of, 162
Symbiotic relationships, **191-209**. *See also* Dominance & submission;
        Interactions between types; Morbid dependency; "Power behind the throne."

Tantrum, 40, 259, 279
Teasing. *See* Sadism.
Television
    character types seen on, 120, 158, 186, 188
    sadism in, 178
Temperament, 130, 363, 371, **430**
    T-score in NPA test, 461-462
Temperaments, 353
    choleric, 136
    melancholic, 53, 64, 104, 129, 130, 136, 190, 309, Pl. 2
    phlegmatic, 63, 98, 130, 136, 309
    sanguine, 92, 136
    theory of "humors" &, 265-266, 353, Pl. 24
Tests, 327-328
    intelligence, 130, 266-267
    personality, 265-266, 358
    NPA personality, 461-472
Tobacco, abuse of, 161
Todd, Mary. *See* Lincoln.
Trail, geographic, 351, 475 ff.
Trait, 237. *See also* Behavioral complex; Character vectors.
Triumphs,
    aggressive-vindictive, **154-156**
        in A type, 97
        in NA type, 114
        in NPA type, 119
        in PA type, 107, 331-334
        sadism &, 171, 331-334
        suicide &, 162
    narcissistic, **155**
    resigned (detachment), **156**
    submissive, **155**, 190, 199, 202
Tutankhamun, Pl. 19
Twin studies, **130**

Twins, identical, 130, 244, 285, 299
Type A & B, 266, 329. *See also* Personality types.

Ulterior motives
    love &, **165-166**
    narcissism &, 93, 150
Unconscious competition, 194, 195
Unconscious mind, 3
Ungulata, 313, 371

van Gogh, Vincent, 129, Pl. 47
Vanity, 371
    in character types, 53, 92, 101, 116
    in orangutan, 310-311
Vindictiveness. *See also* Aggression; Rage, aggressive; Morbid dependency; "Power
                behind the throne"; Sadism.
    absence of, 94-95, 101
    basis in aggression, 38
    in character types, 98, 105, 109, 119, 154-155, 168
    in covert sadism, 175
    in morbid dependency, 202
    in "power behind throne," 193
    Maugham on, 201-202
Violence. *See also* Rages.
    lack of
        in habitancies, 253-259
        in N type, 48-49
        in orangutan, 309-311
    latent, 177
    in NP type, 53, 101
    in NPA type, 54, 116, 119
    in PA type, 55-56, 190, 331-334
    in submissive types, 58-61
    on patient by physician, 182
    in psychoses, 280-281, 283
    reflected in language, 176-177
    sadomasochism &, 172 ff., 331-334
    self-inflicted, 161-162
Voice
    in N type, 48, 92, 94, 96, 283
    in NA type, 109
    in NPA type, 114-115
    in NP type, 90, 100
    in PA type, 105, 108
    in resigned type, 126
    in submissive type, 120, 122
    stentorian, 114

426   INDEX

War
in habitancies, 253 ff., 306
initiated by submissive nation, 320
military strategist, 55
narcissism &, 319-320
potential nuclear, 55, 178, 319-320
sadism &, 171, 319-320
World War II, 163, Pl. 28
Western societies
alcoholism &, 292
character type &, 91, 111
divorce in, 91, 208
precariousness of, 91
schizophrenia &, 283, 308-309
Whale, 312
Wilhelm II (Kaiser), Pl. 37
Wince, perfectionist, 161
Winokur, G., 289, 294
Wolff, P.H., 243
"Workaholism"
in NP type, 52, 99
in NPA type, 115

XA type, 43,45, 372
X-linkage. *See* Gene, X-linkage.

ADDENDUM

Update 1985-2008	**429**
Correspondence with M. Bleuler	**433**
Profiles of NPA Character Types	**438**
Comparison of Narcissism and Aggression	**457**
Inadequacy of DSM-IV Criteria for NPD	**460**
The NPA Personality Test	**461**
Testing of Subjects with NPD	**471**
Population Genetics: "Out of Africa"	**473**
NPA Theory: Original Article (1990)	**481**

UPDATE 1985-2008

Q.   Over twenty years have gone by since the publication of "Toward Self & Sanity." How was it received?

A.   The reception has been mixed. Some individuals have written to us with enthusiasm, to the point of saying that the book has changed their lives for the better. However, the American psychiatric establishment has shown no interest, and our efforts to have the theory published, even in speculative form, were met with stiff resistance from the journal editors. It was finally published in 1990 in the peer-reviewed journal* *Speculations in Science & Technology*, and the full text of that article is included in this volume. The noted Swiss psychiatrist Manfred Bleuler expressed interest in the theory, and his communications are also included below.

Our general impression was that neither the mental health establishment nor the general public, in America, was ready for a genetic theory of personality having possible far-reaching implications.

* Afterwards we served on the editorial board of the journal.

Q.   Have any discoveries in the field of behavioral genetics since 1985 tended to support or disprove the NPA theory?

A.   Progress in behavioral genetics has been slow, but we have seen no information that would cause us to modify the basic ideas of the theory.

Q.   Did you gather any supporting evidence for the validity of the NPA theory?

A.   Our duties in the practice of medicine (critical care) did not permit us to carry out any controlled studies. However, we have examined many family pedigrees, mainly of European Royalty. These pedigrees tend to confirm our initial finding that N and A are recessive traits, with P being Mendelian dominant.

Q.   What were some criticisms of the NPA theory?

A.   Some thought it was too simplistic. Others thought it too complex to understand. Some thought that the character types were presented in too favorable light (especially the narcissistic N type), while others thought the descriptions too pejorative and pessimistic.

Q.   Since you are presenting a serious theory of human behavior, why do you quote such ordinary popular sources?

A.   We would like to emphasize that, for most of us, insight into our behavior will come from awareness of the most ordinary aspects of our daily lives. It is not likely to come from the analysis of the treatises of professors residing in their ivory towers.

Q.   If you were preparing a completely revised edition of the book, what changes would you make?

A.   We would clarify three areas:

- the role of temperament, and how it is a part of personality,

- genetic and environmental factors in the states of submission and resignation, and

- malignant aggression and narcissism.

Q.   Well, how *is* personality involved in social behavior, and what is the difference between personality and temperament?

A.   We consider that, in general, an individual's behavior in a social situation is determined by the individual's *personality*, by the *personalities of other individuals*, and by his (her) *real-life situation* (present-day environment).

An individual's personality is, indeed, a complex entity. It consists of a genetically determined *character type* (the NPA traits), a genetically determined *temperament* ("excitability," or "reactivity" in the Pavlovian sense), and many other *secondary traits* that may have genetic and environmental bases. The environmental bases ("nurture") of the secondary traits are manifold and involve foremost the individual's upbringing by parents and caregivers.

Our theory focuses on the genetically determined NPA character type. Of all of the possible genetic factors in behavior the character type is the easiest to analyze because only three basic traits are involved. The character type is discrete, readily identifiable, and traceable in a family pedigree. Temperament and other secondary traits of personality depend on many genes ("polygenic") and will be much more difficult to quantify.

Consider the simple analogy of behavior with viewing a television set: *character type* is the channel selector, *temperament* is the volume control, and how well the television picture is actually seen depends on the conditions in the room, or *environment*.

Q.   What other factors intrinsic to the individual, other than personality, influence social behavior?

A.   The "basic human drives" in the categories of hunger, thirst, sexuality, territoriality and desire for pleasure may be important. Intelligence and cognitive abilities are important. These would complicate the analysis of a given social situation.

Q.   So knowledge of a person's character type does not allow us to predict his/her behavior?

A.   Not the specific details of behavior. But the genetic NPA character type of an individual is more restrictive than most of us would care to admit.

Q.   Why is our behavior so much of a problem to all of us, especially in love relations? Do we really have a free will?

A.   We do have a free will. But in human behavior, especially in love relations, it is usually a struggle. The struggle is between our cognitive free will (the higher brain functions of thinking, reasoning and learning) and the compulsive aspects of our personality. A good deal of the compulsive aspects stem, of course, from the NPA traits that are genetically inscribed in our character types.

Q.   What would you like to clarify with regard to the states of submission and resignation?

A.   In our original treatment we considered the character vector of *submission* to be primarily environmentally determined. Our analysis of family pedigrees now convinces us that the A– trait is almost always genetically determined. If the A– trait is present in a child, one can almost always find the trait in at least one of the parents. So the aspect of "shyness" associated with the A– trait is unquestionably of genetic origin.

With regard to *resignation*, we now believe that these individuals are of two groups:

- dominant types having the A trait, and

- submissive types.

In Somerset Maugham's description of resigned types who inhabit the South Sea Islands (p. 128) we believe that he was describing mainly submissive types who fled competitive society and adopted solitary lives of detachment.

Thus, referring to our original Figs. 6, 10, 12 and 14, we can add a two-way "pathway" between *Submission* and *Resignation*. In Class II schizophrenia (Fig. 14) this would mean partitioning schizoid individuals undergoing psychosis into two groups, depending on whether the relevant genetic and environmental circumstances were more in the category of submission or resignation.

Q.   And malignant aggression and narcissism?

A.   In the original edition we focused on the malignant aspects of the trait of aggression, namely sadism and the dynamics of the "morbid dependency." But we could equally emphasize the malignant aspects of narcissism, perfectionism, submission or resignation.

There are, indeed, malignant aspects of narcissism (NPD, or "narcissistic personality disorder") that are striking counterparts to those of aggression.* Just as there can exist an aggressive rage and aggressive-vindictive behavior, there can occur the narcissistic rage and narcissistic-vindictive behavior. There are also counterparts to the "morbid dependency" based on narcissism, and to the "power behind the throne" based on an insecure N type or on an authoritarian NP type. There exist not only sociopathic A types but N types as well, and there can occur narcissistic bullying as well as aggressive bullying. Even paranoia can have its roots in narcissism as well as in aggression. And there are other aspects of aggression that have their counterparts in narcissism.** But that, indeed, could be the topic of another book...

* See for example an account by a self-proclaimed "malignant narcissist": Vaknin S. (1999): *Malignant Self-Love*, Narcissus Publications, Skopje.

** See *Comparison of Narcissism and Aggression*, p. 457 ff.

Q.   Well, are you closing on a note of optimism or pessimism?

A.   It costs us nothing to be cautiously optimistic. Permit us to close with the words of Manfred Bleuler (1985), whose humanistic approach to life is one with which we can feel profound sympathy:

> "I think in particular that the consequences of the genetic types can be both: shameful and admirable. We should not neglect that there exists all around us wonderful human doing: the care of a mother to her children, of a husband to his aggressive psychotic wife, of a nurse toward the sick. And I confess that I think admirable heroism plays an important role in human society..."

M. BLEULER

CORRESPONDENCE WITH MANFRED BLEULER
(1903-1987)

In June of 1985 we turned toward Europe in an effort to find someone in the psychiatric community who was interested in our approach. Dr. Bleuler's ideas that schizophrenia was predisposed mainly by "normal traits in unfavorable combination" were consistent with the NPA model, hence it was in this spirit that we contacted him.

Dr. Bleuler was, of course, the son of Eugen Bleuler, one of the true giants of classical psychiatry. (It was the latter who brought the term "schizophrenia" into common use.) It was, thus, with great respect and anticipation that we entered into the correspondence given below.

Alas, our meeting was not to take place: due to personal reasons we were unable to travel at the time, and Dr. Bleuler passed away in 1987.

We all have some regrets in our lives, and among ours is that we were unable to meet this Master of the psychiatric profession, still so active and still maintaining such an inquiring mind after so many decades of service and devotion to what must have surely been the most difficult patients in the world.

PROF. DR. MED. M. BLEULER
BAHNHOFSTRASSE 49
CH-8702 ZOLLIKON/ZURICH   June 21st, 1985

Dear Professor Benis,

Your message of June 13th has reached me today. It has immediately interested and fascinated me and I left other urgent duties in order to study your message and to answer your letter.

The first objection against your theory is clear. The personality is individual, is a world of most different traits which play together to become a personality — how can you think of reducing such a manifold, most coloured and most individual world to a simple genetic basis with 3 genes?

I would answer: the oversimplified image does not represent of course the personality or his disposition but important aspects, important vulnerabilities, important trends in the development. With this idea in mind your theory seems much more acceptable.

As you ask me my advice on the best mode of presentation I should suggest: it seems to me appropriate not to present your new idea as yet as a theory but as an important plan for further research. In order to demonstrate this suggestion I will mention how the summary of your results (page 2) should be more cautiously presented:

You write under the title "immediate results": "Identification of only three basic traits is sufficient to define the genetically determined character type."

Would it not be more likable and more realistic to write: "What arguments exist for the supposition of only three basic traits which are sufficient to define the genetically determined character type?"

I feel certain that your theory gives a new and most promising program for future research. On the other hand you will certainly understand that I feel uncertain in regard to the results of such a research — at least now, as I don't know details of your research results.

I am, of course, particularly interested in the significance of your theory in regard to the genetic background of schizophrenic psychoses. As you know: overlooking my life-long research and my life-long community with schizophrenics and my life-long study of the literature I came to the conclusion: there do exist manifold predispositions for developing schizophrenic psychoses, disposition of the character development, of emotional development and also neurological and neuro-humoral dispositions. They can be summarized as manifold dysharmonies of the disposition for a psychosis which consists in the loss of harmony of the personality (harmony of the Ego, harmony of thinking and feeling.)

This is not a theory but it is a provisional summary of present day knowledge. "Dysharmonic predisposition" is much too vague and up to now of restricted usefulness.

You give me hope for the discovery of more precious concepts of the disharmonic disposition.

Be certain: in spite of my critical remarks I think that your theory — or your research program — is of high importance.

I must excuse me in regard to my awkward English expression: decades have passed since I have lived in USA, and I had few occasions ever since then to exercise writing in English.

> Sincerely yours,
> Bleuler

PROF. DR. MED. M. BLEULER
BAHNHOFSTRASSE 49
CH-8702 ZOLLIKON/ZURICH   July 10th, 1985

Dear Professor Benis,

I have got a great gift from you: the great book "Toward Self and Sanity." I have a high esteem for this book even before I have studied it in details, only by looking it through. I think this book could probably influence very much attitude and thinking of our society.

You want my comments: I am afraid to give them before I have studied your book properly and I am afraid to give them in my awkward English.

I will, however, try to make a comment: It seems to me that in your book is overemphasized what is shameful, distasteful, cruel, antisocial, bad in human doing and being. This, of course, is a modern tendency in describing human history, in poetry, in politics, in journalism. On the other hand it seems to me that in your book has not been described enough, what is good, great, admirable, heroic, wonderful in men and in human doing.

I think in particular that the consequences of the genetic types can be both: shameful and admirable. We should not neglect that there exists all around us wonderful human doing: the care of a mother to her children, of a husband to his aggressive psychotic wife, of a nurse toward the sick. And I confess that I think admirable heroism plays an important role in human society, for instance the heroism of the great psychiatrist's son Bonhoeffer who fought with great bravery against Hitler's dictature and took his execution with calm and dignity.

I want to choose an example which makes it perhaps more clear what I have to suggest: page 198/9 of your book:

> A. says: (in regard to exaggeration) "Yes, of course we are... but, then again, we are not."

I should like to see mentioned another voice in the discussion:

> "B. You are right, Mr. A., you were exaggerating and not exaggerating — but you are always one-sided, you neglect, that the same hereditary tendency can be developed to a poor and shameful character trait, but also in the contrary: it can be a basis for greatness."

For instance one could formulate the paragraph "Exploitation" on page 198 also in the following way:

> "Becoming a master he could feel that he has a right to all of the patient's money and that no fee is too high — but we observe that many (or most?) masters in medicine (and certainly many modest practitioners) ask only for very moderate and honest fee. (An example: in about 1897 the great psychiatrist August FOREL was called for a consultation to Paris: he asked five Franks as his only fee.) In the daily life of a physician there are many occasions to abuse patients sexually. It is wonderful that in comparison to the tremendous number of these occasions real misuse is extremely rare..."

Please forgive my comments and forgive my awkward English. I should like to discuss this problem personally with you: You know this comment does not diminish my appreciation of your work.

Very sincerely yours,

Bleuler

Writing this a great number of examples of human goodness are in my mind: you mention the Zürich clergyman Lavater. Do you know how he died? During the French occupation of Zürich in 1799 Lavater (a physically weak man) saw a French soldier who attacked in a terrible way a woman on the street. Lavater immediately ran to help for the woman, took hold of the criminal — and Lavater was killed.

PROF. DR. MED. M. BLEULER
BAHNHOFSTRASSE 49
CH-8702 ZOLLIKON/ZURICH

6th August, 1985

Dear Professor Benis,

I thank you very much for having sent me your paper "A General Theory of Personality Traits..." I have studied it again with great interest. I think that it will open the way to new and important research. I also hope that it will found a solid found[ation] of my Dysharmony-Concept of schizophrenic psychoses.

Very sincerely yours,

I enclose a short paper* which has just appeared and summarizes some important facts in regard to schizophrenic psychoses. I think that these facts correspond well to your theory.

You remark: "…one tends to focus on psychopathology… in an unfavorable light." This conception has been expressed in an impressive way in a new book: Peter Barham, "Schizophrenia and Human Value," Blackwell Editor, 1984.

———————————————

* Bleuler M. (1985) Widersprüche zwischen Ordnungs-Streben und Wirklichkeit bei der Klassifizierung von endogen-psychotischem Geschehen, in Pflug B., Foerster K. and Straube E., "Pespektiven der Schizophrenie-Forschung," Gustav Fischer Verlag, Stuttgart.

PROFILES OF NPA TYPES

Profiles of NPA Character Types

N type	439
A type	441
NP type	443
NA type	445
PA type	447
NPA type	449
Submissive types: Compliant	451
Submissive types: non-Compliant	453
Resigned types	455
Other Submissive, Resigned & Withdrawn	See Chaps. 5-6

PROFILES OF NPA CHARACTER TYPES

Narcissistic type (N)

N types are typically extroverted, sanguine complexioned, non-perfectionistic, and prone to outlandish posturing and adornment. Low temperament individuals can be soft-spoken, gallant or angelic. High temperament individuals can be charismatic, or intrusive, overbearing and brash. The N type is quite common in aristocratic families and in all lists of famous people, especially in the arts.

Phenotype: N

Genotype: nn, nna

Animal Model: bonobo

Inheritance pattern: An N type must have at least one parent who is either an N or an NP type. Parents who are both N types can have only N or NA children.

Infertility: Increased probability of miscarriages and stillbirths when mated with either an A or PA type.

Rage: Narcissistic rage (mass discharge of parasympathetic nervous system).

Also known as: Narcissist. Sanguine personality. The non-aggressive non-perfectionist. The self-anointed glory seeker.

Complexion: Sanguine, florid, flushed… to blood-red in individuals of light skin color.

Smile: Radiant "gingival" smile, broadly exposing gums and teeth.

Photograph: Looks at camera. Broad charismatic gingival smile. Starry-eyed smile.

Gestures: Deep bow, accompanied by sweeping arm. "Joan of Arc pose" in which the individual's eyes are directed toward the heavens when accepting recognition in the limelight. "Narcissistic arms gesture" in which the arms are extended to the front or sides, with the palms up and the fingers somewhat spread apart. It is a pose often assumed by singers and by religious leaders.

Handwriting: Very variable. May be beautifully well formed with flourishes, but non-perfectionistic. May be illegible scribbling, especially in male.

Narcissistic type (ctd.)

Sexuality: Tendency to promiscuity: high. Tendency to ambivalence in sexual orientation: high.

Population Genetics: "The Sublime Habitancy," having a high prevalence of N types, some NP types, and very few types having the trait of aggression. (Examples: South Sea Islanders, natives of Hawaii, indigenous East Africa).

Susceptibilities: Narcissistic personality disorder (NPD), megalomania, messianism. Attention deficit disorders (ADD, ADHD). Unfocused, undisciplined, disorganized personality. Borderline personality. Pathological liar. Confidence man/woman ("con artist"). Vagabond criminal. Uncontrollable florid rages. Munchausen syndrome. Down's syndrome. Migraine headache. Congenital or rheumatic heart disease. Tuberculosis, leukemia, gastric (stomach) cancer, CVA (stroke). Anorexia nervosa/bulimia. Bipolar depression. Acute manic psychosis. Schizophrenia (especially hebephrenic type).

Pitfalls*: Until their lack of compulsive perfectionism becomes apparent, N types can resemble NP types. High temperament, seductive N types can mimic NA types. Loud N types (especially male) can be confused with NPA types. Sexually profligate NP or NPA– types could possibly be concluded to be N types. Psychotic or bipolar N types can exhibit behavior that resembles aggression or even sadism. Criminal violence in sociopathic N types can lead to mistaken conclusion of A trait.

---

* Possible confusion with other character types.

Aggressive type (A)

A types typically have a non-sanguine complexion, are extroverted, brusque, brash and prone to aggressive arrogance. The female is sometimes denigrated as "masculine." Circumstances can lead this type to be overtly sadistic. Short physical stature is common but not universal. Among illustrious individuals the A type is rare. Disproportionate numbers of A types are found in the categories of "antisocial" or "sociopathic" personality disorder.

Phenotype: A

Genotype: aa, naa

Inheritance pattern: An A type must have at least one parent who is either an A or a PA type. Parents who are both A types can have only A or NA children.

Infertility: Increased probability of miscarriages and stillbirths when mated with either an N or NP character type.

Rage: Aggressive-vindictive rage (mass discharge of sympathetic nervous system).

Also known as: Aggressive. Choleric personality. "The arrogant dynamo." Non-sanguine autocrat.

Complexion: Non-sanguine. Tending toward pallid or sallow in individuals of light skin color. May be milky white.

Smile: Sardonic smirk. Non-gingival, half-open mouthed grin. Grin with short, repetitive laugh.

Photograph: Looks at camera. "Pleased-with-self" grin.

Gestures: Upraised clenched fist, aggressive finger point, haughtily-cocked jaw, intimidating glare.

Handwriting: Non-perfectionistic. Often slurred or bold illegible scrawl.

Sexuality: Tendency to promiscuity: moderate. Tendency to ambivalence in sexual orientation: moderate.

Population Genetics: "The Militant Habitancy," having a high prevalence of A and PA types. (Examples: Middle Eastern subpopulations, Australian Aborigines. See *non-sanguine trails*, pp. 476 ff.)

Aggressive type (ctd.)

Susceptibilities: Attention deficit disorders. Overt sadism, sociopathology. Pathological liar. Antisocial personality disorder. Leader of militant, genocidal movement. Corrupted as absolute ruler... absolutely.

Pitfalls: High-temperament PA types can resemble A types. Accounts of the behavior of NPA types or bipolar N types can resemble A types. Accounts of (pseudo) perfectionist behavior in A type could lead to mistaken conclusion of P trait. A types, in drive for power, may mount the podium, leading to mistaken conclusion of the N trait. "Punky" adornment may lead to mistaken conclusion of the N trait (pseudonarcissism). Psychosis in N or NP types can superficially resemble aggression and even sadism.

Narcissistic-Perfectionist type (NP)

NP types are usually reserved, unaggressive individuals, tending toward a sanguine complexion and a propensity to blush easily. This is the dutiful, aloof "quiet achiever," who is obsessive and compulsive with regard to order, symmetry and neatness, often reflected in handwriting. Despite being unaggressive, these individuals can be very, very stubborn. Low temperament individuals may be described as "rigid," "wooden," "melancholic," or "bovine," while high temperament individuals can be "nervous birds" and may be more prone to intemperate behavior. NP monarchs are sometimes "lambs among the wolves." Tall and lean physical stature is common but not universal.

Phenotype: NP

Genotype: nnP, nnPP, nnPa, nnPPa

Animal Model: orangutan, gorilla

Inheritance pattern: An NP type must have at least one parent who is either an N or an NP type. Two NP types could have children of any character type with the exception of A or PA types.

Infertility: Increased probability of miscarriages and stillbirths when mated with either an A or a PA type.

Rage: Narcissistic rage (mass discharge of parasympathetic nervous system).

Also known as: Obsessive-compulsive personality. Phlegmatic-melancholic personality. Bovine personality. "Nervous bird" personality. "The quiet achiever."

Complexion: Tending toward sanguine or flushed in individuals of light skin color.

Smile: Uncommon sudden warm, radiant sheepish smile, like Cheshire cat appearing in the mist. Sometimes gingival smile (broadly exposing gums and teeth).

Photograph: Looks at camera. Relaxed face; sheepish or radiant "limelight" smile.

Gestures: Sometimes "narcissistic arms" gesture in which the arms are extended in front of the individual, with palms up and the fingers somewhat spread apart. It is a pose often assumed by singers and by religious leaders when praising their gods.

Narcissistic-Perfectionist type (ctd.)

Handwriting: Almost invariably well-formed, with each letter clearly legible. Sometimes striking calligraphic quality.

Sexuality: Tendency to promiscuity: very low. Tendency to ambivalence in sexual orientation: very low.

Population Genetics: "The Punctilious Habitancy," having a high prevalence of NP types and very few A or PA types. (Examples: Switzerland, parts of China).

Susceptibilities: Obsessive-compulsive disorder (OCD). Narcissistic personality disorder (NPD, may be masked by individual's perfectionist trait). Tantrums. "Control freak." Unipolar depression. Bipolar depression (manic-depressive disease). Postpartum depression. Focused, disciplined criminal. Acute episode of paranoia with delusion of being poisoned. Periodic (schizophreniform) psychosis. Schizophrenia. Asperger syndrome, Autism. Epilepsy (seizure disorder). Lymphoma (Hodgkin's Disease), gastric (stomach) cancer, childhood leukemia, migraine headache, rheumatic and congenital heart disease, Down's syndrome, varicose veins/hemorrhoids, CVA (stroke).

Coronary artery disease is rare in the NP type.

Pitfalls: Low temperament individuals can be mistaken for PA types or NPA= types. High temperament individuals can superficially resemble NPA types. Also, NP types (either high or low temperament) in a position of authority can be rigid disciplinarians, inflicting severely cruel punishment, leading to the mistaken conclusion of the A trait (pseudoaggression). The compulsive need for "control" may be misinterpreted as the "thirst for power" characteristic of the A trait. In imminent psychosis the hyperactive behavior of the NP type can mimic aggressive and even sadistic behavior. Not all NP individuals of light skin color have a baseline sanguine (pinkish) complexion: it may tend toward pallor. A gingival, narcissistic smile is striking when seen, but may be rare, especially in melancholic NP individuals.

Narcissistic-Aggressive type (NA)

NA types tend to be sanguine-complexioned, hyperactive extroverts. High temperament individuals, in particular, may be characterized by frantic, frenetic, hypersexual behavior that is not conducive to stable relationships. If the trait of aggression predominates, then the NA type may exhibit sadistic behavior.

Phenotype: NA

Genotype: nnaa

Animal Model: Some chimpanzees

Inheritance pattern: An NA type can have parents of any combination of the other character types. Two NA types can have children of only NA types.

Infertility: No increased probability of miscarriages and stillbirths when mated with any other character type.

Rage: Narcissistic rage (mass discharge of parasympathetic nervous system), or aggressive-vindictive rage (mass discharge of sympathetic nervous system), or combined NA rage.

Also known as: Cyclothymic, histrionic, hysterical or hypomanic-depressive personality. "The ambitious predator." The *prima donna*.

Complexion: Tending toward sanguine or flushed in individuals of light skin color.

Smile: Flashy, glamorous, toothy smile of movie star.

Photograph: Looks at camera. Extroverted, flashy smile.

Gestures: Active or hyperactive gestures. Often seductive body contact in casual social situations.

Handwriting: Variable, non-perfectionistic. Often rounded, elegant letters in female.

Sexuality: Tendency to promiscuity: high. Tendency to ambivalence in sexual orientation: low.

Population Genetics: "The Corybantic Habitancy," having a high prevalence of NA types. (Examples: Brazil, Venezuela, Senegal, Bougainville).

Narcissistic-Aggressive type (ctd.)

Susceptibilities: Attention deficit disorders. Hysteria, hypochondria, fugues.
Hypomanic personality. Histrionic personality disorder. Bipolar, Manic-
depressive disorder. Narcissistic personality disorder (NPD).

Pitfalls: High temperament N types can resemble NA types. Hyperactive, NPA
types can resemble NA types. High temperament, zany, pseudonarcissistic PA
types can superficially mimic NA types. A sullen, depressed NA type can
resemble a PA type.

Perfectionist-Aggressive type (PA)

PA types tend to a non-sanguine complexion. Individuals of low temperament under the best of circumstances can be well-adjusted stodgy, dutiful, socially conscious "solid citizens." However, under the worst of circumstances they can be scheming Machiavellian introverts bent on intrigue with sadistic overtones. Higher temperament individuals can be well-adjusted, somewhat stern or haughty extroverts. If they come to absolute power, then more overt sadistic trends may come to the fore. The PA type is relatively uncommon in the USA and Western Europe.

Phenotype: PA

Genotype: Paa, PPaa, nPaa, nPPaa

Animal Model: baboon

Inheritance Pattern: A PA type must have at least one parent who is either an A or a PA type. Two PA types could have children of any character type with the exception of N or NP types.

Infertility: Increased probability of miscarriages and stillbirths when mated with either an N or an NP type.

Rage: Aggressive-vindictive rage (mass discharge of sympathetic nervous system).

Also known as: Passive-aggressive personality, austere melancholic personality. Paranoid personality. Pseudonarcissistic extrovert. "The chronic criticizer." "The sardonic wit." "The suspicious manipulator." "The Power behind the throne." "The brooding non-sanguine autocrat."

Complexion: Non-sanguine. Tending toward sallow, pallid or milky white in individuals with light skin color.

Smile: Non-symmetric grin or grimace. Mona Lisa "smile." Sardonic smirk. Frozen, toothy but non-gingival grin. Half-open mouthed grin. Grin with short repetitive laugh. Note that the PA type is perfectionistic and aggressive, but non-narcissistic (genetically incapable of the gingival, narcissistic smile which broadly exposes gums and teeth).

Photograph: Usually looks at camera. May pompously look away from camera. Not relaxed. No smile, half smile, tight-lipped sardonic smile, non-symmetric grin or grimace. Non-gingival "frozen smile" showing teeth. Laughs or tries to laugh, showing expressive extroverted countenance.

Perfectionist-Aggressive type (ctd.)

Gestures: Upraised clenched fist, the haughtily cocked jaw, the furrowed brow and strained tight-lipped mouth, and the intimidating glare.

Handwriting: Variable. Usually perfectionistic in female. Often slurred or bold illegible scrawl in males. Sometimes messy corrections. Sometimes bold flourishes.

Sexuality: Tendency to promiscuity: moderate. Tendency to ambivalence in sexual orientation: moderate.

Population Genetics: "The Authoritarian Habitancy," having a high prevalence of PA types, moderate prevalence of A and NPA types, and few N and NP types. (Examples: Western Russia, Serbia. See *non-sanguine trails*, pp. 476 ff.).

Susceptibilities: Paranoia. Pathological "power behind the throne." Paranoid personality disorder, covert sadism, serial criminal. Childhood onset diabetes. Schizophrenia (paranoid type).

Pitfalls: High temperament PA types can mimic A types, or even zany NA types. Low temperament PA types in non-confrontational circumstances can mimic disciplinarian NP types. Unusual adornment in sexually ambivalent PA types may lead to mistaken conclusion of the N trait (pseudonarcissism).

Narcissistic-Perfectionistic-Aggressive type (NPA)

NPA types tend to be sanguine-complexioned, overbearing maternalistic or paternalistic extroverts. The voice is LOUD and the eye contact intense. They tend to be "conventional" in their dress and behavior. Their greatest vulnerability is the tendency to explosive rages, often followed by forced affability or apologetics. The female is sometimes denigrated as "masculine."

Phenotype: NPA

Genotype: nnPaa nnPPaa

Animal Model: Some chimpanzees

Inheritance pattern: Two NPA types can have children of only NPA or NA types.

Infertility: No increased probability of miscarriages and stillbirths when mated with any other character type.

Rage: Narcissistic rage (mass discharge of parasympathetic nervous system), or aggressive-vindictive rage (mass discharge of sympathetic nervous system), or combined NA rage.

Also known as: Explosive personality. Managerial-autocratic personality. The overbearing achiever. The sanguine autocratic tyrant.

Complexion: Tending toward sanguine or flushed in individuals of light skin color.

Smile: Warm paternalistic or maternalistic smile.

Photograph: Looks at camera. Relaxed smile.

Gestures: Active or hyperactive gestures. Intense eye contact with loud, penetrating voice. Often non-seductive body contact in casual social situations.

Handwriting: Variable. Often perfectionistic but sometimes grossly illegible.

Sexuality: Tendency to promiscuity: low to moderate. Tendency to ambivalence in sexual orientation: very low.

Population Genetics: "The Demonstrative Habitancy," having a high prevalence of NPA and NA types. (Examples: Mediterranean subpopulations, Colombia).

Narcissistic-Perfectionist-Aggressive type (ctd.)

Susceptibilities: Overbearing need for control. Explosive personality disorder.
Megalomania. Narcissistic personality disorder (NPD).

Pitfalls: In absolute power the NPA type can be a cruel disciplinarian, even a
tyrant, resembling the behavior of the A type. Hyperactive, promiscuous NPA
types can resemble NA types. Loud N types (especially male) can be confused
with NPA types.

SUBMISSIVE TYPES

Narcissistic-Perfectionistic Compliant Submissive type (NPA=)

Origin: *Compliant* (*C*) submissive types are those in whom the trait of aggression is profoundly suppressed by genetic and/or environmental factors (notation A=).

[In *non-compliant* (*n-C*) submissive types the trait of aggression is partially, but not profoundly, suppressed (notation A−).]

In these types the trait A is profoundly suppressed, so that whatever the circumstances they tend not exhibit aggressive behavior in social situations. They can exhibit the aggressive A-rage, but this is very rare indeed. Compliant submissive types tend to be very introverted and seek a life style involving little responsibility and much protection. Sexually promiscuous individuals are very vulnerable and have a strong tendency to masochistic behavior.

Phenotype: NPA=

Genotype: nnPaa-, nnPPaa-, nnPa-a- or nnPPa-a-. In addition, unusual cases of **nnPaa** or **nnPPaa** (phenocopies based on environmental conditioning).

Inheritance pattern: The allele of submission **a-** is expressed as a dominant with respect to the allele of aggression, **a.**

Rage: Narcissistic rage (mass discharge of parasympathetic nervous system), or aggressive-vindictive rage (mass discharge of sympathetic nervous system), or combined NA rage. In most compliant submissive individuals the aggressive-vindictive rage is rarely seen.

Also known as: NP-like compliant. Quiet achiever submissive. Depressive, masochistic personality. "The shrinking violet."

Complexion: Tending toward sanguine or flushed in individuals of light skin color.

Smile: Warm smile when at ease. Otherwise nervous smile. .

Photograph: Uncomfortable before camera in unfamiliar settings.

Gestures: Reserved and tentative.

Handwriting: Neat and legible, as a slave writing for his masters.

Sexuality: Tendency to promiscuity: low. Tendency to ambivalence in sexual orientation: high.

Narcissistic-Perfectionistic Compliant Submissive type (ctd.)

Population Genetics: "The Introspective Habitancy" (Example: New Zealand, Anglo-Saxon subpopulations). *See* Non-compliant submissive types.

Susceptibilities: Shyness. Masochism. Sadomasochistic "morbid dependency" as the dependent partner. Social phobia. Reactive depression. Dependent, avoidant personality disorder. Periodic (schizophreniform) psychosis. Relatively low risk of schizophrenia in NPA= and NA= types. (Higher risk of schizophrenia in PA= and A= types). *See also* Non-compliant submissive types.

Pitfalls: Some NPA= individuals can superficially resemble NP or N–P types, while NA= types can resemble N or N– types.

SUBMISSIVE TYPES

Narcissistic-Perfectionistic Non-compliant Submissive type
(NPA–)

Origin: *Non-compliant* (*n-C*) submissive types are those in whom the trait of aggression is partially suppressed by genetic and/or environmental factors (notation A–).

[In *compliant* (*C*) submissive types the trait of aggression is profoundly suppressed (notation A=).]

In these types the trait A is partially suppressed, but depending on the circumstances they can exhibit aggressive behavior in social situations. Most individuals tend to be non-confrontational. They can exhibit the aggressive A-rage, but this is unusual. Low-temperament individuals tend to be introverted. Higher temperament individuals can be "nervous extroverts," often with speech tremor or hesitancy.

Phenotype: NPA–

Genotype: nnPaa-, nnPPaa-, nnPa-a- or nnPPa-a-. In addition, unusual cases of **nnPaa** or **nnPPaa** (phenocopies based on environmental conditioning).

Inheritance pattern: The allele of submission **a-** is expressed as a dominant with respect to the allele of aggression **a**.

Rage: Narcissistic rage (mass discharge of parasympathetic nervous system), or aggressive-vindictive rage (mass discharge of sympathetic nervous system), or combined NA rage. In non-compliant submissive individuals the aggressive-vindictive rage may be rarely seen.

Also known as: NP-like non-compliant. Agitated achiever submissive. Reactive depressive personality. "Type A" personality.

Complexion: Tending toward sanguine or flushed in individuals of light skin color.

Smile: Relaxed social smile rarely seen. Nervous smile.

Photograph: Usually uncomfortable before camera, sometimes to the extent of phobia.

Gestures: Agitated, inconsistent and tentative.

Narcissistic-Perfectionistic Non-compliant Submissive type (ctd.)

Handwriting: Usually legible but "jerky" quality, with large variations in size of letters. Sometimes slurred. Sometimes messy corrections. (May be highly perfectionistic in PA– type.)

Sexuality: Tendency to promiscuity: moderate. Tendency to ambivalence in sexual orientation: high.

Population Genetics: "The Introspective Habitancy," having a high prevalence of NPA– types, and moderate prevalence of NP and NPA types. (Example: Anglo-Saxon subpopulations).

Susceptibilities: Attention deficit disorders. Sadomasochistic "morbid dependency" as either the dominant or dependent partner. Shyness. Public speaking phobia. Social phobia. Avoidant personality disorder. Speech hesitation, stuttering (stammering), spasmodic dysphonia. Reactive depression. Narcissistic personality disorder (NPD, in dominant role). Coronary artery disease. Hypertension. Adult-onset diabetes. Thyroid disease. Rheumatoid arthritis, lupus, inflammatory bowel disease. Periodic (schizophreniform) psychosis. Relatively low risk of schizophrenia in NPA– and NA– types. (Higher risk of schizophrenia in PA– and A– types.)

Pitfalls: Unstressed NPA– individuals can resemble NP types. High temperament N types and NP types can resemble NA– types and NPA– types, respectively. Sexually profligate submissive types can have behavior that could be confused with that of N types. Borderline N– and N–P types can resemble submissive types.

RESIGNED TYPES

Narcissistic-Perfectionistic-Aggressive Resigned type (NP–A)

Origin: *Resigned* types are those in whom the trait of aggression is partially suppressed by genetic and/or environmental factors after maturity (notation –A).

[In non-compliant and compliant *submissive types* the trait of aggression is partially suppressed by genetic and/or environmental factors *before* maturity (notation A– and A=).]

In resigned types the trait A is suppressed in a former healthy, mature dominant type or in a former non-compliant submissive type. Because of environmental factors, especially "stress," the individual becomes detached from competitive social interaction, or "abdicates," in order to devote himself (herself) to more serene activities, sometimes feeling the inner satisfaction of "splendid isolation."

Phenotype: NP–A

Genotype: nnPaa, nnPPaa (former NPA dominant types) and **nnPaa-, nnPa-a-, nnPPaa-** and **nnPPa-a-** (former NPA– submissive types)

Inheritance pattern: *See* NPA dominant type and NPA– non-compliant submissive type.

Rage: Narcissistic rage (mass discharge of parasympathetic nervous system), or aggressive-vindictive rage (mass discharge of sympathetic nervous system), or combined NA rage. In most resigned types the rages are rarely seen.

Also known as: NP-like resigned. Quiet achiever resigned.

Complexion: Tending toward sanguine or flushed in individuals of light skin color.

Smile: Smiles easily: former NPA types. Relaxed social smile rarely seen or nervous smile: former NPA– types.

Photograph: Relaxed: former NPA types. Camera-shy: former NPA– types.

Gestures: Reserved.

Handwriting: Usually neat and legible.

Sexuality: Variable according to life situation. Wary of strong attachments.

Resigned types (ctd.)

Population Genetics: Etiology of resigned types is highly dependent on environmental factors.

Susceptibilities: Former submissive types: *See* Compliant, Non-compliant submissive types. Former dominant types: *See* NPA and NA types above.

Pitfalls: Unless details of the individuals' lives are known, descriptions of resigned types NP–A and N–A can mimic N–P and N– types, respectively.

NARCISSISM & AGGRESSION

Comparison of Traits of Unbridled Narcissism and Aggression
(unbridled = no P trait in character type)

Note: the descriptive categories are trends only; notable exceptions may exist.

	___narcissism___	___aggression___
Phenotype:	narcissistic N type	aggressive A type
Genetics:	Mendelian recessive	Mendelian recessive
Genotype:	**nn** or **nna**	**aa** or **aan**
Autonomic nervous system:	parasympathetic	sympathetic
Primary neurotransmitter:	cholinergic	adrenergic
Rage:	narcissistic rage (tantrum)	aggressive-vindictive rage
Goal:	glory	power
Style:	charisma, vanity	militancy, coercion
Slogan:	"moving toward people"	"moving against people"
Sociability scale (hierarchy):	deferential	"pecking order" (dominance/submission)
Superficial behavior:	charming, gallant	serious, imperious
Requires others:	for adulation, support	for convenience, service
Responsive to:	flattery, "limelight"	deference, submission
Entourage:	sycophantic parasites; disciples	subjugated slaves; enforcers
Common perception:	benevolence	malevolence
Strong points:	creativity, sociability	quick action, energetic industriousness
Weak points:	frivolity, nebulousness	brusqueness, brutishness
Stature:	average to tall; soft, paunchy	short to average; solid, wiry
Complexion (light skin color):	sanguine, florid	non-sanguine, dull, pallid
Smile:	radiant, gingival smile	grin, smirk

	narcissism	***aggression***
Flushing, blushing:	common	rare
Tearfulness:	common	rare
Adornment, Dress:	colorful, garish	drab, disheveled, "punky"
Gestures:	expansive arms, bowing	finger point, clenched fist
Oratory:	pontificating, charismatic, unctuous	personal, confrontational
Handwriting:	non-perfectionistic to ornamental	non-perfectionistic to messy
Music:	soaring, melodious, theatrical; ballad	rhythmic, loud; "rap"
Sports, objective in:	applause, recognition	obliterate, injure opponent
Psychopathology:	narcissistic personality disorder (NPD)	sadomasochism
Interpersonal pathology: **[Proponent]:**	narcissistic co-dependency [various authors]	"morbid dependency" [Karen Horney]
Sociopathology:	"borderline", vagabond, "con artist"	"antisocial personality"
Confrontational style:	intrusive, persuasive	coercive, threatening
Reprimand style:	irritable; scolds, quarrels	pugnacious; berates, castigates
When challenged:	vociferous, withdrawing	violent, vengeful
Histrionics:	tearful, diffuse	directed, threatening
Vindictiveness, violence:	spontaneous, unpredictable	unrelenting, sadistic
Bullying:	demonstrative, demeaning	coercive, intimidating
Pathological lying:	engaging, persuasive	abrupt, dismissive
Taboos:	"following the rules"	assenting to a weaker position
Tattoos:	ornamental themes	violent themes
Recreational drugs:	hallucinogens	narcotics
Sexuality:	seductive, persistent	coercive, intimidating

	narcissism	*aggression*
Inversion:	pseudoaggression	pseudonarcissism
Infertility (miscarriage, etc.):	with A, PA types	with N, NP types
Diabetes:	maturity-onset	juvenile-onset
Schizophrenia:	hebephrenic	paranoid
Acute psychosis:	tantrum-like; "poisoning delusion"	confrontational, violent
Paranoia:	acute, episodic	chronic
Society:	open, ingenuous, naive	closed, paranoid, cryptic
Religion:	colorful, musical, expressive	monotone, regimented
Conqueror (war):	vainglorious, inclusive	barbaric, draconian
Habitancy, archetype:	Polynesia, Hawaii (Sublime Habitancy)	Middle East, Australian Aborigine (Militant Habitancy)
Examples, historical:	Eleanor of Aquitaine Kaiser Wilhelm II	Wladyslaw Gomulka Charles Manson
Absolute ruler, archetype:	Henry VIII	Robert Mugabe

Bridled Narcissism and Aggression
(bridled = P trait is present character type)

	narcissism	*aggression*
Perfectionist type:	obsessive-compulsive (NP)	paranoid aggressive (PA)
Power behind throne: **Example, archetypal**	NP type Sophie of Bavaria (Plate 42)	PA type Nancy Reagan, R.B. Cheney
Habitancy:	Switzerland, China (Punctilious Habitancy)	Russia, Serbia (Authoritarian Habitancy)
Examples, historical:	Catherine of Aragon Queen Victoria	Charles de Gaulle Joseph Fouché
Absolute ruler:	Caesar Augustus Tsar Nicholas II	Vladimir Putin Slobodan Milosovich

INADEQUACY OF DSM-IV

Inadequacy of American Psychiatric Association Criteria
for Narcissistic Personality Disorder (NPD)

Fulfillment of DSM-IV criteria for NPD

criterion number	*narcissistic N type*	*aggressive A type*
#1. grandiose self-importance....	yes	yes
#2. fantasies of unlimited success, power...	yes	yes
#3. believes that he or she is "special"...	yes	yes
#4. requires excessive admiration...	yes	no
#5. sense of entitlement...	yes	yes
#6. is interpersonally exploitative	yes	yes
#7. lacks empathy...	yes	yes
#8. envious of others...	yes	yes
#9. arrogant, haughty behavior...	yes	yes

Comment: There is great overlap between unbridled narcissism and aggression in the DSM-IV criteria for NPD. We conclude that these criteria are of minimal value in differentiating the two traits.

Reference:

Frances A. *et al.* (1994): *DSM-IV: Diagnostic and Statistical Manual of Mental Disorders*, 4th Edition, American Psychiatric Assn., Washington D.C., p. 661.

THE NPA PERSONALITY TEST

Q.   What is the NPA personality test?

A.   It is a 50 question online test whose purpose is to determine one's NPA profile. The output of the test is the individual's NPA character type, as well as three scores that reflect submissiveness (S), temperament (T) and focus (F).

Q.   What is the S score?

A.   The S score is a measure of *anxiety, depression* and/or *submissiveness* in social relations, on a scale of 0-100. If the S score is greater than 20-30 then it becomes more likely that the traits N and/or A are not fully expressed, either because of genetics or environment. Thus, the S score is used to identify two major categories of character type: (1) *dominant types*, and 2) *submissive and borderline types.*

If the S score is in the range 20-60, then a minus sign (–) is added (A–, or *non-compliant submissive* types).

If the S score is > 60, then a double minus sign (=) is added (A=, or *compliant submissive* types).

The test does not explore the exact reasons for a high S score. The most common reason in healthy, mature individuals is suppressed aggression (submissive types, as noted above) and suppressed narcissism (narcissistic withdrawn types). The S-score may also be elevated in individuals having a tendency to bipolar behavior.

Q.   What is the T score?

A.   The T score is a measure of *temperament*, again on a scale of 0-100. Introverted, reserved individuals will tend to score low on this scale, while highly extroverted, volatile individuals will tend to score high.

The temperament T score is as follows:

0-10 = low: phlegmatic
10-20 = low: reserved
20-30 = moderate
30-60 = high: reactive
60-100 = high: volatile

Q.   What are typical ranges of the T score for the various NPA character types?

A.   They are given below. One should keep in mind that any biological quantity may exhibit "outliers." These are values that because of an unusual factor, or an unusual combination of "usual" factors, lie outside of the normally accepted ranges.

Presence of the P trait tends to reduce the temperament T score. The lowest scores are seen in NPA= and borderline types. The highest scores are seen in NA types. The types having the widest variability of scores are also the NA types.

<u>dominant types</u>
N   20-40
NP   5-25
NA   20-80
NPA   20-40
PA   10-30
A   30-60

<u>submissive types</u>
NA–   10-50
NA=   5-20
NPA–   10-25
NPA=   5-15

<u>borderline types</u>
low   0-20

Q.   What is the F score?

The F score, for *focus,* is a measure of the degree of organization of an individual's personality, again on a scale of 0-100. Introverted or reserved individuals with analytical tendencies will tend to score high on this scale, while less focused, expansive or practical individuals will tend to score low.

The focus F score is as follows:

0-20 = low: diffuse, expansive
20-70 = moderate: reflective, pragmatic
70-100 = high: contemplative, analytical

Q.   What are typical ranges of the F score for the various NPA types?

A.   They are given below.

Individuals lacking the P trait (perfectionism) are unlikely to score higher than a score of 60. The highest scores are seen in NPA= and borderline types, as well as NP and PA dominant types. The lowest F scores are seen in individuals who lack the P trait.

<u>dominant types</u>
N   0-50
NP   60-100
NA   20-40
NPA   30-50
PA   60-90
A   20-40

<u>submissive types</u>
NA–   0-50
NA=   0-50
NPA– 60-90
NPA= 70-100

<u>borderline types</u>
variable: depends on presence of P trait

Q.   Is there a relationship between the NPA test and the results of other tests, such as the Enneagram or the Myers-Briggs?

A.   Since the NPA approach focuses on specific genetic traits, there is no simple correlation between the NPA test and the results of other tests. We consider the NPA traits as primary with respect to other secondary aspects of an individual's genetically determined personality.

Q.   How is the test analyzed by computer?

A.   Evaluation of any kind of test is not straightforward, as one must consider two separate aspects of testing that are somewhat antagonistic to each other: the sensitivity and specificity of the test. *Sensitivity* is a measure of the frequency of false negatives, while *specificity* is a measure of false positives. (For example, if you are an N type and the test states that you are not, that is a false negative. If the test says that you are an N type but you are not, that is a false positive).

The test was scored by the averaging of three different statistical methods employing "square scores," correlation coefficients and a technique similar to Chi square comparisons. For example, a "square score" for a given NPA type was determined from the differences of the subject's responses from the "expected values" for that type. The "square scores" for each type were then transformed to relative probability values with the use of the reciprocal function. The effect of the squaring technique was to amplify responses concordant with the most probable NPA type. The probabilities of NPA character type and of the three traits are displayed on the bar graphs.

The *S score* was calculated from questions emphasizing submissiveness, social anxiety and depression. The S-score was determined as a "square score" for the S category of questions, divided by the maximum possible score times 100. Four categories of S-score are considered: (1) less than 20 (dominant types), (2) between 20-30 (transition range), (3) between 30-60 (moderately elevated), and (4) greater than 60 (high).

The temperament *T score* was calculated from questions that emphasized dominant behavior in N, NA, NPA and A character types: individuals who answer the selected questions will tend to score higher on the T scale. The S and T scores are mathematically independent, the two scores being calculated from separate subsets of the list of fifty questions.

The focus *F score* was calculated from questions that emphasized behavior in the NP and PA character types. The basis of the F score was a set of questions separate from those used in the computation of the S and T scores.

The bar graph giving the probabilities of the three NPA traits is most applicable to the six dominant types. If the S score is elevated (>20) then the estimate of the probability of traits N, P and A is not theoretically accurate, since the traits N and/or A may have been suppressed by genetic or environmental factors.

Values of probability generated by the statistical techniques are inherently dependent on (1) the particular questions that were chosen for the test, and (2) the varied ways that the questions will be interpreted and responded to by various subjects. Hence, the probability values finally produced are relative ones only. The values are relevant only to this particular test, and they do not have any other significance in a quantitative sense.

Q.   What are the questions of the NPA test?

A,   A sample test is given below. The subject took the test anonymously on the internet and submitted the results to us for the computerized analysis. The subject, a male and parent of an autistic child, tested as a dominant NP type with an S score below 20:

NPA PERSONALITY TEST

This is a 50-question test to assist in the determination of your NPA Character Type. The charts below will display when all questions are answered.
Instant results!

All questions are answered on a 0 to 4 scale, as follows:

> *4 = "I agree" or TRUE or Yes!*
> *3 = "I somewhat agree"*
> *2 = Neutral*
> *1 = "I somewhat disagree"*
> *0 = "I disagree" or FALSE or No!*

<u>*Suggestions*</u>*: Take a bit of time before answering. Imagine the situations depicted. Think about yourself rather than about the test. Try to see yourself as others see you. Read the entire list of questions before starting the test.*

1. If you succeed, would you like to bask in the limelight of public acclaim?	0

2. Are you a "strong silent type" who rarely attempts to smile and is good at staring people down?	0

3. Is this you: talkative, impatient, much sex appeal, smiles and laughs easily, having a sharp-edged personality?	0

4. When things must get done, are you liable to push people aside and ask questions later?	0

5. With minor stress, do you have a tremulous voice or hesitancy in speech?	0

6. At a quiet party, would you welcome a newcomer with a hearty yell across the room?	0

7. Is this you: quiet, friendly, meticulous, extremely organized, aloof, nice smile, careful handwriting, strong and stubborn?

2

8. Does your laughter tend to the high-pitched, hysterical side?

0

9. Do you tend to be suspicious and confrontational, rather than accepting other people's motives and actions?

2

10. In a touching, emotional situation, would you be unsympathetic to the "weeping weaklings" around you and head for the door?

0

11. In school were you an earnest, introspective student who was hesitant to speak up in class, dying a thousand deaths waiting to be called upon?

0

12. In relaxed social situations, do you speak with a loud, carefully-controlled voice and make intense eye contact with others?

2

13. Are you attracted to activities that are slow moving and require careful attention to detail?

4

14. In school or college would you be attracted to cheerleading, singing, drama <u>or</u> the dance?

0

15. On the stage would you be natural as the villain: cool, calculating, suspicious, vindictive, a stern colorless face, and a dagger at hand?

0

16. Do you have the potential to be a well-organized, dynamic speaker who would never read to an audience from a prepared text?

2

17. Could you be described as hyperactive, unembarrassable, and attracted to good food, sex, travel and excitement?

0

18. Is it possible that you will tangle with the law for vindictive, belligerent, violent behavior?

0

19. When "wronged" by someone, do you typically reply with "the silent treatment" rather than being vocal?

0

20. Do you sometimes "get into a world of your own", become depressed and think that you could survive as a hermit?

2

21. In photographs do you <u>and</u> at least one family member show a broad charismatic smile?

0

22. Could this be you: loud, unrestrained, unsmiling, colorless complexion, tendency to bully, unfocused, unruly, a real brute?

0

23. Are you a nice person who is easily taken advantage of, or bullied, in work or love relationships?

3

24. Is this you: sociable but not sexy, rather loud voice, conventional lifestyle, maternalistic or paternalistic, dynamic, a managerial outlook on life, a bulldog?

1

25. Do these terms apply to you: extroverted, unaggressive, unfocused, warm, vain, sexy, unconventional: a teddy bear or a baby doll?

0

26. When excited does your voice easily become piercing (male) or shrill (female)?

0

27. When you have a seething rage, does the blood drain from your face, giving you a frightening, evil appearance?

0

28. Do people say, much to your annoyance, that you have the "clean gene", or that you are "obsessive-compulsive"?

4

29. Are you rather slow to be drawn into intimate or sexual relationships?

2

30. Could this be you: reserved, well organized, aloof, tending to be sarcastic or cynical, a grin rather than a smile, a coiled spring capable of pouncing on someone?

4

31. Do you get extremely nervous or anxious about speaking in public, even in benign situations?

0

32. Are you an extroverted, active person who has no interest at all in social events or friendly people?

0

33. Do you become more upset than most people if there is a lack of order, neatness or cleanliness?

0

34. When angered, could you be described as tearful, hysterical, and liable to lash out with talons bared?	0

35. If someone were slow at doing something, would you immediately intervene and say with a forceful voice, "Let me help you with that!"	0

36. At a party can you become really carried away while telling people about your interests and expertise?	4

37. Do you generally lack self-confidence or have feelings of free-floating anxiety or inferiority?	0

38. Would you do well in the imagined role of a stern, pallid-faced Roman emperor (empress) capable of harsh judgments?	0

39. Could you picture yourself as the TV anchorperson: likeable, reserved smile, very organized, very focused, forceful voice: a Wolf Blitzer or a Christianne Amanpour?	0

40. Does colorful adornment, e.g., fancy clothes, rings or accessories of gold, come naturally to you?	0

41. Are you shy in social situations, e.g., small groups, parties, dinners?	0

42. If someone were casually to lay a hand on you, is it likely that you would immediately respond with a punch in the face?	0

43. Are you rather likely to have many short-lived love affairs in which you treat your weaker partner rather cruelly?	0

44. Are you rather slow and very stubborn in getting a task well done?	3

45. Are you (potentially) a good leader because of your dynamic style, your talents for organization, as well as your ease in social situations?	3

46. Do you go at lengths to avoid anxiety-producing confrontations with other people?	2

47. Is your humor reserved, biting and sarcastic, rather than open and jocular?	0

48. Are you more hyperactive and sexually provocative than other people, a vampire in your dreams?	0

49. At a social occasion, would you be at ease in musical, theatrical or clownish situations?	0

50. In sports or games are you be likely to become involved in physical fights in which you really intend to injure, if not kill, your opponent?	0

By selecting "I agree", I acknowledge that (1) this test is to be used for my personal use only, (2) the information herein may, or may not, be applicable to my personal life, and (3) NPA theory involves copyrighted materials.	*I agree*

Graphical results: dominant NP type with S score < 20 :

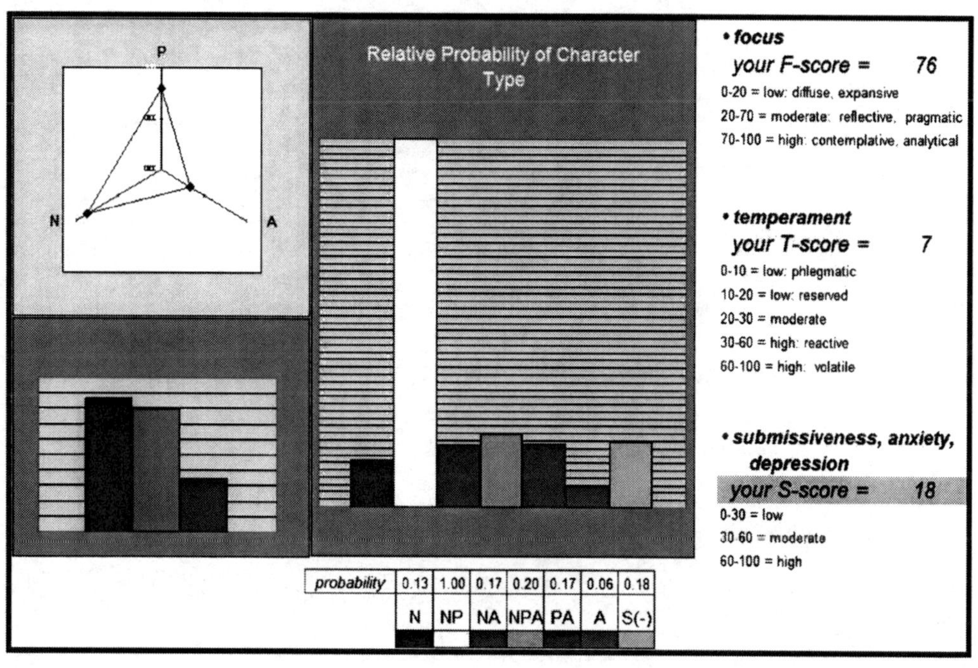

probability	0.13	1.00	0.17	0.20	0.17	0.06	0.18
	N	NP	NA	NPA	PA	A	S(-)

TESTING NPD

Testing of Subjects with NPD: Sam Vaknin takes the NPA test

In NPA theory narcissistic personality disorder, or NPD, is "malignant narcissism" in the same sense that sadism is "malignant aggression." In recent years aspects of NPD have been greatly popularized by Sam Vaknin, a self-proclaimed "narcissist" and author of the book *Malignant Self Love: Narcissism Revisited.* We asked Sam Vaknin if he would take the NPA personality test, and he graciously agreed.

For comparison, here are the results of the NPA test for a typical narcissistic N type:

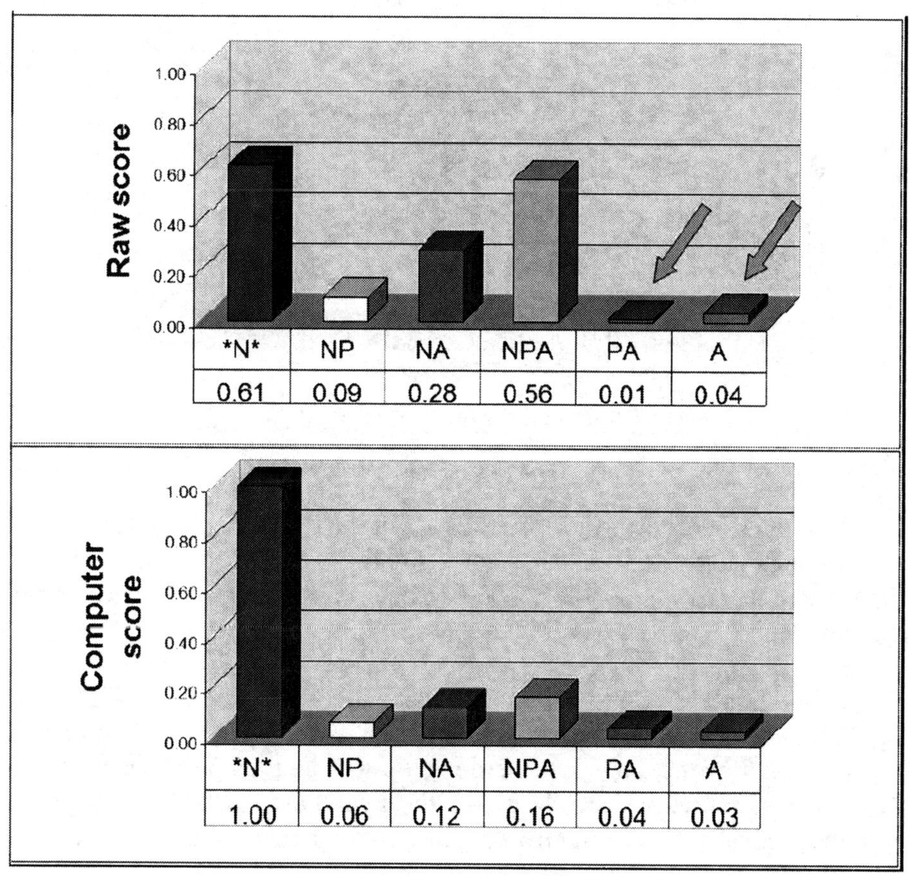

The raw scores are shown in the top panel, and the computer score (probability) in the lower panel. Note that this N type has very low scores in the categories of non-sanguine aggression, i.e., in NPA types lacking the N trait (arrows).

In contrast, here is Sam Vaknin's test:

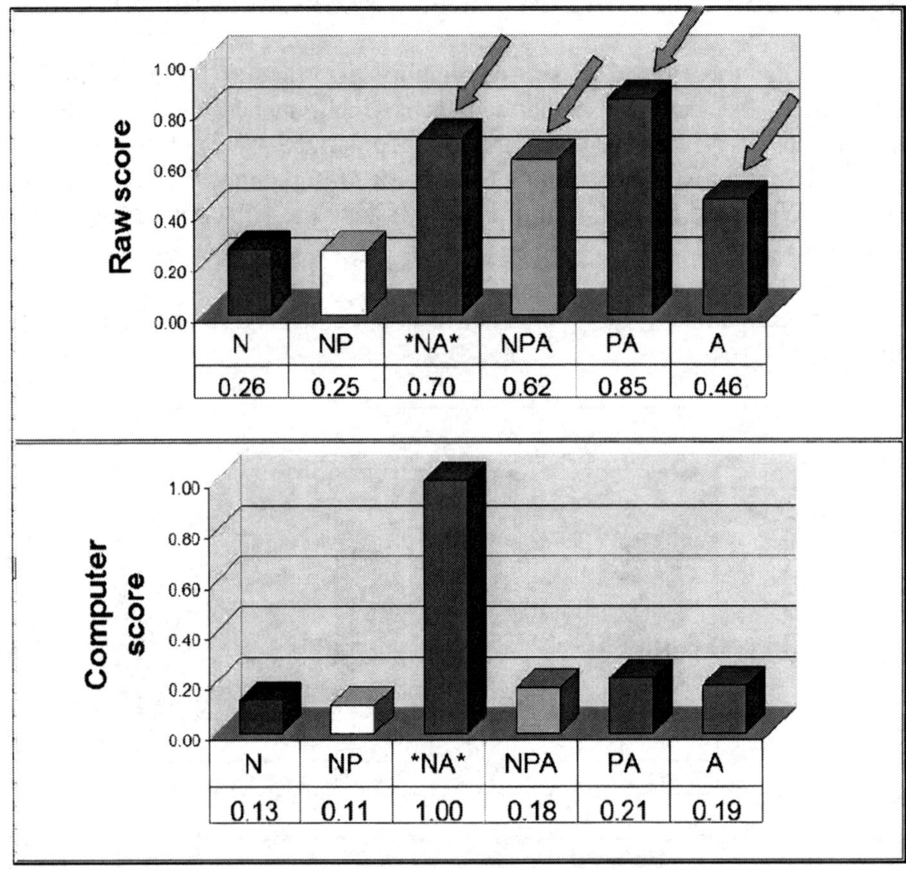

Note that here the results are very different. All of the NPA categories having a component of the trait of aggression are elevated (arrows), while those lacking the trait are very low. The computer interpretation (lower panel) is narcissistic-aggressive NA type, rather than N type.

In addition, 31 other subjects (23 male and 8 female) with NPD took the NPA test online anonymously and submitted the results to us. (We do not have information on how the diagnoses were made). Of the total 32 subjects, 26 tested as NA types. Hence, we are drawn to the hypothesis that NPD, as presently interpreted clinically, has as significant genetic component of aggression.

Conclusion: it may turn out that many cases of NPD are more correctly NAPD, or narcissistic-aggressive personality disorder.

POPULATION GENETICS

Population Genetics: Geographical Distribution of NPA Character Types with Reference to Ancestral Migration from Africa

Overview of NPA personality theory

Three genetic traits

The NPA personality theory of genetically determined character types posits that three separate autosomal loci determine the traits of narcissism (N), perfectionism (P) and aggression (A). Consequently, various combinations of the traits lead to various *character types.* The traits N and A are posited to be related to branches of the autonomic nervous system, with the trait P being a mediator of the N and A traits. According to NPA theory, in any individual: (1) either the trait N or A, or both, must be present, and (2) the trait P may or may not be present. As a complicating factor the traits N and A may be partially expressed. Character types having full expression of the trait N or trait A are denoted as *dominant* types, while types having the partially expressed trait A (termed A–) are denoted as *submissive* types.

On the basis of archetypal family pedigrees we posited that the mechanism of transmission of the NPA traits is autosomal, that of traits N and A being Mendelian recessive while that of P being Mendelian dominant. The unusual mechanism of transmission for N and A, as high-frequency *recessive* traits, leads to the further hypothesis that these loci code for repressors of the traits, and that an inactive repressor would lead to a "release of inhibition" allowing expression of that trait.

Character types

Confining the discussion to phenotypes in which the three traits are fully expressed, NPA theory allows for the following *dominant character types*:

N   narcissistic
A   aggressive
NA   narcissistic-aggressive
NP   narcissistic-perfectionist
PA   perfectionistic-aggressive
NPA narcissistic-perfectionistic-aggressive

Sanguine and non-sanguine types

Character types are denoted sanguine or non-sanguine depending on the presence or absence of the trait N, respectively. Of special interest are the non-sanguine types: A and PA.

Aggressive and non-aggressive types

Character types are denoted aggressive or non-aggressive depending on the presence or absence of the trait A. Of special interest are the non-aggressive types: N and NP.

Subspecies and hybrids

Since the traits N and A are recessive, the mating of a group of pure-bred non-sanguine types with a group of pure-bred non-aggressive types would not produce viable progeny (all zygotes would lack both traits N and A). Here, "pure-bred" denotes that neither group contains individuals who are genetic carriers of the other recessive trait. Thus, in the context of NPA theory, non-sanguine types (A and PA) and non-aggressive types (N and NP) could be considered to be two separate subspecies. However, if either group contained carriers of the other recessive trait, then mating between the two groups could yield NA and NPA types in the progeny. In this sense, NA and NPA dominant types can be considered to be "hybrid" types.

Consequences of NPA genetics

The NPA genetic model, consisting of one dominant trait and two recessive traits, has its own peculiarities and subtleties. The genetic mechanisms underlying the model lead to the following generalizations:

- A non-sanguine child must have a non sanguine parent, or in equivalent terms, "non-sanguinity" is a dominant trait. Thus, a non-sanguine type cannot arise *de novo* in a geographic location where all individuals are sanguine types.

- A child of non-aggressive type must have a non-aggressive parent, or "non-aggressiveness" is a dominant trait. Thus, an individual of non-aggressive type cannot arise *de novo* in a geographic location where all individuals have the trait of aggression.

- The "hybrid" NA and NPA types could arise *de novo* in a population in which neither type existed. These types could, in principle, arise in any geographic location, even one where all individuals are non-sanguine.

The above generalizations assume progeny where no new mutations have occurred.

Note the clear distinction between the trait of *aggression* and that of *non-sanguine aggression*. The trait of aggression is a recessive trait, while non-sanguine aggression is transmitted effectively as a dominant trait. This is because in order for the trait of aggression to be transmitted to a child, both parents must either have the trait or be a carrier of the recessive **a** gene, while in order for the trait of non-sanguine aggression to be transmitted to a child, in addition at least one parent must lack the trait N.

Geographic NPA trails

Focusing on an individual having a particular *dominant trait*, we define an ancestral "geographic trail" as set of connected geographic points corresponding to the location of the birth of parents in successive preceding generations where the trait was passed from parent to child. We presume that for any individual with a dominant trait one could trace the geographic trail for that trait deep back into the individual's ancestral population, even to the origin of a causative mutation.

In the context of NPA theory three types of continuous ancestral geographic trails are relevant:

1. non-sanguine trails (relevant to A or PA types)

2. non-aggressive trails (relevant to N or NP types)

3. P trails (relevant to types having the P trait)

With regard to the geographic mapping of the NPA traits, the following points are relevant for any individual, whether living or long deceased:

- A non-sanguine individual must have a non-sanguine trail.

- A non-aggressive individual must have a non-aggressive trail.

- An individual with the P trait must have a P trail.

- An individual of the NPA character type has only a P trail.

- An individual of the NA character type has no trail at all.

- The traits N and A, being recessive, have no defined trails.

Non-sanguine trails: "Cholericus"

The distribution of non-sanguine types in Eurasia and Australasia is as yet not documented. In concordance with the "Out of Africa" concept, we hypothesize that all non-sanguine individuals alive today have non-sanguine ancestral trails that lead to a single founding individual who lived in Africa (or in the Mideast) about 60,000 years ago. Following the ancient theory of humors, we call this individual "Cholericus." Cholericus could have been male or female, but the individual must have been of either the A or PA type. Descendents of Cholericus, thus, might be found among present-day Mideasterners, North Africans, Eastern Europeans and Russians, Indians and Pakistanis, Vietnamese, as well as among Australian Aborigines, Fijians and individuals who have immigrated to distant lands.

Non-aggressive and P trails: "Sanguineous" and "Perfectus"

In analogy with the above concept of "Cholericus" we can similarly denote two single individuals to whom all living non-aggressive types and perfectionist types could be traced. Here, by "non-aggressive" it is meant an N or NP type, and by "perfectionist" it is meant any type having the P trait. For non-aggressive types we can call the founding individual "Sanguineous," and for types having the P trait we can call the founding individual "Perfectus." Again, Sanguineous and Perfectus could have been either male or female. Sanguineous must have been either an N or NP type, while Perfectus must have been an NP, PA or NPA type.

Descendents of Sanguineous and Perfectus are assumed to have wide distributions over the world.

Origins of the distribution of NPA types

If one accepts that the distribution of NPA traits worldwide is not uniform, and that some regions have very different preponderances of NPA types in relation to other regions, the natural question arises: "how did this come about?"

To approach this question we rely heavily on recent data that has proven, to the exclusion of reasonable alternative theories, the "Out of Africa" concept of ancestral human migration. The Genographic Project of the National Geographic Society [2-4] has been in the forefront of quantifying the ancestral migrations in the precise terms of mitochondrial DNA and Y-chromosomal markers. These data show unequivocally that all modern humans can be traced to single common male and female ancestral individuals who lived in Africa during the Middle Paleolithic Era. The distributions of the prevalences of various mitochondrial

DNA and Y-chromosomal markers has led to maps showing the probable paths of ancestral migrations. The main results of the Genographic study were: (1) the migrations out of Africa were very recent in terms of human evolution, and (2) it was possible to deduce probable paths taken by the migrating subpopulations.

Applying the usual concepts of population genetics, the present-day distribution of the NPA traits (and, consequently, of NPA character types) could be dependent on the following conditions and mechanisms:

1. *Initial conditions*, or the geographical distribution of genotypes present in our African ancestors at the time of the start of the "Out of Africa" migrations.

2. *Mutation*, especially in the NPA genes or in genes having direct interaction with the NPA genes, and effects of recombination

3. *Genetic drift, founder effects, migrations* and *immigrations*.

4. *Natural selection*, especially according to adaptability of a particular NPA type to conditions of climatically induced changing environment.

5. *Sexual selection*, including *assortative mating*, especially according to NPA character type, and *inbreeding*.

It can be seen from the above that the deduction of the origins of the present day distribution of types is a complex problem, with probably all of the above mechanisms being relevant at some times and in some regions.

Mutations in NPA loci and Assortative Mating

With regard to mutations, we point out that a mutation in the putative repressor of the N or A trait could eventually lead to the appearance of that trait in a subpopulation. Similarly, a mutation at the P locus could lead to individuals lacking the P trait in the affected subpopulation.

With regard to assortative mating, it is well known that humans mate assortatively for many reasons, one of which is on the basis of physical similarity [1]. A striking physical feature often present in individuals of the N type is the ability to display a "gingival smile," broadly exposing teeth and gums (see Chapters 5 and 6). We posit that this "narcissistic smile" would be a powerful driving force for assortative mating between N types, and possibly between other character types having the N trait. The Masai tribe and others of East Africa are exemplified by individuals who show flamboyance of dress and behavior, and who also typically display the gingival smile. From photographs of these individuals, one notes the very high prevalence of a marked upper incisor gap (UIG). It is possible that this wide gap between the front teeth developed in conjunction with assortative mating based on the gingival smile and became a powerful bonding signal of attractiveness and affection. In some of the African tribes the UIG is even manually accentuated by the ritualized filing of a triangular notch between the upper incisors. Finally, we have noted that the gingival smile and the UIG are often present together in present day non-African individuals, and the simplest hypothesis would be that this association has a deep ancestral origin.

Role of NPA traits in major events of the Paleolithic and Neolithic Eras

It is possible that the NPA approach may be useful in the interpretation of some of the major events occurring during those critical times when modern humans first began to migrate from Africa and began to populate all of the major land masses. Although the details of the events will always be conjectural, we offer the following hypotheses:

Birth of art: We posit that the first works of cave art during the Middle Paleolithic Era, from their imaginative and non-perfectionistic features, were done by N types.

Neolithic era: The Neolithic era, with its emphasis on perfectionism in tool making, was primarily the result of the activity of NP types.

Agrarian revolution: Modern NP types have been observed to be drawn to farming. Hence, we posit that the agrarian revolution with its roots in the "Fertile Crescent" of the Mideast [2,3] was also primarily the result of the activity of NP types. The "Fertile Crescent" of the Mideast was, then, the nidus for a growing expanse of NP types, who then migrated to the north to Central Europe and Scandinavia, and to the east toward China and Korea. NP types ancestral to the present day San and Mbuti peoples would have also been prevalent in East Africa at the time of the early migrations.

Eastward migration in Asia: The analysis of the distribution of NPA types in present-day eastern Asia, and in particular Mongolia, may cast light on the dynamics of the eastward migration in Asia during settlement of that region.

Settlement of Pacific Islands: From the observation that N types are especially frequent in some of the Polynesian Islands and Hawaii, we posit that it was adventurous N types who originally populated these islands. It is possible that both the broad potential for imaginative exploits and religious-like fervor contributed to the motivation of these individuals to undertake dangerous, if not imprudent, voyages. From indications of probable A trait in New Guinea, Bougainville, Vanuatu, Fiji and finally in New Zealand, we infer that individuals with the A trait were also instrumental in the settlement of some of the islands in this area.

Melanesian origins: The nidus for the trait of non-sanguine aggression (aggression without the presence of the N trait) was likely in the Mideast, and we have suggested that all non-sanguine types would have geographic trails beginning in a single founding individual, denoted "Cholericus." According to NPA genetics, non-sanguine trails leading to Melanesian islands, such as Fiji, and possibly aboriginal Australasia as well, would have been due to either A or PA types, or both. Some possible non-sanguine ancestral trails are shown by the paths in Figure 1 below:

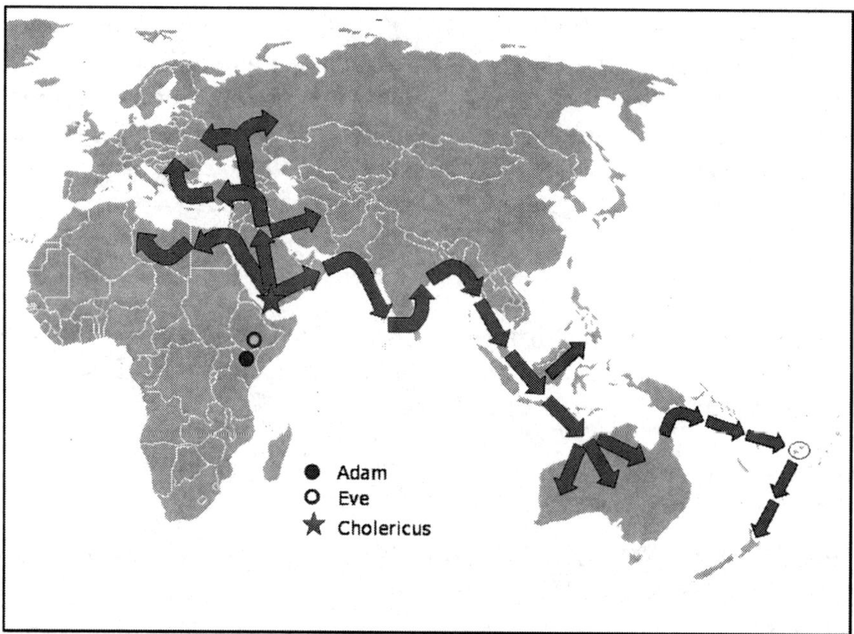

Figure 1.   Possible non-sanguine trails emanating from a single founding individual, Cholericus, consistent with the "Out of Africa" ancestral migrations. According to NPA theory, at least some non-sanguine A and PA types should be found in present-day indigenous peoples along the major trails. The locations of ancestral Y-chromosomal "Adam" and m-DNA "Eve" are also shown.

Settlement of Americas: From indications of the absence of the A trait (and in particular, non-sanguine A trait) in North and South American indigenous peoples, we posit that it must have been N and NP types who were the migratory founders of the New World.

Settlement in Arctic regions: Indications that present-day clans of the Arctic regions are either primarily N types or primarily NP types should make the comparisons of their lifestyles a fruitful object of study. Our anecdotal observations are that Arctic hunters requiring perfectionistic tools tend to be of the NP type, while nomadic reindeer herders tend to be of the N type.

Future genetic studies

The first priority would be to identify the NPA loci by linkage studies done by analysis of family pedigrees. The A locus would be of particular interest, as partial expression of the A trait (denoted by A– in NPA theory) is relatively common and may be related to common diseases, such as coronary artery disease and HLA-associated diseases (Chap. 11, above).

Next of interest would be population studies of the type being done in the Genographic Project: one would look for association and linkage disequilibrium between the NPA genes and known alleles of interest, for association with diseases, and, of course, for association with known mitochondrial DNA and Y-chromosomal markers. The concept of ancestral NPA trails may prove to be useful in the geographic mapping of NPA genotypes and phenotypes.

References

1. Alvarez, L. (2005): Narcissism guides mate selection: Humans mate assortatively, as revealed by facial resemblance, following an algorithm of "self seeking like." *Evolutionary Psychology 2*, 177-194.

2. Wells, Spencer (2006): *Deep Ancestry: Inside the Genographic Project*, National Geographic Society, Washington D.C.

3. Wells, Spencer (2002): *The Journey of Man: A Genetic Odyssey*, Princeton University Press, Princeton, New Jersey.

Internet site

4. *The Genographic Project*, National Geographic Society, https://www3.nationalgeographic.com/genographic/atlas.html

NPA PERSONALITY THEORY
Original Publication (1990)

A Theory of Personality Traits Leads to a Genetic Model for Borderline Types and Schizophrenia

Anthony M. Benis, Sc.D., M.D.

Mount Sinai School of Medicine,
One Gustave L. Levy Place, New York 10029

[*Speculations in Science & Technology* (1990) Vol. 13, No. 3, pp. 167-175]

<u>Summary in Plain English</u>

We present here an original theory of human personality traits based on traditional genetics. We identify three separate traits, each being dependent on the expression of a single gene. We call the traits "behavioral complexes" because each trait expresses itself in a spectrum of related behavioral characteristics. The three traits are: aggression (A), narcissism (N) and perfectionism (P). The traits of aggression and narcissism are fundamental to the human personality structure, and each is associated with a rage reaction: the classic aggressive-vindictive rage ("flight or fight" reaction), and the not so well-known narcissistic rage ("tantrum of vanity"). The trait of perfectionism is well known in the context of "obsessive-compulsive" behavior. The latter trait is present in some personality types as a modifier trait, but it is not essential to human social behavior.

The resulting model based on only the traits N, P and A appears at first sight overly simple. However, it is complex enough to generate all of the classic personality disorders of the psychiatric literature.

The traits N and A are special, in that a functioning individual must have at least one of the two traits underlying his/her personality structure. Individuals who have only one of the traits (N or A) are particularly vulnerable in the context of schizophrenia. That is, if the single trait that the individual possesses is not allowed to flourish, for example because of environmental constraints, then the individual will be incapacitated in social relations. In a *schizophrenic type* the single trait is profoundly suppressed so that the individual has symptoms of psychosis (delusions and hallucinations). In a *borderline type* the trait is partially suppressed so that the individual is schizoid or withdrawn (but not psychotic), yet is still identified as having difficulties in interpersonal relations.

Our approach leads to a model of schizophrenia based on two separate genetic loci (corresponding to the traits N and A), with three variants of each gene (alleles) at each locus. We have found some evidence that the gene for the trait of aggression is linked to a well-known gene (HLA) on chromosome 6.

We note that our model predicts that schizophrenia is a heterogeneous entity, being composed of several distinct types. It predicts that the genetic structure underlying schizophrenia is not distinct "abnormal genes," "genetic defects," or "inborn errors of metabolism." Rather, the basis of schizophrenia is very frequently-occurring genes that are present in many "normal" personality types, but happen to occur in an unfavorable combination in the schizophrenic or borderline individual. Our approach is consistent with a so-called "diathesis-stress" model in which clinically evident schizophrenia is triggered by environmental stresses imposed on a genetically vulnerable individual.

<u>Outline</u>

Summary

Introduction

Model Based on Three Genetic Traits

Genetics and environment

Traits of aggression, narcissism and perfectionism
Aggression (A)
Narcissism (N)
Perfectionism (P)

Origin of the character traits

Aggression: Complete and Incomplete Expression (Dominant, Submissive and Resigned Types)

1) Dominance: Dominant Character Types

N type
A type
NA type
NP type
PA type
NPA type

2) Submission: Submissive Types, including Borderline and Schizophrenic Types

Non-compliant types

A– type
PA– type
NA– type
NPA– type

Compliant types

A= type
PA= type
NA= type
NPA= type

3) Resignation: Resigned Types, including Schizophrenic Types

–A type
P–A type
N–A type
NP–A type

Narcissism: Incomplete Expression (Non-aggressive Withdrawn Types)

> N– trait
> N= trait
> –N trait

Discussion: Genetics of character traits

Acknowledgements

References

Table 1

Table 2

A Theory of Personality Traits Leads to a Genetic Model for Borderline Types and Schizophrenia

Anthony M. Benis, Sc.D., M.D.

Mount Sinai School of Medicine,
One Gustave L. Levy Place, New York 10029

[*Speculations in Science & Technology* (1990) Vol. 13, No. 3, pp. 167-175]

Summary

A genetic model of basic personality traits is presented, proposing three "behavioural complexes," or traits, which are the expression of major pleiotropic genes. The traits are: aggression, narcissism and perfectionism. The traits of aggression and narcissism represent the dual nature of human ambition, and each is associated with a rage reaction. The trait of perfectionism is associated with obsessive-compulsive behaviour. The resulting model is parsimonious in that it generates all of the classic personality disorders and borderline types of the psychiatric literature. The genetic predisposition to schizophrenia lies in lack of expression of the traits of aggression and narcissism, leading to a model of two loci, with three alleles at each locus. The gene for the trait of aggression may be linked to the HLA loci. The model proposes that schizophrenia is a heterogeneous entity, being composed of several distinct types. It is suggested that diathesis-stress models of schizophrenia based on the lack of expression of natural character traits are worthy of further exploration.

Introduction

Although the domains of psychology and psychiatry have flourished during the twentieth century, no objectively testable model of the human personality has been advanced. Indeed, since the time of Freud, the approaches to personality have been qualitative, or at best empirical, in nature [1-9]. The lack of prior success in the adequate identification of basic character traits, and their origins, indicates that new approaches to this question are necessary.

Model Based on Three Genetic Character Traits

Genetics and environment

Although it is universally accepted that both genetic and environmental factors comprise personality, no model based on the combined action of these factors has been put forward. To begin, we make the assumption that although both factors are operative, it is — based on studies with identical twins [10] — clearly the genetic, or structural, factors that first need to be identified.

Traits of aggression, narcissism and perfectionism

Horney [11] advanced the concept that at maturity there exist at least three expansive character types, namely the "arrogant-vindictive," the "narcissistic" and the "perfectionistic." Extending these ideas, we posit that the human character rests primarily on the existence of three major traits: aggression (A), narcissism (N) and perfectionism (P). Each of these traits is assumed to exist as a "behavioural complex," being the expression of a single major pleiotropic gene (a gene determining several related characteristics). Our interpretation of the three traits is as follows:

Aggression (A): The behavioural trait of aggression is observed to be the most labile of the three. The stereotypic acts associated with this trait involve body posturing, gestures, and eye contact of intimidation and deference, with individuals having this trait continually competing with each other on a scale of dominance and submission. The trait of aggression corresponds to a striving for *power* over one's environment, hence it is one main component of competitiveness or ambition. In a pejorative connotation the trait may reveal itself in the context of sadism or sadomasochism. The facial expression is non-sanguine, i.e., tending toward sallowness or pallor in individuals of light skin colour. The hallmark of the trait of aggression is a mass discharge of the sympathetic nervous system: the "flight or fight" response or the aggressive-vindictive rage. During the expression of this rage, the facial complexion of pallor is accentuated.

Narcissism (N): The trait of narcissism is noted to be less labile than that of aggression (where individuals may be constantly altering their character states on a scale of dominance and submission). The stereotypic acts associated with the trait include self-flaunting body posturing, expansive arm gestures, bowing, instinctive self-adornment, and a natural attraction to the limelight of personal recognition. Individuals having only this trait (of the three) are competitive but non-aggressive in their strivings for recognition. The trait corresponds to a striving for *glory* in one's environment, hence it is the second main component of human ambition. In a pejorative connotation, the unbridled trait of narcissism may reveal itself in the context of conceit, exhibitionism, vanity or messianism. An associated facial expression includes the radiant gingival smile (broadly exposing gums and teeth). The facial complexion tends toward blood-red or ruddy. Hallmarks of the trait include blushing, flushing, and a second type of mass discharge of the autonomic (parasympathetic) nervous system: the narcissistic rage of defence and withdrawal. During expression of this rage the normally sanguine complexion becomes even more florid.

Perfectionism (P): The trait of perfectionism is not a basic drive of ambition and is not associated with a rage reaction. Rather it is a mediator of the unbridled drives of aggression and/or narcissism. The stereotypic acts associated with the trait of perfectionism are obsessiveness, compulsiveness, repetition, and the maintenance of neatness, order and symmetry. A clue to the nature of the trait lies in the compulsive, repetitive mannerisms of autistic children and some adult schizophrenic individuals [12-14]. The behavioural pattern is often ritualistic and the speech characterised by echolalia. It will be posited later that certain autistic and schizophrenic individuals are those in whom the two components of ambition, i.e., aggression and narcissism, have been suppressed by genetic or environmental factors, either congenitally, in childhood, or after maturity, thus revealing in the individual a primitive state of perfectionism.

Origin of the character traits

The most vigorous proponent for the traits A, N and P in interpersonal behaviour was Horney [11,15-20], who considered that the traits have environmental origins, being the result of an individual's desperate search for dominance in the context of a stifling upbringing. The present model — in ascribing the traits to genetic origins — emphasises even more the physiognomic attributes that tend to exist concurrently with the behavioural ones. Thus, when divorced from the trait of narcissism, the trait of aggression is found to be associated with a non-sanguine complexion, with an incapacity to express the warm smile of recognition, and with the non-florid aggressive-vindictive rage. In contrast, the trait of narcissism is associated with a sanguine complexion, the radiant gingival smile, and the florid narcissistic rage.

Aggression: Complete and Incomplete Expression (Dominant, Submissive and Resigned Types)

We now introduce a category of basic states of personality, or character vectors, which take into account the quality and degree of expression of the trait of aggression, A. Again, extending the views of Horney [11], three character vectors are defined, namely *dominance, submission* and *resignation*. We use the term "vector" because dominant character types tend to project themselves above other individuals, submissive types tend to place themselves beneath others, and resigned types tend to retreat away from others.

1) Dominance: Dominant Character Types

In the character vector of dominance, the traits A and N, if present at all, are fully expressed. Therefore, allowing for the various combinations of the traits N, P, and A, we obtain the following dominant character types:

Narcissistic (N)

Aggressive (A)

Narcissistic-aggressive (NA)

Narcissistic-perfectionistic (NP)

Perfectionistic-aggressive (PA)

Narcissistic-perfectionistic-aggressive (NPA).

These phenotypes are summarised below. They are also listed in Table 1 (p. 496), together with the corresponding personality states that have been described in the literature. Note that an ambitionless, pure perfectionist (P) phenotype does not exist as a viable dominant character type.

Narcissistic (N) type: The narcissistic (N) type is found in the writings of Horney [11,15] and others (Table 1). In our view, this is the equivalent of the "sanguine" character type described by the ancients. The important attributes of this type are: expansiveness but unaggressiveness, non-perfectionism, a tendency to flamboyant self-adornment, a natural attraction to the limelight, the gingival smile of recognition, and the florid narcissistic rage. In extreme forms this type appears as a self-anointed visionary, a proselytising evangelist or a messianic personality.

Aggressive (A) type: The aggressive (A) type corresponds to Horney's arrogant-vindictive type [11] and to her concept of "moving against people [16,17]. In our interpretation, this is the classic "choleric" character type of antiquity. The main attributes of this type are: unbridled arrogance, instinctual vindictiveness, non-perfectionism, no tendency to self-adornment, a wry or sardonic grin in place of a gingival smile, and the pallid-complexioned aggressive-vindictive rage. In extreme forms this type appears as a sadistic personality, as an extroverted paranoid personality, or as the so-called antisocial or sociopathic personality.

Narcissistic-aggressive (NA) type: The narcissistic-aggressive (NA) type may be regarded as a composite of the previously described narcissistic and aggressive types. Horney described the essence of this character type, in the female, in an article, "The overvaluation of love: a study of a common present day type" [18]. The main attributes of this type are: a sanguine complexion, synergistic merging of unbridled narcissism and aggression, hyperactivity, non-perfectionism, a tendency toward extreme self-adornment, exhibitionism in the limelight, a "flashy" extroverted smile, a tendency toward hypersexuality, and the capacity to exhibit the narcissistic, aggressive-vindictive or combined narcissistic-aggressive rages. In extreme forms this character type appears as the hypomanic, histrionic or hysterical personality.

Narcissistic-perfectionist (NP) type: The attributes of the narcissistic-perfectionist (NP) type were described by Horney [11] in her exposition of the "perfectionist type." In our view, this encompasses the classic "phlegmatic" type known to the ancients. The main qualities of this type are: a tendency toward a sanguine complexion, industriousness, orderliness, an intense sense of duty, unaggressiveness, stubbornness, negativism, a tendency to ruminate, perfectionistic rather than unbridled self-adornment, an uncommonly seen gingival smile of recognition, and the capacity to exhibit the florid narcissistic rage. In extreme forms this character appears as the obsessive-compulsive personality.

Perfectionist-aggressive (PA) type: The perfectionistic-aggressive (PA) type is alluded to by Horney [16] in her mention of aggressive types who function in the capacity of a "power behind the throne," that is, personages who utilise intellectual qualities and planning rather than overt aggression to achieve their aims. In our view, this is the classic non-sanguine, austere "melancholic" personality of the ancients. The principal qualities of this type are: a non-sanguine complexion, passive aggressiveness, dour perfectionism, vigilance, manipulativeness, a proud bearing, haughty reservedness, a calculated vindictiveness, a lack of an innate tendency to self-adornment, a sardonic grin, and the pallid-complexioned aggressive-vindictive rage. In extreme forms this is the passive-aggressive, rebellious-distrustful, or ruminating paranoid personality.

Narcissistic-perfectionist-aggressive (NPA) type: The narcissistic-perfectionistic-aggressive (NPA) type was not explicitly described by Horney, although she did note [11,15] that the three traits can coexist in the same individual. The main attributes of this type are: a sanguine complexion, a loud voice, dynamism with a tendency to be overbearing, bombastic garrulity, intense eye contact, a strong sense of duty, a bent toward conventional values, unpretentious self-adornment, an outgoing smile of moderate intensity, and the capacity to exhibit the narcissistic, aggressive, or explosive narcissistic-aggressive rages.   In the extreme cases this individual is the managerial-autocratic or explosive personality.

2) Submission: Submissive Types, including Borderline and Schizophrenic Types

In the character vector of submission the trait of aggression is not fully expressed. We define two gradations of submission: *non-compliance*, in which the individual is basically submissive but is easily activated to an energetic state of aggression, and *compliance*, in which the individual tends to remain in a profound state of submission.

The state of submission is most often the result of a congenital, inherited, incomplete expression of the gene for the trait A. It may also be induced during the juvenile period on the basis of environmental constraints to character development.   That is, phenocopies (based on environmental factors) of a genetically disposed submissive state may exist.   Also, like dominant types having full expression of the trait A, submissive types may exhibit the aggressive-vindictive rage. However, the threshold for the triggering of such a rage may be higher.

Non-compliant submissive types — We denote the state of non-compliance by A–, obtaining the following phenotypes:

Aggressive borderline (A–)

Perfectionistic-aggressive borderline (PA–)

Narcissistic non-compliant (NA–)

Narcissistic-perfectionistic non-compliant (NPA–)

Compliant submissive types — We denote the state of compliance by A=, obtaining the following phenotypes:

Aggressive schizophrenic (A=)

Perfectionistic-aggressive schizophrenic (PA=)

Narcissistic compliant (NA=)

Narcissistic-perfectionistic compliant (NPA=)

The NPA– non-compliant type corresponds to active, motivated, non-confrontational individuals whose baseline personality tends toward submissiveness, as described by Horney in her discussion of inverted sadistic behaviour [17]. In the therapeutic setting, these individuals are found over the spectrum of "Type A," dependent, and phobic-anxious personalities (Table 1). The NA– type is a non-perfectionistic, active individual exhibiting pronounced narcissistic behaviour. In the therapeutic setting this is a cyclothymic or dependent histrionic personality (Table 1).

The compliant types NA= and NPA= correspond to more profoundly submissive individuals, having more pronounced tendencies toward masochistic behaviour (Table 1). They correspond to Horney's compliant "self-effacing" personality [11] and to her concept of "moving toward people" [19].

Since the above-mentioned non-compliant types possess the trait N, they tend to have sanguine complexions. Among the remaining types, we note categories of non-sanguine, potentially aggressive borderline types (A– and PA–) and schizophrenics (A= and PA= types).

By definition, the borderline types possess only one of the traits of ambition (N or A) and it is only partially expressed; the schizophrenic types have only one of the traits, and it is profoundly suppressed.

3) Resignation: Resigned Types, including Schizophrenic Types

In the character vector of avoidance, or resignation, the trait of aggression is stunted after maturity because of environmental constraints. Unlike submissive types who readily involve themselves in the relative competition of dominance and submission (and sometimes sadomasochism), resigned types remain relatively detached from such activities and only with difficulty can be stressed to an energetic state of aggression. However, like submissive types, resigned types can be induced into the aggressive-vindictive rage.

Denoting the state of resignation by –A, we obtain the following resigned types:

Aggressive schizophrenic (–A)

Perfectionistic-aggressive schizophrenic (P–A)

Narcissistic resigned (N–A)

Narcissistic-perfectionistic resigned (NP–A)

The resigned types having the narcissistic trait correspond to detached individuals, as described by Horney [11]. She considered that "moving away from people" [20] was a maladaptive response that could develop as a growing individual struggled toward maturity. The NP–A type tends to have strong perfectionistic tendencies, while the N–A type is more labile (Table 1).

Again, the above resigned types tend to have sanguine complexions. Among the remaining types, we note once more a category of non-sanguine, potentially aggressive schizophrenics. Here, however, the attainment of state –A or P–A represents a dramatic "schizophrenic break" after maturity from a premorbid state of dominance (A or PA).

Narcissism: Incomplete Expression (Non-aggressive Withdrawn Types)

We focus on the dominant types N and NP, who lack the trait A. In analogy with the previous discussion of the partial expression of the trait A, we define the following states of incomplete expression of the trait N, as shown in Table 1: N–, N= and –N. Again, these states may be rooted either in heredity or in environmental constraints.

In summary: Our analysis, presented in detail elsewhere [21] shows that the model predicts two major categories of schizophrenia, namely 1) that in which the trait A is genetically absent and the trait N is suppressed, and 2) that in which the trait N is genetically absent and the trait A is suppressed.

Thus, our model leads to an analysis of a character structure that exists when naturally existing character traits are only marginally expressed. This generates a diathesis-stress model for schizophrenia, i.e., incorporating aspects of both genetics and environment.

Discussion: Genetics of character traits

On the basis of archetypal examples, the model assumes that in their full expression the traits A and N are transmitted according to autosomal recessive inheritance, with the trait P being transmitted in the autosomal dominant mode. For the traits A and N, we assume the existence of three alleles, corresponding to full expression, partial suppression, and total suppression of the traits. Thus, for trait A these alleles are denoted by **a**, **a-** and **a~**, respectively, and the corresponding alleles for the trait N are **n**, **n-** and **n~**. For the trait P, we postulate two alleles, **p** and **p~**, corresponding to full expression or total absence of the trait P, on the assumption that the trait is always transmitted with complete penetrance.

We propose that the alleles **a~** and **n~** control the production of inhibitors of the traits A and N at the level of the central nervous system, and the alleles **a-**

and **n-** produce partial inhibition. Thus, we are led to posit that the alleles **a~** and **n~** are dominant, alleles **a** and **n** are recessive, with alleles **a-** and **n-** occupying intermediate positions (Table 2).

A result of autosomal recessive inheritance for the traits A and N is that only matings of individuals of the A or PA category with those of the N or NP category could produce "ambitionless" progeny, totally lacking both the N and A traits. Such an event would be lethal to the offspring (miscarriage or stillbirth), or if the offspring did survive intra-uterine life, they would be placed in the "failure to thrive" category.

The present model does not mean that all individuals can be readily stereotyped by simple caricatures, as depicted in Table 1. First of all, besides the three posited genes, many other "modifier" genes very likely influence personality. In particular, human temperament in the Pavlovian sense — a measure of an individual's level of activity, or excitability — probably has a genetic basis [1,6,8]. Thus, for each character type we allow for the possibility of a spectrum of temperaments having a genetic basis. Secondly, environmental influences are posited to play an important role, the model being not inconsistent with the concept that some maladapted character types at maturity are rooted in stressful environmental influences during childhood. Finally, the many adaptive modes of coping with the stresses of life means that in the final analysis the real life situation — that is, environment — is the key to the individual's actual behaviour.

As an example of a type of approach to the identification of the genes for the traits N, P and A, we mention the supposition that the **a** locus lies on the short arm of chromosome 6, closely linked to the HLA (histocompatibility) group of genes. This follows from several observations. First, several conditions associated with the HLA loci (e.g. rheumatoid arthritis, ulcerative colitis, Graves' disease) are associated with an incapacity to express fully the trait of aggression [21-23]. Second, there is evidence for the existence in the HLA region of a single gene predisposing to a wide variety of endogenous disorders [21-24]. Finally, there have been reports that some schizophrenic and affective disorders may be associated with the HLA loci [25-27]. Thus, it is possible that a well-known genetic marker for the trait A already exists.

Acknowledgements

The author thanks J. Mendlewicz, M.D. for his encouragement and J.H. Rand, M.D. for his critical reading of the manuscript.

References

1. Buss A.H. and Plomin R. (1984). *Temperament: Early Developing Personality Traits*, Erlbaum, Hillside, New Jersey.

2. Cloninger R.C. (1987). A systematic method for clinical description and classification of personality variants, *Arch. Gen. Psychiatry 44* 573-588.

3. Eysenck H.J. (1980). *A Model for Personality*, Springer-Verlag, New York.

4. Freud S. (1896, 1962). "Heredity and the aetiology of the neuroses," in *Early Psycho-Analytic Publications*, Hogarth, London.

5. Leary T. (1957). *Interpersonal Diagnosis of Personality*, Ronald, New York.

6. Millon T. (1981). *Disorders of Personality: DSM-III, Axis II*, Wiley, New York.

7. Stone M.H. (1980). *The Borderline Syndromes*, McGraw-Hill, New York.

8. Strelau J. (1983). *Temperament, Personality, Activity*, Academic Press, New York.

9. Widiger T.A., Trull T.J., Hurt S.W., Clarkin J. and Frances A. (1987). A multidimensional scaling of DSM-III personality disorders, *Arch. Gen Psychiatry 44* 557-63.

10. Bouchard T.J. (1984). "Twins reared together and apart: What they tell us about human diversity," In: Fox S.W. (ed.), *Individuality and Determinism*, Plenum, New York.

11. Horney K. (1950). *Neurosis and Human Growth*, Norton, New York.

12. Leonhard K. (1979). *The Classification of the Endogenous Psychoses*, Irvington Press, New York.

13. Rimland B. (1964). *Autism*, Appleton-Century-Crofts, New York.

14. Rutter M. (1978). "Diagnosis and definition," In: Rutter M. and Schopler E. (eds.), *Autism*, Plenum, New York.

15. Horney K. (1939). "The concept of narcissism," In: *New Ways in Psycho-analysis*, Norton, New York.

16. Horney K. (1943). "Moving against people," In: *Our Inner Conflicts*, Norton, New York.

17. Horney K. (1943). "Sadistic trends," In: *Our Inner Conflicts*, Norton, New York.

18. Horney K. (1937, 1967). "The overvaluation of love: A study of a common present day type," In: *Feminine Psychology*, Norton, New York.

19. Horney K. (1945). "Moving toward people," In: *Our Inner Conflicts*, Norton, New York.

20. Horney K. (1945). "Moving away from people," In: *Our Inner Conflicts*, Norton, New York.

21. Benis A.M. (1985, 2008). *Toward Self and Sanity: On the Genetic Origins of the Human Character*, Psychological Dimensions Publishers, New York.

22. Moos R.H. (1977). Personality factors associated with rheumatoid arthritis, *J. Chronic Dis. 17* 41-55.

23. Weiner H. (1977). *Psychobiology and Human Disease.* Elsevier, New York.

24. Christy M., Mandrup-Poulsen, T. and Nerup J. (1984). Genetic markers in insulin-dependent (type I) diabetes mellitus, *Ann. Clin. Res. 16* 53-63.

25. Gottesman L.I. and Shields J. (1982). *Schizophrenia: The Epigenetic Puzzle*, Cambridge University Press, Cambridge.

26. Mendlewicz J. and Shopsin B. (1979). *Genetic Aspects of Affective Illness*, SP Medical and Scientific Books, New York.

27. Weitcamp L.R., Stancer H.C., Persad E., Flood C. and Guttormsen S. (1981). Depressive disorders and HLA: A gene on chromosome 6 that can affect behavior, *N. Engl. J. Med. 305* 1301-06.

28. Kraepelin E. (1919, 1971). *Dementia Praecox and Schizophrenia*, Krieger, Huntington, New York.

29. Paré A. (1634). "Of Humours," In: (1968) Johnson T. (Ed.), *The Collected Works of Ambroise Paré*, Milford House, Pound Ridge, New York.

TABLE 1

Principal character types of model: Comparison with literature

Character types of model	Personality states from literature*
Dominant	
N Narcissistic	narcissistic [1,3,5,8] sanguine [6]
A Aggressive	aggressive vindictive [1] aggressive-sadistic [3] paranoid-aggressive [5] antisocial [5,8] choleric [6]
NA Narcissistic- aggressive	competitive-hypersexual [1] histrionic [5] hysterical-hypomanic [8]
NP Narcissistic- perfectionistic	perfectionistic [1] responsible-hypernormal [3] compulsive [5] phlegmatic [6] obsessive [8]
PA Perfectionistic- aggressive	power behind the throne [1] rebellious-distrustful [3] passive-aggressive paranoid [5,8]

NPA Narcissistic-
    perfectionistic-
    aggressive
	managerial autocratic [3]
	explosive [8]

---

* [1]Horney (refs. 11,15-20); [2]Kraepelin (ref. 28); [3]Leary (ref. 5); [4]Leonhard (ref. 12); [5]Millon (ref. 6); [6]Paré (ref. 29); [7]Rimland (ref. 13); [8]Stone (ref. 7).
==

Submissive

Non-compliant: juvenile onset

A–	Aggressive borderline	schizoid-schizotypal [5] schizoid [8]
PA–	Perfectionistic-aggressive borderline	schizoid-schizotypal [5] schizoid [8]
NA–	Narcissistic non-compliant	dependent histrionic [5] cyclothymic [8]
NPA–	Narcissistic-perfectionistic non-compliant	inverted sadistic [1] dependent [5] phobic-anxious [8]

Compliant: juvenile or maturity onset

A=	Aggressive schizophrenic	paranoid schizophrenic [2]
PA=	Perfectionistic-aggressive schizophrenic	catatonic or paranoid schizophrenic [2], paraphrenia systematica [2], affect-laden paraphrenia [4]
NA=	Narcissistic compliant	self-effacing masochistic [1] dependent histrionic [5], depressive, masochistic, hypomanic [8]
NPA=	Narcissistic-perfectionistic compliant	self-effacing masochistic [1] docile dependent [3] schizoid dependent [5] depressive masochistic [8]

Resigned

Maturity Onset

–A	Aggressive schizophrenic	paranoid schizophrenic [2]
P–A	Perfectionistic-aggressive schizophrenic	catatonic or paranoid schizophrenic [2]
N–A	Narcissistic resigned	resigned [1], avoidant [5]
NP–A	Narcissistic-perfectionistic resigned	resigned [1], avoidant [5]

Non-Aggressive (Narcissistic withdrawn)

Juvenile onset

N–   Narcissistic   schizoid-schizotypal [5]
      borderline   schizoid [8]

N-P   Narcissistic   schizoid-schizotypal [5]
      perfectionist   schizoid [8]
      borderline   borderline autistic child [7]
                             idiot savant [7]

Juvenile or maturity onset

N=   Narcissistic   hebephrenic schizophrenia [2]
      schizophrenic

N=P   Narcissistic-   catatonic, hebephrenic or acute-
      perfectionistic   onset paranoid schizophrenia [2,4]
      schizophrenic   autistic child [7]

Maturity onset

-N   Narcissistic   hebephrenic schizophrenia [2]
      schizophrenic   expansive paraphrenia [4]

-NP   Narcissistic-   catatonic, hebephrenic or acute-
      perfectionistic   onset paranoid schizophrenia [2,4]
      schizophrenic

TABLE 2*

Possible genotypes corresponding to phenotypes of model

<u>Phenotype</u>	<u>Genotype</u>
<u>Aggression (A)</u>	
Absence of A	a~a, a~a-, a~a~
A	aa
A-	aa-, a-a-
A- (phenocopy)	aa
A= (environmentally induced)	aa, aa-, a-a-
-A (environmentally induced)	aa, aa-, a-a-
<u>Narcissism (N)</u>	
Absence of N	n~n, n~n-, n~n~
N	nn
N-	nn-, n-n-
N- (phenocopy)	nn
N= (environmentally induced)	nn, nn-, n-n-
-N (environmentally induced)	nn, nn-, n-n-
<u>Perfectionism (P)</u>	
Absence of P	p~p~
P	pp, pp~

** Table 2 was omitted from the published version because of space considerations.*

CPSIA information can be obtained at www.ICGtesting.com
Printed in the USA
LVOW11s2002151013

357047LV00021B/806/P